CHINA'S DILEMMA

CHINA'S DILEMMA

ECONOMIC GROWTH, THE ENVIRONMENT AND CLIMATE CHANGE

Ligang Song and Wing Thye Woo (eds)

ANU

THE AUSTRALIAN NATIONAL UNIVERSITY

E PRESS

Asia Pacific Press

Brookings Institution Press

SOCIAL SCIENCES ACADEMIC PRESS (CHINA)

ANU

E PRESS

Co-published by ANU E Press and Asia Pacific Press
The Australian National University
Canberra ACT 0200 Australia
Email: anuepress@anu.edu.au
This title available online at http://epress.anu.edu.au/china_dilemma_citation.html

CHINA BOOK
INTERNATIONAL

Co-published with SOCIAL SCIENCES ACADEMIC PRESS (CHINA)
under the China Book International scheme. This scheme supports
co-publication of works with international publishers.

National Library of Australia Cataloguing-in-Publication entry

Title:	China's dilemma : economic growth, the environment and climate change / editors Ligang Song ; Wing Thye Woo.
ISBN:	9780731538195 (pbk.) 9781921536038 (pdf.)
Notes:	Includes index. Bibliography.
Subjects:	Economic development--Environmental aspects--China. Climatic changes--China. Energy consumption--China. China--Economic conditions. China--Environmental conditions.
Other Authors/Contributors:	Song, Ligang. Woo, Wing Thye.
Dewey Number:	338.900951

Cover design: Teresa Prowse
Cover photo: Jason Lyon. iStockphoto, File Number: 2831996

Contents

Tables

ix

Figures

Symbols used in tables

..   not available
n.a.   not applicable
-       zero
.   insignificant

Abbreviations

ADB	Asian Development Bank
APEC	Asia Pacific Economic Cooperation
ASEAN	Association of Southeast Asian Nations
BIS	Bank for International Settlements
BP	British Petroleum Global Limited
BTU	British thermal unit
CASS	Chinese Academy of Social Sciences
CGE	computable general equilibrium
CPI	consumer price index
CWIM	China Water Institutions and Management
EIA	Energy Information Administration
EKC	Environmental Kuznets curve
FDI	foreign direct investment
FIE	foreign-invested enterprises
FTA	Free Trade Agreement
gce	grams of coal equivalent
GDP	gross domestic product
GIOV	value of gross industrial output
GTAP	Global Trade Analysis Project
HRS	Household Registration System (*hukou*)
HRS	household responsibility system
ICT	information and communication technology
IEA	International Energy Agency
IGCC	integrated gasification combined-cycle
IMF	International Monetary Fund
IO	input-output (analysis)
IPCC	Intergovernmental Panel on Climate Change
MCL	minimum cost of living
MFA	Multifibre Arrangement
MPC	marginal propensity of consumption
Mtoe	million tonnes of oil equivalent
MWH	McKibbin–Wilcoxen hybrid approach
NBS	National Bureau of Statistics
NGO	non-government organisation
NAFTA	North American Free Trade Agreement

NDRC	National Development and Reform Commission
NERI	National Economic Research Institute
OECD	Organisation for Economic Co-operation and Development
OEF	Oxford Economic Forecasting model
PBC	People's Bank of China
PHH	pollution heaven hypothesis
PPP	purchasing power parity
PRC	People's Republic of China
REER	real effective exchange rates
RMB	renminbi
SCE	standard coal equivalent
SCE	state-controlled enterprise
SEPA	state environmental protection administration
SITC	Standard International Trade Classification
SME	small and medium enterprise
SOE	state-owned enterprise
tcae	tonnes of carbon equivalent
TFP	total factor productivity
TVE	township and village enterprise
WGE	waste-gas emissions
WHO	World Health Organization
WTO	World Trade Organization
WUA	water user associations

Contributors

Prema-chandra Athukorala — Research School of Pacific and Asian Studies, The Australian National University, Canberra

Iain Bain — School of Economics, College of Business and Economics, The Australian National University, Canberra

Bao Qun — School of Economics, Nankai University, Tianjin

Cai Fang — Institute of Population and Labour Economics, Chinese Academy of Social Sciences, Beijing

Cheng Fang — Economic and Social Development Division, Food and Agriculture Organization of the United Nations, Rome

Yuanyuan Chen — School of Economics, Nankai University, Tianjin

Du Yang — Institute of Population and Labour Economics, Chinese Academy of Social Sciences, Beijing

Ross Garnaut — Research School of Pacific and Asian Studies, The Australian National University, Canberra.

Jane Golley — Crawford School of Economics and Government, The Australian National University, Canberra

Xiaodong Gong — Research School of Pacific and Asian Studies, The Australian National University, Canberra

Health and Mortality Transition in Shanghai Project Research Team — Zhongwei Zhao, Xizhe Peng, Yuan Cheng, Xuehui Han, Guixiang Song, Feng Zhou, Yuhua Shi and Richard Smith

Stephen Howes — Crawford School of Economics and Government, The Australian National University, Canberra

Xiulian Hu — Energy Research Institute, Beijing

Jikun Huang — Center for Chinese Agricultural Policy of the Chinese Academy of Sciences, Beijing

Yiping Huang — Citigroup Global Markets Asia Limited, Hong Kong and Crawford School of Economics and Government, The Australian National University, Canberra

Qiuqiong Huang — Department of Applied Economics, University of Minnesota, Twin Cities

Frank Jotzo — Research School of Pacific and Asian Studies, The Australian National University, Canberra

Kejun Jiang — Energy Research Institute, Beijing

Sherry Tao Kong — Research School of Social Sciences, The Australian National University, Canberra

Shi Li — Department of Economics, Beijing Normal University, Beijing

Justin Yifu Lin — China Center for Economic Research, Peking University, Beijing and World Bank, Washington, DC

Warwick McKibbin — School of Economics, College of Business and Economics, The Australian National University, Canberra

Dominic Meagher — Crawford School of Economics and Government, The Australian National University, Canberra

Mark Rosegrant — Environment and Production Technology Division, International Food Policy Research Institute, Washington, DC

Scott Rozelle — Freeman Spogli Institute for International Studies, Stanford University, California

Peter Sheehan — Centre for Strategic Economic Studies, Faculty of Business and Law, Victoria University, Melbourne

Xunpeng Shi — Crawford School of Economics and Government, The Australian National University, Canberra.

Ligang Song — Crawford School of Economics and Government, The Australian National University, Canberra

Fiona Sun — Centre for Strategic Economic Studies, Faculty of Business and Law, Victoria University, Melbourne

Rod Tyers — School of Economics, College of Business and Economics, The Australian National University, Canberra

Jinxia Wang — Centre for Chinese Agricultural Policy, Institute of Geographical Sciences and Natural Resource Research, Chinese Academy of Sciences, Beijing

Xiaolu Wang — National Economy Research Institute, China Reform Foundation, Beijing

Peter J. Wilcoxen — The Maxwell School, Syracuse University, New York

Wing Thye Woo — Brookings Institution, Washington, DC, University of California–Davis and Central University of Finance and Economics, Beijing

Stanley Wood — Environment and Production Technology Division, International Food Policy Research Institute, Washington, DC

Xin Meng — Department of Economics, Research School of Social Sciences, The Australian National University, Canberra

Nobuaki Yamashita — Department of Economics and Finance, La Trobe University, Melbourne

Liangzhi You — Environment and Production Technology Division, International Food Policy Research Institute, Washington, DC

Lijuan Zhang — Center for Chinese Agricultural Policy of the Chinese Academy of Sciences, Beijing

Acknowledgments

The China Economy and Business Program gratefully acknowledges the financial support for China Update 2008 from the Australian government's development agency, AusAID, and the assistance provided by Rio Tinto through the Rio Tinto-ANU China Partnership.

1

China's dilemmas in the 21st Century

Ligang Song and Wing Thye Woo

Thirty years of reform (1978–2008) have turned China into one of the largest and most dynamic economies in the world. China, however, faces three significant and profound challenges towards the end of the first decade of the twenty-first century: First, to maintain continued high growth amid global financial turbulence, the slow-down of the major economies abroad, and some rising socially destabilising tensions such as growing income inequality; second, to bring its growth path in line with environmental sustainability; and third to manage the rising demand for energy to moderate oil price increases and to placate heightening domestic and international concerns about global warming. This book, in three parts, offers some analyses as to how China could confront these challenges and discusses some of the key implications for China and the world.

Part I of the book deals with the first challenge of maintaining rapid and sustainable growth. It covers issues such as whether it is likely that China will fall into stagflation amid the slow-down of the US economy and rising costs of production. What are the likely implications of American and European financial shocks for Chinese economic performance? Why is it important to look at the US–China bilateral trade imbalances in the context of global production fragmentation? What are the conditions for achieving equitable and sustainable growth in China; and what are the key lessons that can be learned from China's experience of 30 years of reform?

According to Yiping Huang in Chapter 2, China faces an enormous challenge in sustaining its rapid growth and maintaining macroeconomic stability; but stagflation is not likely to materialise in China any time soon. He attributes this precarious future situation to three factors: deceleration of the US economy, elevated international oil prices and gradual cost normalisation in China. His model does not support the recent claim of some China experts that the Asian economies, and China in particular, have decoupled themselves from the US economy. Huang's model estimates that a 1 percentage-point slow-down in the US economy could lower Asian economic growth by 1.1 percentage points and Chinese growth by 1.3 percentage points.

Huang emphasises that, apart from the increase in demand for energy and rising marginal costs of oil production, financial investment demand (which outweighs physical investment demand by a large margin) could have contributed significantly to the oil-price jump that slowed growth and raised inflation. Subsidies for fuel are not advisable because they cause distortion and inefficiency without relieving demand pressure. China already suffers from too many price distortions—notably those in the factor markets (that is, in the labour, capital and land markets)—that reduce production costs and ensure high profits, and hence sustain high investment growth. Because the reduction of these distortions through cost normalisation is an inevitable structural adjustment, China from now on will face stronger than ever inflation pressure.

Since exports make up more than 40 per cent of China's gross domestic product (GDP) and most of its exports are directed to European and North American markets, negative financial shocks in those regions might be expected to hamper China's economic growth. Rod Tyers and Iain Bain in Chapter 3 use a dynamic model of the global economy to analyse this issue. They find that a rise in North American and European financial intermediation costs retards neither China's GDP nor its import growth in the short run because these shocks set in train some mitigating factors. Specifically, the temporary flight of savings from Organisation for Economic Cooperation and Development (OECD) countries into investments in China raises China's domestic aggregate demand to offset export decline, therefore maintaining output level. This capital inflow also forces China's central bank to choose between appreciating the renminbi (RMB) more quickly and allowing faster inflation: the choice so far has been RMB appreciation, which has boosted imports.

The widening bilateral trade imbalances between the United States and China have become a contentious political issue in the United States. Premachandra Athukorala and Nobuaki Yamashita argue in Chapter 4 that the policy debate about US–China trade relations has been based wrongly on the conventional notion of horizontal specialisation, in which trade takes place in the form of final goods. This assumption ignores the continuing process of global production sharing in which both countries are increasingly engaged. The share of parts and components in US exports to the other East Asian countries is much higher than US exports to China; and their share in US imports from China are remarkably low compared with the figures for the other East Asian economies, as well as the global average. When components are netted out, it becomes evident that China is specialising in labour-intensive niches within otherwise skill-intensive sectors. Furthermore, the continuing process of production fragmentation has increasingly blurred the difference between merchandise trade and services trade. Athukorala and Yamashita's conclusion is that any analysis that overlooks the exports of new production-related services relating to the process of global production sharing could overstate the magnitude of the US–China trade imbalance by a wide margin. Therefore, the widely held view that China's rapid market penetration of the US economy is driven by unfair trade practices needs to be re-examined in the light of the fact that the two economies are deeply interconnected and interdependent within global production networks.

Increasingly excessive income inequality is one of the most undesired outcomes of China's 30-year-long reform. The key question one could ask is whether China should let efficiency be the focus of primary distribution, and then redistribute income through the fiscal system and other government measures. Justin Lin in Chapter 5 argues that this course of action would be unproductive, if not counterproductive. In his opinion, equality and efficiency can be achieved simultaneously by adopting a strategy for development that follows underlying comparative advantage. This strategy leads to a high rate of capital accumulation, which, in turn, automatically upgrades production technologies. The outcome is a spontaneous shift from labour-intensive industries to relatively capital and technology-intensive industries. Lin believes that this production shift will always move income distribution in favour of labour. Because this strategy requires a competitive market system to work, China must deepen the reform of its financial sector, unwind the distortion of resource prices, reduce state monopolies, emphasise education, improve the quality of Chinese institutions and maintain macroeconomic stability.

China experienced a historically unprecedented scale of rural-to-urban migration in the past 30 years. More than 126 million rural workers have moved from the low-productivity agricultural sector to high-productivity urban sectors, contributing significantly to China's productivity improvement and increases in total output. Current knowledge about China's migration is, however, grossly inadequate primarily because of a lack of comprehensive data for how the conditions of the migrants and their families have changed over time. In Chapter 6, Xiaodong Gong, Sherry Tao Kong, Shi Li and Xin Meng make an important attempt to fill this knowledge gap by using the data from the first comprehensive survey of rural–urban migration in China collected in 2007. The study results provide information on the individual and family characteristics of migrants, their jobs and job-related welfare, income and poverty dynamics, living conditions and the broader well-being measures of migrants such as physical and mental health conditions, and the health and education of migrants' children.

Xiaolu Wang in Chapter 7 undertakes a rethinking of Chinese economic transformation in the past 30 years. He looks at the reform approaches in Russia and China, and questions whether the conditions of reform could lead to a process of Pareto improvement—namely, to make one person better off without making someone else worse off. Wang argues that most of the reforms in China have led to improvement in all social groups—at least for the majority of people—and therefore Chinese reforms have generally achieved a process of Pareto improvement. In his opinion, a key factor in determining this outcome is that the reform policy must take the public interest into consideration, which can be accomplished through policy debates and practical experiments. Agricultural reform, price-system reform and ownership reform are cited as examples to showcase why the Chinese approach has been successful.

It is important, however, to be cognisant of the fact that the structural conditions in China in 1978 were vastly different from those in Eastern Europe and the former Soviet Union, and hence that most of the Pareto-improvement outcomes in China that Wang identified were not replicable in transitional economies that were highly urbanised and over-industrialised. In China, marketisation of its economy enabled the movement of underemployed rural labour into more productive industrial jobs in the new non-state sector. In Russia, however, marketisation by Russia's fiscally strapped government meant substantial reduction in subsidies to heavy industry and the movement of workers from these into light industries and services in the newly legalised private sector. In short, economic reform in China meant unleashing the economic development process and, in Russia, it meant implementing economic restructuring; the former process is much more likely to produce more Pareto-improvement outcomes than the latter process.

Just as Premier Zhou Enlai famously said that it was still too early to judge the success of the French Revolution of 1789, it would certainly be premature to trumpet the superiority of China's reform policies. After all, as Wang notes, the task of reform is not yet accomplished. The success of future reform for China's long-run growth and development will depend crucially on whether China is capable of building governmental institutions that are transparent, efficient and accountable, and supported by an effective legal system.

Part II of this book seeks to address the following key questions: how much does China's economic growth contribute to global greenhouse gas emissions? How can China escape from the dilemma of achieving high growth and environmental protection? What is the political economy of emissions reduction in China? What are the environmental consequences of foreign direct investment in China? What are the impacts of climate change on the Chinese economy, particularly on sectors such as agriculture? Will water shortages impose constraints on economic growth in China? How much does air pollution increase mortality in China?

The world has entered a period of exceptionally fast economic growth— with rapid economic development especially in China, followed by India and many other low-income countries—which could be called the 'Platinum Age' (Garnaut and Huang 2007). 'There are reasonable prospects for growth rates in the vicinity of 10 per cent per annum—or even higher for a while—to continue for a considerable period and for growth rates to remain high until average Chinese productivity levels and living standards are approaching the range of industrialised countries in the late 2020s' (Garnaut 2008:3). This rapid economic growth goes hand in hand with increasing resource use and pressure on the environment, including the build-up of greenhouse gases and the resulting climate change. The combination of China's large, rapidly growing economy and its carbon intensity means that in the coming years China will have an influence on greenhouse gas emissions unmatched by any other country.

Ross Garnaut, Stephen Howes and Frank Jotzo in Chapter 8 provide 'business-as-usual' projections for China's and the world's carbon dioxide emissions to 2030, by which time China will be responsible for 37 per cent of global emissions. Fortunately, China has recently announced goals of reducing the energy intensity of the economy and increasing the share of low-emissions energy, and is starting to put a range of ambitious climate change policies in place. If implemented, these policies will limit emissions growth to well below the business-as-usual trajectory. China's goal of limiting the growth in energy use to half the growth rate of the economy could become the basis for a near-term emissions target that will allow it to play a leading role in international discussions. The authors argue, however, that strong domestic policy action in China is likely to happen only if there is strong action in other major countries, especially the United States. They also believe that a global mechanism that ensures comprehensive emissions pricing (through a carbon tax or an emissions trading scheme) would be the most economically efficient way of achieving the projected emissions growth targets.

Warwick McKibbin, Peter Wilcoxen and Wing Thye Woo point out that under 'Platinum Age' growth rates, GDP per capita in China could catch up with that in Western Europe, Japan and the United States by 2100. This outcome could be undermined, however, by a possible fallacy of composition created by environmental constraints. The present atmospheric carbon dioxide concentration level is 380 parts per million (ppm), and it is increasing by 2 ppm annually. If, as commonly believed, significant irreversible environmental damage occurs with carbon dioxide concentrations above 560 ppm, there is the terrible possibility that either China will not catch up or it will catch up with the rich countries in 2100 only because of environmentally induced income declines occurring in these countries. For the world, the challenge is how to slow carbon dioxide emissions adequately to provide the breathing space required in which to develop alternative fuels that are greenhouse gas-free. For China, a key challenge is how to minimise the negative growth impact from limiting greenhouse gas emissions.

McKibbin, Wilcoxen and Woo use the G-cubed model to compare the merits of three market-based carbon dioxide reduction mechanisms: a domestic carbon tax, an international cap-and-trade scheme and the McKibbin–Wilcoxen hybrid (MWH) approach. They find the MWH approach to lower GDP growth the least in the short and middle term; however, as the only long-term solution is likely to be shifting to non-fossil fuel energy, the authors stress that the market-based carbon dioxide reduction mechanism must be combined with an ambitious program to accelerate the development of green technology. In their opinion, such a program would probably have a higher chance of success if some important parts of it were based on international collaboration.

Cai Fang and Du Yang in Chapter 10 attribute China's environmental problems to its growth pattern, which is characterised by: 1) a reliance on large amounts of factor inputs (for example, capital) rather than on productivity; 2) a concentration on those heavy industries (for example, steel) that are energy and pollution intensive; and 3) rapid economic growth driven by strong pro-growth central and local governments. Taking these characteristics into account, Cai and Du make two points about effective ways to reduce pollution. First, if foreign pressure is the sole factor pressing China to deal with its environmental problems, emissions reductions will be small because of the lack of incentives to carry out emissions reduction policies. Second, the political economy of emissions reduction relates to the different reactions of the affected parties, including the central government, local governments, enterprises and households. The key is to make relevant regulations incentive compatible among all stakeholders. For example, different policy packages for different regions in terms of emissions reduction would be needed given the existence of a considerable heterogeneity of sulphur dioxide emissions among different regions.

Cai and Du believe that the effectiveness of any emissions reduction policy lies eventually in the endogenous demands for change in growth patterns and for a cleaner environment because 'there is now a great deal of evidence supporting the view that rising incomes affect environmental quality in a positive way' (Copeland and Taylor 2004:66). They think that anti-emissions policy packages could be incentive compatible with the motivation of local governments for development and the behaviour of enterprises only after China has moved to the stage at which economic growth is generated mainly by productivity growth.

China has been the largest destination for foreign direct investment (FDI) in the world and the environmental consequences of FDI in China have attracted increasing attention, with growing concerns about the deterioration of China's environment. Qun Bao, Yuanyuan Chen and Ligang Song in Chapter 11 find that there is an inverted-U curve relationship between FDI and pollution, suggesting that the continued flow of FDI will eventually reduce pollution emissions. It also implies, however, that FDI inflows into the richer regions will reduce pollution while FDI flows into the poorer (mostly inland) regions will worsen their environment. Bao, Chen and Song propose that the poorer provinces should be treated in the same manner as developing countries in fulfilling their global obligations to reduce emissions. These poorer provinces can be given financial means and technological support to enable them to comply with the toughened governmental emissions regulations. The authors conclude that it is also in the interest of inland regions to reduce emissions even though pollution abatement efforts will involve additional costs because much environmental damage is irreversible.

China is growing food to feed 1.3 billion people and there is increasing concern about whether climate change will attenuate Chinese food security. This issue is particularly acute because the amount of arable land has been declining due to rapid urbanisation and industrialisation. Liangzhi You, Mark W. Rosegrant, Cheng Fang and Stanley Wood in Chapter 12 investigate the climate's contribution to Chinese wheat-yield growth (after controlling for physical inputs). They find that the gradual increase in growing-season temperatures in the past few decades has had a measurable effect on wheat productivity. A 1 per cent increase in wheat growing-season temperature reduces yield by about 0.3 per cent. Overall, the rising temperatures from 1979 to 2000 cut wheat-yield growth by 2.4 per cent. They conclude that this negative impact will probably become worse with accelerating climate change in the future.

Water shortages appear to pose an immediate environmental threat to China's continued high economic growth. The present water situation is already fairly critical because of the uneven distribution of water and lower than normal rainfall in the past 15 years. Right now, about '400 of China's 660 cities face water shortages, with 110 of them severely short' (The Straits Times 2004). Jinxia Wang, Jikun Huang, Scott Rozelle, Qiuqiong Huang and Lijuan Zhang in Chapter 13 report that there is a water crisis in northern China because 70 per cent of the villages they surveyed are facing increasing water shortages. Agricultural production in 16 per cent of these villages is severely constrained by a shortage of water. Their survey also shows that 10 per cent of the villages using ground water in the past decade have seen their water tables falling at an alarming rate of about 1.5 metres annually. The chapter identifies two problems in responses to the looming water crisis. First, the measures adopted by the government, such as better water management, are not being implemented effectively. Second, farmers do not always respond in ways in which they save water. The common cause of the two problems is that governments have not created the incentives for farmers to save water. The authors believe that the institutionalisation of water pricing and water-use rights policies would be more effective and much cheaper than the South to North Water Transfer Project being undertaken by the government.

Rapid economic growth and expansion of urban areas have led to considerable changes in the living environment for an increasing proportion of the population in China, which in turn could greatly influence improvements in public health and socioeconomic development. The Health and Mortality Transition in Shanghai Project Research Team has conducted a detailed study of health and mortality transition in Shanghai from the late 1950s to the beginning of the twenty-first century. Their findings in Chapter 14 show that air pollution in Shanghai has had a notably negative impact on population health and mortality, particularly in the case of sulphur dioxide emissions. The study also finds that the health impacts of air pollution are more observable in cold months when the level of pollution concentration is markedly higher than in warm months. The encouraging sign is that the Shanghai municipal government has made some progress in controlling air pollution in recent years. The concentration of particulate matter reached its peak in 2003 and has been declining since. The level of nitrogen dioxide has also displayed a trend of slow decrease despite a rapid increase in the number of motor vehicles in the city. The level of sulphur dioxide has, however, increased slightly in recent years.

Part III of the book focuses on energy use and the environment. It addresses issues such as: what is the relationship between China's energy consumption and the environment? How much does household energy consumption contribute to carbon dioxide emissions? Can China's coal industry be reconciled with environmental protection? Must China choose between rapid growth and acceptable environmental outcomes?

Kejun Jiang and Xiulian Hu in Chapter 15 detail China's position and policies for dealing with environmental challenges. In the short run, emission-mitigation policies will be implemented, mainly increasing energy efficiency through technological progress and the development of renewable and nuclear energy. In the long run, China's policies will focus increasingly on reducing greenhouse gas emissions (through measures such as a carbon tax) and adapting to climate change. Given China's heavy dependence on the use of coal, it will work closely with other countries to develop a new generation of clean-coal technologies. China will also readjust its economic structure towards producing goods that are less energy and resource intensive.

Much of the Chinese literature on energy consumption focuses on the industrial sector, with household energy consumption a less researched area. Using urban household survey data, Jane Golley, Dominic Meagher and Xin Meng in Chapter 16 find that the total energy requirements of households in China are substantially higher when their indirect energy requirements are added to their direct energy consumption. Poor households are more emissions intensive than better-off households because of their heavy dependence on coal, suggesting that policies aimed at raising the income growth of poor households would also reduce emissions of greenhouse gases. The study identifies a linear relationship between per capita income and energy demand, highlighting the need for the government to promote ways to reduce the emissions intensities of the goods consumed by households.

Coal provides nearly 70 per cent of China's primary energy consumption and is the single most important source of pollution in China. Xunpeng Shi in Chapter 17 reports that it is estimated that 85 per cent of the sulphur dioxide and 60 per cent of the nitrogen oxides emitted into the atmosphere in China come from the combustion of coal. The expectation is that the share of coal in China's primary energy consumption will remain unchanged in the next 20 years. Shi provides some positive answers to the question of whether China's coal industry can be reconciled with environmental protection. There is a declining trend in waste-gas emissions because of the fall of the overall emissions intensity, which has been helped by the application of clean-coal technologies such as coal washing and dust precipitation. Furthermore, increases in energy and carbon prices will accelerate the pace of adopting clean-coal technologies, making it possible for the coal industry to develop together with improvements in the environment.

Despite its low per capita emissions, China has now passed the United States as the largest emitter of carbon dioxide, and these emissions continue to rise rapidly (International Herald Tribune 2008). Peter Sheehan and Fiona Sun in Chapter 18 examine two questions: 1) whether China has to choose between rapid economic growth and stabilising emissions; and 2) whether there are implementation paths within the current approach by the government that could enable China to realise the twin objectives of maintaining economic growth and stabilising emissions within 25–30 years. Their model simulations show that there could be some realistic options in which China can reduce energy use and emissions with continued rapid development. Success in achieving this objective depends on whether China can effectively shift the pattern of economic activity from energy-intensive areas (for example, specific forms of heavy industry) to industry and service sectors that are knowledge intensive and rely less on energy and other resource inputs; and stimulate the adoption of advanced technologies, processes and practices that are energy efficient and more environmentally benign. Achieving such an outcome is in the interests of China and the international community, and it is therefore important for industrialised countries to adopt strong measures to support this policy implementation process in China.

It is therefore encouraging to learn that 'prices have already reached levels at which investments in many alternative energy sources are profitable' (Garnaut and Song 2006:393). 'Seeking new sources of energy through technological innovations and international economic and technological cooperation would…be an important long-term solution to the dilemma of maintaining the sustainability of economic growth while protecting the environment' (Song and Sheng 2007:245).

It is too early, however, to be confident that such optimism is justified. The uncomfortable reality remains for China that unless ecological balance is restored within the medium term, environmental limitations could choke off further economic growth.[1] The uncomfortable reality for the rest of the world is that the negative consequences of large-scale environmental damage within a geographically large country are seldom confined within that country's borders. The continued march of China's desertification initially brought more frequent sand storms to Beijing and then, beginning in April 2001, sent yellow dust clouds not only across the sea to Japan and Korea but across the ocean to the United States. China's environmental management is a concern not only for China's welfare, but for global welfare.

Proper management of the environment has now become critical if China is to continue its industrialisation process. The expurgated version of a 2007 World Bank report said that 'about 750,000 people die prematurely in China each year, mainly from air pollution in large cities' (Financial Times 2007a);[2] and a 2007 OECD study estimated that 'China's air pollution will cause 20 million people a year to fall ill with respiratory diseases' (Financial Times 2007b). Pan Yue, the deputy head of China's State Environmental Protection Agency, summed up the present situation in China very well.

> If we continue on this path of traditional industrial civilization, there is no chance that we will have sustainable development. China's population, resources [and] environment have already reached the limits of their capacity to cope. Sustainable development and new sources of energy are the only road that we can take (Kynge 2004).

The environment is an important area in which China could help to build a harmonious global system. Specifically, China should be mobilising international consensus to form an international research consortium to develop ways to burn coal cleanly because China is now building a power station a week and is hence able to facilitate extensive experimentation on such prototype plants. If successful, this global cooperation on clean-energy research will unleash sustainable development in China as well as in the rest of the world.

Notes

1   The environment is of course not the only serious obstacle to continued high growth in China. Two other serious obstacles that have a high probability of appearing are its outmoded governance structure, which exacerbates social tensions, and its trade friction with the United States and the European Union, which threaten a breakdown of the post-World War II multilateral free-trade system. See Woo (2007) for a discussion of these two challenges to China's growth.

2   Some 350,000 to 400,000 people died prematurely from air pollution in Chinese cities—300,000 from poor air quality indoors and 60,000 (mostly in the countryside) from poor-quality water.

References

Copeland, B.R. and Taylor, M.S., 2004. 'Trade, growth, and the environment', *Journal of Economic Literature*, XLII:7–71.

Financial Times, 2007a. '750,000 a year killed by Chinese pollution', *Financial Times*, 2 July.

——, 2007b. 'OECD highlights Chinese pollution', *Financial Times*, 17 July.

Garnaut, R. and Huang, Y., 2007. 'Mature Chinese growth leads the global Platinum Age', in R. Garnaut and L. Song (eds), *China: linking markets for growth*, Asia Pacific Press and ANU E Press, The Australian National University, Canberra:9–29.

Garnaut, R. and Song, L., 2006. 'Rapid industrialisation and market for energy and minerals: China in the East Asian context', *Frontiers of Economics in China*, 1(3):373–94.

Garnaut, R., 2008. 'Will climate change bring an end to the Platinum Age?', *Asian-Pacific Economic Literature*, 22(1):1–14.

International Herald Tribune, 2008. 'China increases lead as biggest emitter of carbon dioxide', *International Herald Tribune*, 13 June. Available from http://www.iht.com/articles/2008/06/13/business/13emit.php

Kynge, J., 2004. 'Modern China is facing an ecological crisis', *Financial Times*, 26 July.

Song, L. and Sheng, Y., 2007. 'China's demand for energy: a global perspective', in R. Garnaut and L. Song (eds), *China: linking markets for growth*, Asia Pacific Press and ANU E Press, The Australian National University, Canberra:225–47.

The Straits Times, 2004. 'China may be left high and dry', *The Straits Times*, 3 January.

Woo, W.T., 2007. 'The challenges of governance structure, trade disputes and natural environment to China's growth', *Comparative Economic Studies*, 40(4):572–602.

2

Will China fall into stagflation?

Yiping Huang

The new policy challenge

The growing risk of stagflation has perhaps been one of the most important macroeconomic surprises around the world since the beginning of the year. China is not excluded from the risk. According to the National Bureau of Statistics (NBS), real gross domestic product (GDP) growth moderated from 11.9 per cent in the fourth quarter of 2007 to 10.6 per cent in the first quarter of 2008. Between mid 2006 and late 2007, the growth of export volumes fluctuated largely within the 20–30 per cent range. After that, however, export growth tanked (Figure 2.1). The growth of industrial production showed a similar trend, especially when adjustments were made for changed producer prices.

Meanwhile, inflation rates continued to reach new highs. During much of 2005 and 2006, the headline consumer price index (CPI) stayed well below 2 per cent. From the beginning of 2007, however, it began to rise steadily (Figure 2.2). In February 2008, the CPI reached a new 11-year high of 8.7 per cent. Rising inflation was a result initially of spikes in pork prices, but soon it spread to other food markets, with the overall food CPI exceeding the 20 per cent level in early 2008. The non-food CPI has been largely stable so far; however, non-food CPI could be underestimated as the government controls domestic fuel prices. Producer prices accelerated to 8.2 per cent in May 2008 from below 3 per cent a year ago.

Figure 2.1 **Growth of volumes of Chinese exports and imports,**
January 2005–May 2008 (per cent per year/year)

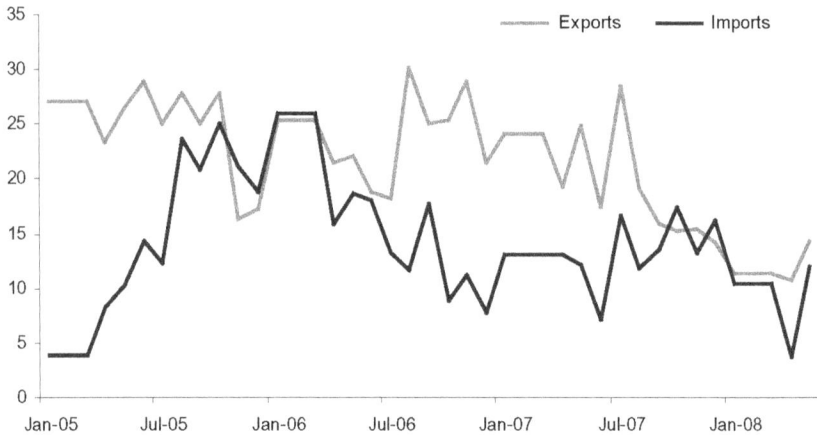

Note: In drawing the chart, I have averaged year-on-year growth rates for January, February and March and smoothed out the effect of the Chinese New Year holiday.
Sources: CEIC Data Company and Citi.

Figure 2.2 **China's headline CPI, food CPI and non-food CPI,**
January 1999–May 2008 (per cent per year/year)

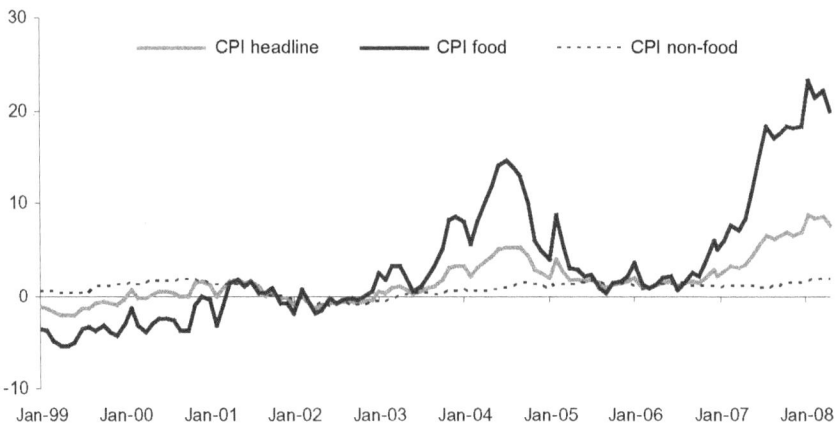

Sources: CEIC Data Company and Citi.

The combination of slowing growth and rising inflation poses serious challenges for Chinese macroeconomic policymaking. Deterioration of the US economy, the continuing surge of oil prices, liberalisation of domestic factor markets, the upcoming Olympic Games and unexpected events such as the recent snowstorm and earthquake further complicate policymakers' decisions.

In fact, the Chinese government's macroeconomic policy stance has already swung notably during the past months. At the beginning of December 2007, the Central Economic Work Conference identified two top policy priorities for 2008: fighting overheating and controlling inflation. As the economic conditions in the United States deteriorated sharply in early 2008, however, the State Council added an additional goal for its macroeconomic policies: preventing a slide of the economy. From May 2008, the environment shifted again. The near-term outlook for the United States improved steadily, with many investors believing that the worst might be over. In the meantime, inflation rates continued to ratchet up. The Chinese authorities again toughened their stance for battling inflation problems.

The heated policy debate has not ended. The 'hawks' advocate aggressive policy tightening, including sharp appreciation of the currency. They argue that inflation, regardless of its composition, is a monetary phenomenon; therefore, only aggressive monetary tightening can tame inflation. The 'doves', on the other hand, warn about the downside risks to the economy and call for some tolerance of higher inflation. They point, in particular, to already declining corporate profits and collapsed equity prices. It follows, then, that continuous aggressive tightening will not only kill the growth engine, it could escalate social and financial risks.

In this chapter, I argue that the scenario of stagflation is unlikely to materialise in China any time soon. Development in three areas could, however, reinforce the challenge of slowing growth and rising inflation in the coming year: deceleration of the US economy, elevated international oil prices and gradual cost normalisation in China. We are likely to see growth shifting to a slightly slower gear—about 10 per cent—while inflation will ratchet up to higher levels, of about 5–7 per cent, in the coming year or two.

In this environment, policy decisions will likely continue to be difficult. With elevated inflation pressure, however, the central bank will probably maintain its tightening bias, with a focus on direct liquidity management. Steady appreciation of the renminbi should also continue. In the meantime, fiscal policy could turn more expansionary to support post-disaster reconstruction and to facilitate industrial upgrading. If an expansionary fiscal policy successfully boosts domestic demand and reduces the external surplus, it could in the end help ease pressures on the currency and inflation.

US growth and the decoupling thesis

Since the outbreak of the sub-prime crisis, US economic activity has slowed steadily. The annualised quarter-on-quarter real GDP growth rate fell from close to 5 per cent during the third quarter of 2007 to below 1 per cent in the next two quarters. Economists are still divided about whether the US economy will fall into deep recession, but most agree that US economic growth is likely to stay well below the trend level for at least several quarters. This should have important implications for the outlook for the Chinese economy in the coming year.

To some economists, slowing growth in the United States but steady expansion of the Chinese economy in 2007 provided convincing evidence of the decoupling thesis. The decoupling proposition suggests that as Asian economies grow and economic interactions between them deepen, the relative importance of the US economy for Asia will decline.

The decoupling thesis is an exaggeration of real economic relations, to say the least. If anything, Asia and China's economic linkages with the United States strengthened, rather than weakened, during the past 10 years. China, in particular, became a much more open economy. The export share of GDP rose from 18.6 per cent in 1997 to 36.1 per cent in 2007, while the share of United States-bound GDP increased from 3.3 per cent to 6.9 per cent in the same period (Figure 2.3).

The fact that Chinese growth did not soften alongside slowing US growth in 2007 was probably more of a special situation than a general rule. First, the slow-down in the United States in 2007 was concentrated in the housing sector; non-housing activities continued to grow at 2.5 per cent. The situation has now changed and the growth slow-down in the United States in 2008 is occurring mainly in the non-housing sector, especially in consumer spending and business investment. Second, softening of United States-bound exports was offset by European Union-bound exports, when the euro strengthened significantly. This will, however, be difficult to repeat; the European economy already shows increasing weakness and the euro is under pressure to weaken, at least in the coming quarters.

It is probably also premature to count on growing intra-regional trade to support Asian economic growth. Although intra-regional trade grew exponentially after China joined the World Trade Organization (WTO) in 2001, the majority of this—at least 70 per cent—was trade in intermediate goods. The key destinations for Asia's finished-goods exports are the G3 economies (the United States, the European Union and Japan, 61 per cent), Asia (21 per cent) and the rest of the world (18 per cent). This implies that Asian domestic

Figure 2.3 **Shares of total exports and United States-bound exports in China's GDP, 1993–2007** (per cent)

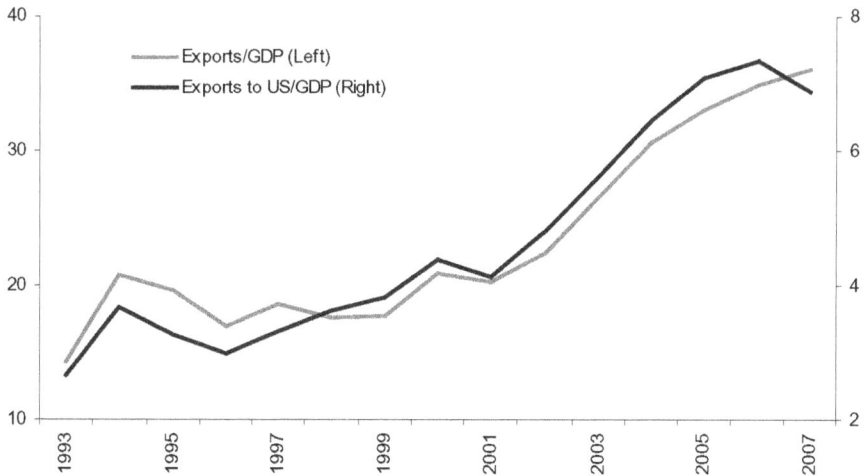

Sources: CEIC Data Company and Citi.

demand is still not significant enough to support regional economic growth, should the US economy slow sharply. Asia and China could become decoupled from the United States and other industrial economies when regional domestic demand is large enough, but that is at least 10 years away.

I examine the likely impacts of a US slow-down on Asian economies through model simulations applying the Oxford Economic Forecasting (OEF) model. The results suggest that a 1 percentage-point slow-down in the US economy could lower Asian economic growth on average by 1.1 percentage points and reduce Chinese growth by 1.3 percentage points (Figure 2.4). The real changes are likely to be smaller as the model cannot endogenously generate a policy response to support growth. In China, for instance, the government will most likely employ fiscal stimuli if external demand weakens significantly.

Another mechanism that could offset growth moderation in China is the likely change in capital flows. In a world faced with slowing US growth and rising US financial risk, capital inflows to China could accelerate, as evidenced by changes during the first two quarters of 2008. Unfortunately, however, China's economic constraint is not a lack of capital; rather, too much capital inflow could further increase domestic liquidity and therefore add further pressure to domestic inflation.

Figure 2.4 **Model simulation: impacts of a 1 per cent slow-down of the US economy** (per cent)

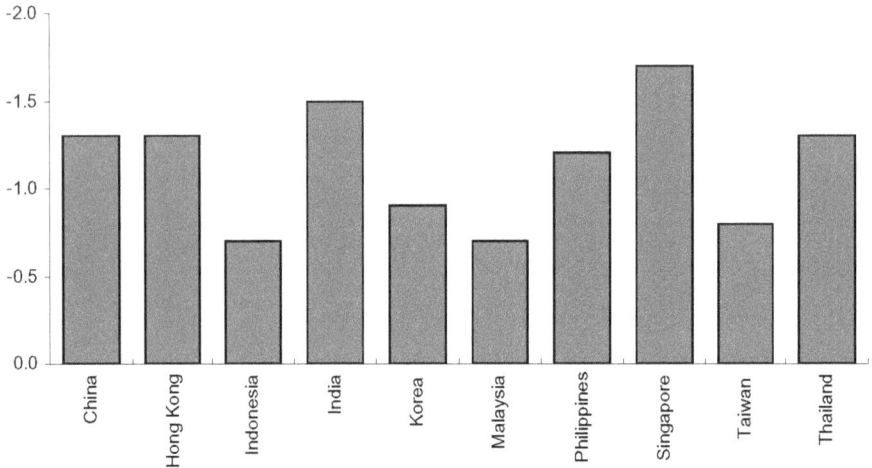

Source: Simulation results applying the Oxford Economic Forecasting model.

Figure 2.5 **Nominal and real oil prices, January 1959 – June 2008** (US$/barrel)

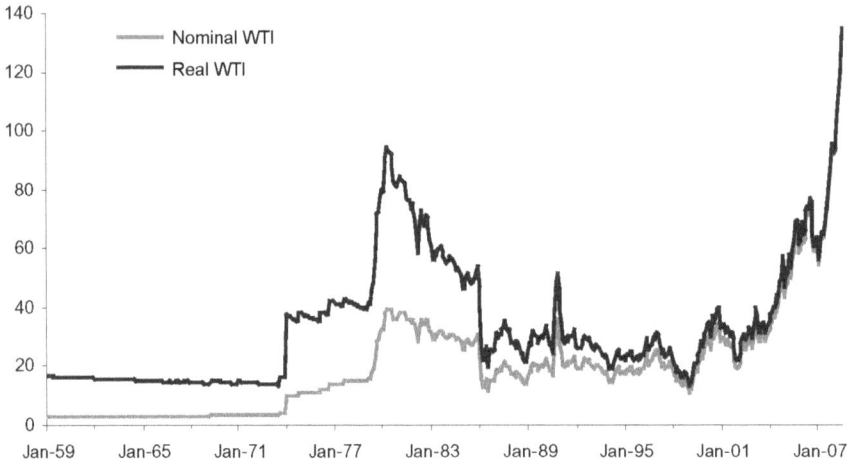

Source: Citi.

The new oil price shock

One of the biggest macroeconomic surprises during the past year was the combination of moderating global economic growth and rising oil prices. The West Texas Intermediate (WTI) oil price rose from less than US$67.50 per barrel in June 2007 to US$135/barrel in June 2008 (Figure 2.5). Even in real terms, the current WTI price already exceeded the previous peak during the 1980 oil crisis.

There are a number of hypotheses for why oil prices increased rapidly although the global economy showed softness. The first hypothesis highlights the growing importance of the emerging market economies in international oil markets. In 2008, for instance, the emerging market economies are likely to account for all net increases in global oil demand. The second hypothesis points to rising marginal costs of oil production. As it becomes more costly to produce oil, prices have to go up regardless of the conditions of demand. A third hypothesis reveals a new contribution from financial investment demand, compared with real economic demand, in an environment with a depreciating US dollar. Financial investment usually accounted for about 16 per cent of global oil demand; today, this share is already 60 per cent.

The first two hypotheses explain why oil prices have been on a long-term upward trend, but they fail to account for the doubling of oil prices during the past year. The third hypothesis—that is, financial demand outweighing physical demand—probably plays a dominant role in the latest episode of oil price increases. If that is the case, however, oil prices could correct in the coming year, especially if the US dollar rebounds steadily. Even if corrections do happen, prices could stay at elevated levels. The good old days of oil prices of $20/barrel are probably long gone. Oil prices averaged $72/barrel in 2007 and $110/barrel during the first six months of 2008.

High oil prices could have significant impacts on the Chinese economy. According to simulation results applying the OEF model, a $10/barrel increase in the oil price could lower China's real GDP growth by 0.4 percentage points and lift its CPI by 0.2 percentage points. The real observed effects, especially the growth effect, were, however, smaller than the model's predictions.

One important reason for the smaller than predicted oil price impacts is the declining share of oil in production over time. Falling oil intensity of GDP is a global trend as technologies improve. In China, efficiency gain through reform efforts further reinforces this trend. During the 30 years of economic reform, China's oil intensity improved by about 70 per cent (Figure 2.6). As a result, current oil expenditure's share of nominal GDP is still significantly lower than the peak in 1980, despite the recent surge in oil prices. This means that GDP growth elasticity of oil prices changes over time, while model predictions are based on historical correlations.

Another important reason for the smaller than expected oil price impacts is the domestic fuel price subsidy; therefore, changes in international oil prices do not necessarily affect domestic prices for end users. This effectively lessens impacts on GDP and inflation. The domestic economy is not, however, entirely insulated. Oil and electricity shortages are a common and growing phenomenon. In 2007, when international oil prices averaged US$72/barrel, China's fiscal subsidy to oil prices was about US$8.2 billion, or equivalent to 0.2 per cent of that year's GDP. The uncompensated losses of oil refiners could be even greater.

China is not the only economy in Asia that subsidises domestic fuel prices. In India, Indonesia and Malaysia, fiscal subsidy to oil prices accounted for 1–1.5 per cent of GDP in 2007. All these countries adjusted domestic prices in May or June in order to reduce the fiscal burden and to improve energy efficiency. The rising oil prices already generated serious consequences for the Chinese economy. Fiscal burdens grew significantly. If oil prices are sustained at the

Figure 2.6 **Share of oil expenditure in GDP and oil intensity of GDP, 1976–2007**

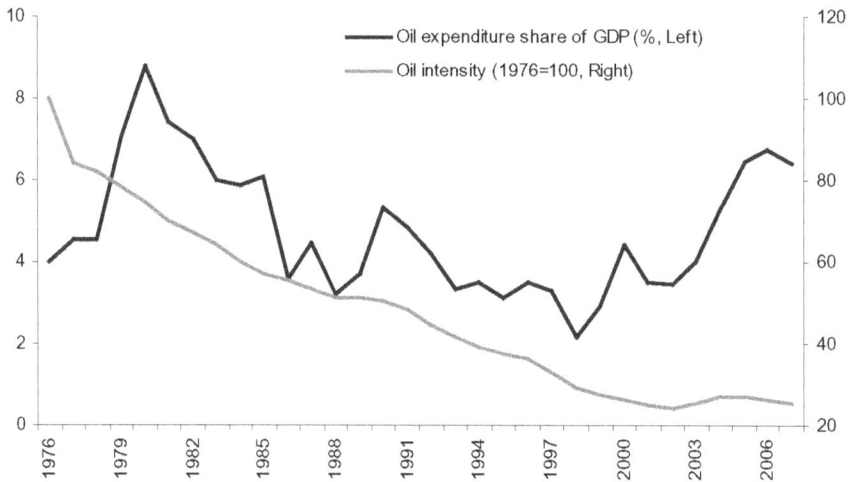

Source: Citi.

US$130/barrel level, the explicit and implicit subsidies could amount to 1-2 per cent of GDP. The government is, however, already stretching its expenditure on disaster relief and reconstruction. More importantly, price controls also cause serious inefficiency problems and distortions to economic structure. Lower energy costs already attract lots of energy-intensive industries to China, such as steel mills and aluminium smelters—industries in which China does not have a comparative advantage. China is therefore really subsidising the global industry. Lack of demand response in China, in turn, added further pressures to international oil markets, especially as China was already a key player. Between 2003 and 2007, China accounted for about one-third of the net increase in global oil demand.

In mid June, the government finally announced to lift gasoline prices by 16.8 per cent, diesel prices by 18.1 per cent and electricity tariff by 4.5 per cent. The upward adjustment of domestic fuel prices will probably add 0.3 percentage point to inflation in the coming months.

These moves were relatively modest compared with the doubling of the international prices during the past months. In particular, as the policy change does not establish new mechanism linking domestic prices to international prices, distortions remain and likely will grow again if international oil prices continue to rise. But they were important first steps in reducing domestic distortions. It's possible that the authorities may adjust domestic oil prices in the coming year.

The process of cost normalisation

Fuel price increases are only a small part of the broad cost adjustment currently under way in China. The average wage, for instance, has been growing at close to 20 per cent year on year during the past year (Figure 2.7). Average land prices also rose by more than 15 per cent. The base lending rate is at 7.47 per cent—nearly 1 percentage point higher than a year ago. The real average bank lending rate is already above 9 per cent. Many companies are already feeling the rising cost pressures and some have been forced to close.

Cost distortion has been one important feature of Chinese economic institutions during the reform period. After 1978, the authorities proceeded to rapidly liberalise the goods market. The factor markets, however, remain highly distorted. For instance, policies on labour mobility are still highly restrictive. Land belongs to the public and the government controls its allocation. State-owned entities dominate the allocation of capital. In other words, there are not yet well-developed free markets for production factors. As a result, prices for labour, capital and land are still significantly depressed. According to my rough

Figure 2.7   **Growth of land prices and average wages in China, 2001–2008** (per cent per year/year)

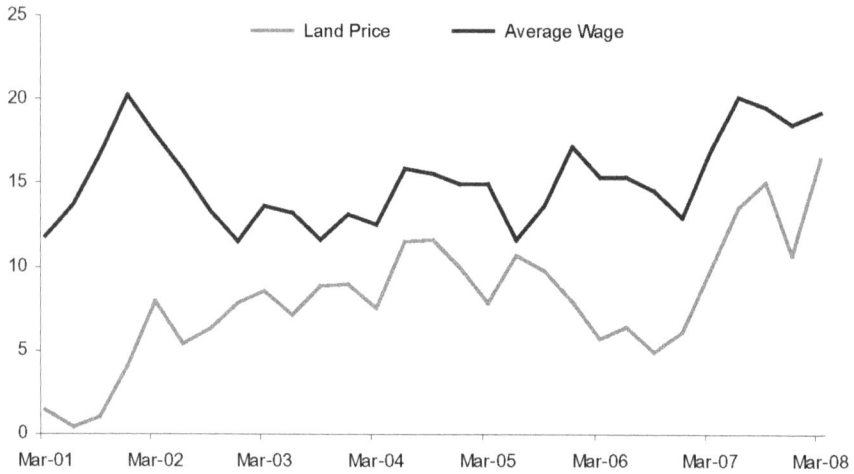

Sources: CEIC Data Company and Citi.

Table 2.1   **Estimation of China's cost distortion** (RMB billion)

Item	Cost	Key assumption
1. Labour	203	Assuming the new labour law is fully enforced
2. Land	154	Assuming 20 per cent of land sales revenue in 2006
3. Energy	1,632	RMB1,572 billion price difference; RMB60 billion in resource taxes
4. Capital	337	Assuming a 2 percentage-point hike in policy rates after financial liberalisation
5. Environment	1,080	Without considering the RMB287.4 billion per annum maintenance fee
6. Others	429	Equivalent to the amount of export tax rebates in 2006
Total	3,835	15.5 per cent of 2007 GDP

Source: Citi.

estimation, cost distortions probably totalled RMB3,835 billion in 2007—or 15.5 per cent of GDP (Table 2.1).

While cost distortions were not all the result of deliberate policy decisions, they were consistent with the policymakers' clear objective of promoting production and economic growth. By depressing opportunity costs for most factors, distortions in factor markets reduce production costs and ensure high profits in production. This supports continuation of unusual investment growth. Distortions in factor markets, therefore, are like subsidies to producers and investors, which fuel extraordinary growth of the Chinese economy.

The broad production subsidy regime also caused a few problems and risks. First, economic growth becomes increasingly investment dependent because of high profits from production and strong incentives for investment. Second, depressed production costs boost exports and attract foreign direct investment. Third, income distribution deteriorates quickly. Owners of enterprises capture vast production profits. In fact, these have developed into serious risks in China, threatening the sustainability of its rapid growth. This suggests that such a regime of cost distortion cannot continue forever.

In fact, important changes have already begun to take place. The recently introduced Labour Contract Law provides better protection of workers' rights, including their job security and social welfare benefits. Labour costs could rise sharply once the new law is rigorously implemented. Energy prices are still controlled by the State, but the government now intends to grant greater roles for the market mechanisms. Capital costs have also started to rise, as a result of the tightening of domestic liquidity and appreciation of the currency. Finally, the authorities have stepped up efforts to protect the environment, responding to the recent sharp deterioration of water, land and air quality.

This is what I describe as cost normalisation. The whole process could take decades to complete, but rises in production costs could again add pressure to growth and inflation.

Policy considerations

Policy outlooks could be very uncertain in an economic environment in which growth slows but inflation rises. Policymakers might become more hawkish when inflation becomes a prominent risk but might turn more dovish when there is significant pressure on growth. We need to have a good understanding of several critical questions in order to gauge the likely policy trend in the coming year.

First, is China's inflation a temporary price adjustment or more structural in nature? Policy actions should be more restrained if it's the former but can be more aggressive if it's the latter. The official documents so far still use the phrase 'structural price increases'. I am sympathetic to the view that this round of inflation was triggered initially by food prices, but almost every major episode of high inflation during the reform period began with rising food prices. With loose monetary conditions, the effects of increases in food prices could spread quickly to other sectors. Tightening of monetary policy should therefore continue in order to control inflation.

Second, is the Chinese economy overheating? Many economists attribute rising inflation to such a problem. Real GDP growth averaged more than 10 per cent during the past five years and reached 11.9 per cent in 2007. High growth does not, however, necessarily equate with overheating. Overheating occurs only when the economy grows at a pace faster than domestic resources can support. China recorded current account surpluses for 14 consecutive years. In 2007, the current account surplus reached 10.8 per cent of GDP. This means that China saved domestic resources for investment overseas; therefore, China's inflation risk probably stems mainly from the liquidity condition instead of the overheating problem.

Third, will the US slow-down drag down the Chinese economy significantly? In the case of a deep and protracted US recession, China's exports could suffer badly. This might not only slow economic growth, it could create serious overcapacity problems in China, which would, in turn, ease China's inflationary pressure. Although the US economy still faces significant uncertainty, its near-term outlook has improved compared with two months ago. This should help limit the downside risks for the Chinese economy.

Finally, will the Olympic Games be an obstacle to the tightening policy? Economists have come to agree that the boom and bust cycle associated with the Olympic Games, often observed in many countries, might not materialise in China. The games could, however, still serve as an important psychological factor for investors and policymakers. While it is unlikely to stop the tightening policies if inflation remains a major macroeconomic risk, policymakers will probably be cautious in determining the timing of such policies. Decisive policies are therefore more likely to be implemented after the games than before them.

China began its current round of tightening policy in early 2007, when the People's Bank of China (PBC) first raised its base policy rates by 27 basis points. In subsequent months, the PBC hiked its policy rates another five times (Figure 2.8). Meanwhile, the central bank also adopted a series of measures to directly manage the liquidity conditions, including reserve requirements, open market operations and credit control.

The focus of the tightening policies probably shifted recently. The PBC has not hiked its policy rate since the beginning of 2008; it has, however, stepped up efforts to control liquidity. By the end of June 2008, the commercial banks' reserve requirement had reached 17.5 per cent. Direct credit control also began to affect commercial banks' loan growth (Figure 2.9). According to the PBC, the real average lending rate rose to more than 9 per cent in the second quarter of 2008, compared with about 8 per cent at the end of the year and the base lending rate of 7.47 per cent.

The effects of the tightening measures on broad liquidity conditions have been less than clear. The simple excess liquidity indicator—subtracting the growth of industrial production from the growth of broad money (M2)—suggests that the liquidity conditions probably loosened again from the beginning of 2008 (Figure 2.9). This probably confirms that the central bank is still lagging behind the market. More importantly, it likely reflects the burdens created by massive capital inflows.

Unfortunately, no one can satisfactorily explain the nature and composition of capital inflows. Between January and April 2008, foreign exchange reserves rose by US$228 billion, while the sum of the trade surplus and utilised foreign direct investment explained only US$93 billion (Figure 2.10). Some economists attribute the gap between the two to 'hot money'. Hot money is, however, probably not the proper term to describe recent increases in capital inflow,

Figure 2.8 **Central bank benchmark lending rate and commercial bank base lending rate, January 1991–May 2008** (per cent)

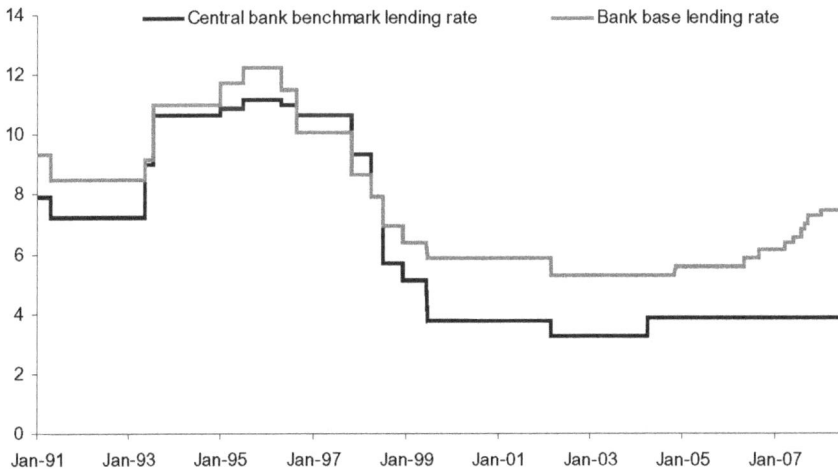

Sources: CEIC Data Company and Citi.

Figure 2.9 **Excess liquidity and bank credit, January 1998–May 2008**
(per cent per year/year)

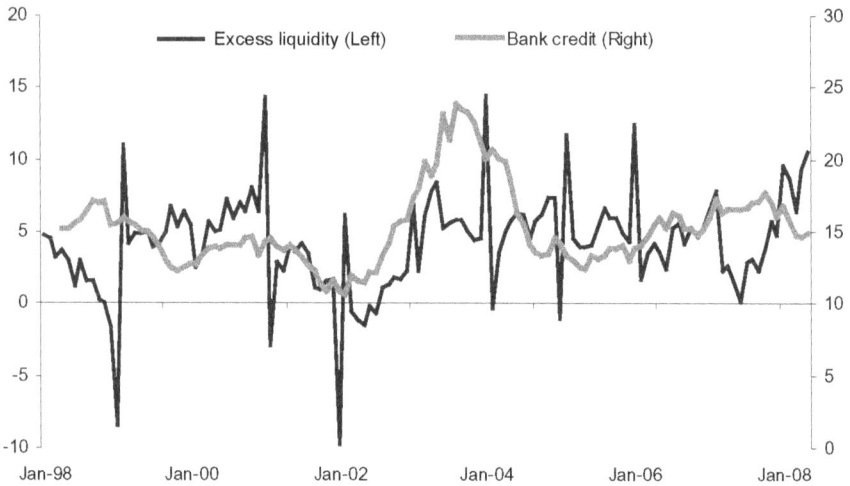

Sources: CEIC Data Company and Citi.

Figure 2.10 **Monthly increase in foreign reserves, the trade surplus and utilised foreign direct investment, January 2007–May 2008**
(US$ billion)

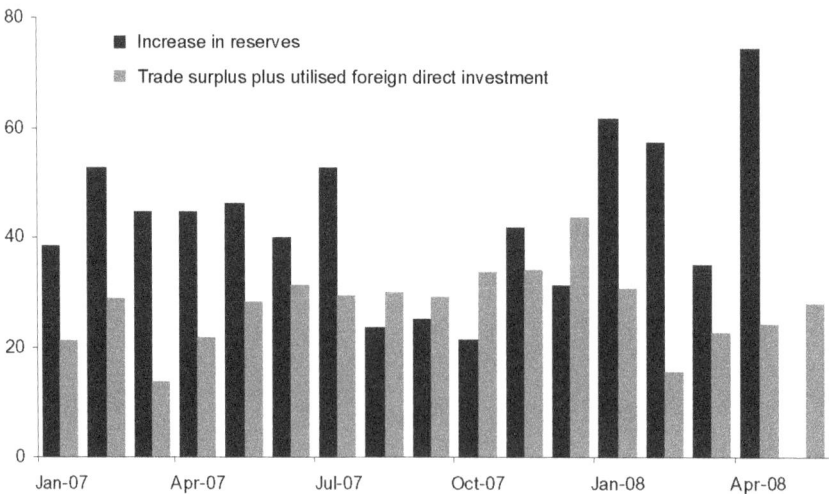

Sources: CEIC Data Company and Citi.

Figure 2.11 **The renminbi's nominal effective exchange rate and bilateral exchange rate against the US dollar, October 2006–June 2008**

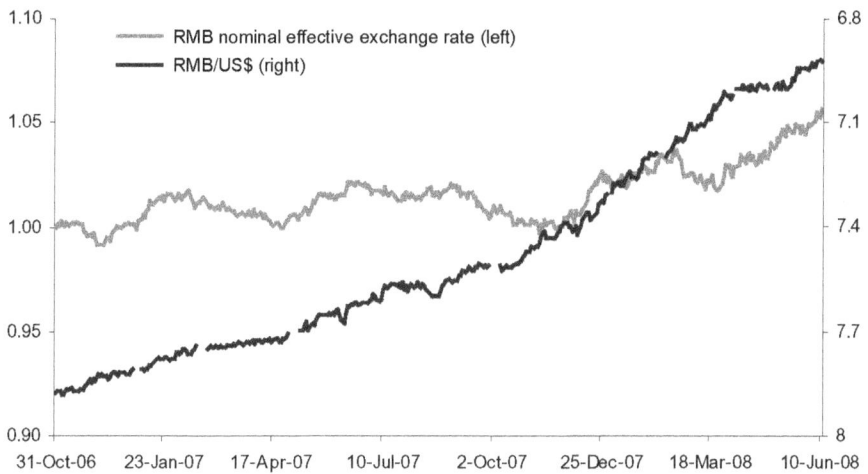

Sources: CEIC Data Company and Citi.

Figure 2.12 **Shanghai A share stock-market: PE ratio and index, January 2001 – May 2008**

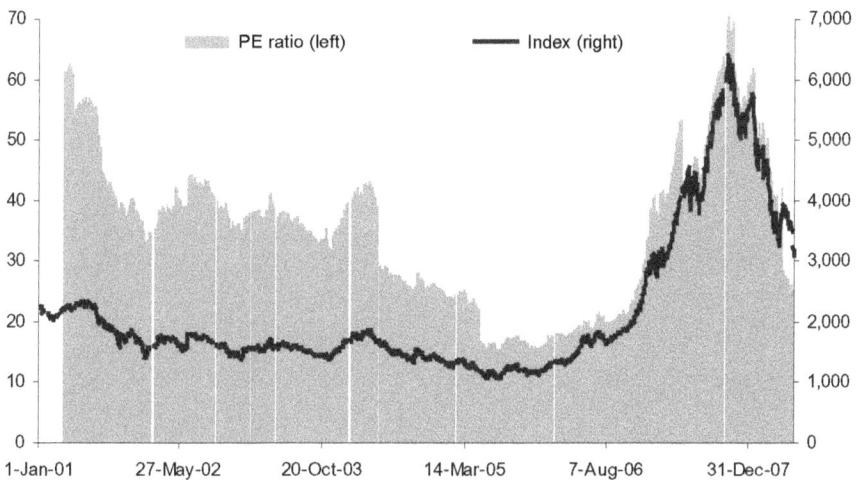

Sources: CEIC Data Company and Citi.

27

as much of it will likely stay within China for relatively long periods. There is, however, no denying that capital inflows surged in recent months.

There were probably many reasons why capital inflows increased dramatically during the past months. One reason could be the financial crisis in the United States. In a volatile international capital market, China becomes a safe heaven. Another reason might be the rapid appreciation of the renminbi, which encouraged expectations of more currency gains in the near future. During the first quarter of 2008, the annualised monthly pace of the renminbi's appreciation against the US dollar averaged 15–20 per cent—the fastest pace since the exchange rate policy reform in July 2005 (Figure 2.11). In April, this pace decelerated sharply to about zero, before picking up again to about 10 per cent in early June.

Concluding remarks

Like most other economies in the world, China currently faces the extraordinary challenge of slowing growth and rising inflation. Some continuing changes, especially the slow-down of the US economy, elevated oil prices and the normalisation of domestic costs, are likely to reinforce this challenge in the coming months. The probability of stagflation, however, remains extremely remote. Real GDP growth will likely moderate slightly; however, barring major surprises, China should be able to comfortably maintain growth of about 10 per cent in the next five years. The earthquake in early May in Sichuan was a human tragedy, but its direct economic cost should be limited.

Inflation rates, however, could move up, although the food-driven CPI should moderate further in the coming months. The low-inflation phase of the past 10 years, supported by relocation of factories and labour and associated productivity gain, is probably over. The anticipated broad-based increases in costs of factors including labour, land, capital, energy and the environment imply that the economy should face inflationary pressures that are stronger than ever. More and more policymakers will also adopt the idea of the need to tolerate high inflation. Given that China is still a developing country and a transitional economy, it is reasonable to expect it to maintain 5 per cent inflation. This could open the door to major inflation problems, as the central bank is always behind the curve.

Tight monetary policy is likely to continue, as long as inflation remains a key risk. The central bank will, however, probably continue to focus on liquidity management. It stopped hiking its policy rates for fear of attracting more capital inflows. Currency appreciation will probably maintain an annual pace of 5–10 per cent in the coming year or two, despite recent dramatic fluctuations in market expectations. At the end of the day, policymakers will have to strike a balance between the need to rebalance the economy and the need for smooth structural adjustments. Job losses remain the top concern for the government. For the same reason, the probability of one-off revaluation or devaluation looks extremely low. If, however, the global financial risks recede quickly in the coming months, the reform of the exchange rate policy and liberalisation of the capital account could accelerate significantly.

Currently, the Chinese authorities employ a combination of tight monetary policy and neutral fiscal policy. While the purpose of tight monetary policy is to control inflation, the intention of neutral fiscal policy is to prevent downside risks to the economy. Fiscal policies could turn more expansionary in the second half of 2008 or in 2009. First, the government will probably need to spend at least RMB250 billion—or 1 per cent of GDP—every year in the next three years for post-earthquake reconstruction. Second, the government needs to play a more active role in facilitating structural adjustment, including helping technological upgrading and re-employment. Finally, active fiscal spending could boost domestic demand, lower the external account surplus, reduce capital inflow and ease inflationary pressure.

There is a misunderstanding among many economists and policymakers that China needs to reduce investment. One important reason for this is the risk of over capacity. While overcapacity risks existed in certain industries at certain times, there is no evidence of an economy-wide overcapacity problem. There is still huge investment potential in areas of urban development, the resources sector and environmental protection. In fact, a large current account surplus means China needs to expand its domestic demand in order to rebalance the economy. China needs less but better quality investment.

Finally, 2008 could be the first year since 2001 in which the Chinese economy sees a major downturn in growth. This could point to significant increases in financial risks. The stock index has already declined by 50 per cent from its peak in late 2007. Housing prices are also becoming less stable. The banking sector has already experienced major reform steps during the past six years, but the new institutions have not had any stress tests. A 1998 East Asian-type financial crisis is unlikely; however, any significant increases in financial risks could still lead to losses of large amounts of financial assets.

3

American and European financial shocks

Implications for Chinese economic performance

Rod Tyers and Iain Bain

China's exports amount to almost half of its gross domestic product (GDP), with most of these directed to Europe and North America, so China can expect that negative financial shocks in these regions might retard its growth. Mitigating factors, however, include the temporary flight of North American and European savings into Chinese investment and some associated real exchange rate realignments. These issues are explored using a dynamic model of the global economy. A rise in North American and European financial intermediation costs is shown to retard neither China's GDP nor its import growth in the short term. Should the Chinese government act to prevent the effects of the investment surge, through tighter inward capital controls or increased reserve accumulation, the associated losses would be compensated for by a trade advantage since its real exchange rate would appreciate less against North America than against the rates of other trading partners. The results therefore suggest that, as long as the financial shocks are restricted to North America and Western Europe, China's growth and the imports on which its trading partners rely are unlikely to be hindered significantly.

During the past decade, reference to China in the financial and academic press has lauded its growth performance but tended to emphasise its exchange rate regime and its controversial current account surpluses with the United States and the European Union (see, for example, Tung and Baker 2004; Bernanke 2006; Lardy 2006; McKinnon 2006; Xiao 2006; Callan 2007; Woo and Xiao 2007; Tyers et al. 2008). After the downturn in the US housing market in 2007, however, and the associated credit squeeze in the United States and Europe, attention shifted to the 'decoupling' issue: whether China's comparatively rapid expansion could be sustained in the face of slow-downs in Organisation for Economic Cooperation and Development (OECD) countries.[1] It appeared that the credit squeeze would bring the oft-anticipated 'hard correction' to the imbalance constituted by the extraordinarily large US current account deficit and that the US dollar would sink, even relative to the renminbi (RMB) (Edwards 2005; Obstfeld and Rogoff 2005; Roubini and Setser 2005; Eichengreen 2006; Krugman 2007). How would this affect China's economic performance?

With exports amounting to almost half of its GDP and most of these directed to Europe and North America, China can expect that negative financial shocks in those regions might retard its growth.[2] Since China assembles manufactured components from elsewhere in Asia and the Pacific,[3] the extent of its 'decoupling' is the key to wider regional performance. In the short term, one mitigating factor is the transitory flight of increased amounts of the world's savings into Chinese investment (McKibbin and Stoeckel 2007b). The Chinese government might, however, oppose this on volatility grounds, via the strengthening of inward capital controls; yet even if the additional financial capital is kept out of China, mitigation remains possible since substantial real exchange rate realignments are likely and these could advantage China in the short term. In the long run, a rising consumption share (Lardy 2006; Kuijs 2006; Kuijs and He 2007; Azziz and Cui 2007) and the redirection of investment within China to its services sector, where considerable potential remains for a productivity catch-up (Ma 2006), will underpin China's growth.

In this chapter, these issues are explored collectively using a dynamic model of the global economy. The model simulates the real effects of shocks that take the form of transitory rises in region-specific interest premiums in North America and Western Europe, combined with increases in investment financing costs in both regions. The key effects of these shocks are for the real net rates of return on North American and Western European investment to fall while the yields demanded by financiers increase. Investment falls in those regions and real wage rigidity ensure that unemployment rises, GDP growth slows and import demand falls in both regions—at least temporarily. The focus of the analysis is, then, on factors influencing China's growth performance in the face of these shocks.

The genesis of the North American and European slow-down

The story of the slow-down is frequently told with a focus on US monetary policy, starting with the succession of monetary expansions by the US Federal Reserve after the stock-market corrections of 2000 and the demand contraction of late 2001 (Figure 3.1).[4] The federal funds rate fell 5 percentage points by 2002 and a further percentage point by 2003–04 (Federal Reserve Board of Governors). So the story goes, this unilateral easing inspired a housing bubble, which burst in 2007, unravelling packaged mortgage investments that had, apparently, been priced in a manner that relied on continued housing price inflation (BIS 2007). Linked with this story is the extraordinary blow-out of the US current account deficit since 2000, via the effects of the housing bubble on the US private saving rate. The growth of private wealth, combined with low borrowing rates, tended to boost consumption during this period, requiring that US investment be financed from foreign, rather than US, savings (Edwards 2005; Eichengreen 2006). It stands to reason, then, that the raising of short-term rates by the Federal Reserve during 2004–06 by at least 4 percentage points (Figure 3.1) would eventually prick the housing bubble and that the US economy would have a hard or soft landing, with either outcome redressing the current account imbalance (Krugman 2007).

Figure 3.1 **US short and long Treasury bond rates, December 1998 – May 2009**

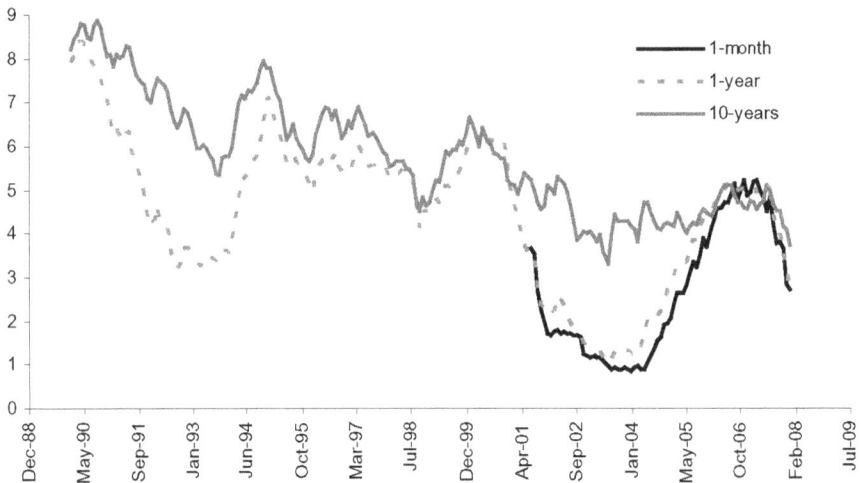

Note: Market yield in per cent per annum on US Treasury securities, quoted on an investment basis.
Source: Federal Reserve Board of Governors. Federal Reserve Board of Governors, Washington DC. Available from wwwfederalreserve.gov..

These linked stories ignore, however, the considerable role of the surge in the growth of the 'emerging economies', and particularly China, since the late 1990s, and the simultaneous yet independent information technology (IT) related boom in US productivity. The growth surge in emerging economies improved the US terms of trade in this period, raising US imports and increasing domestic price competition.[5] Its effect on the US price level can be inferred from the decline in the Chinese bilateral real exchange rate with the United States, shown in Figure 3.2. While US producer prices showed a rising trend during 2000–05, the US dollar prices of an ever-expanding supply of Chinese goods were falling. The deflationary force yielded was bolstered by the IT-related US productivity boom, which began in the early 1990s and continued through to 2006 (Table 3.1).[6] Along with the negative shocks associated with the US stock-market correction and the 11 September 2001 terrorist attacks, the collective deflationary force was considerable, justifying the observed monetary easing on inflation targeting grounds alone. Even with this monetary expansion, a temporary deflationary effect is seen clearly in Figure 3.2 from the decline in the US producer price during 2001. Of course, the coincidence of US asset price inflation with product price deflation was bound to create a crisis of priority in 2000–03. With asset price targeting always controversial, it is not surprising that the Federal Reserve gave priority to the control of product price deflation, thus keeping annual consumer price index (CPI) changes in the positive range (Figure 3.3).

Later in the period, the transitional economies' growth surge caused a global commodity price boom, with oil prices reaching unprecedented highs (Figure 3.4), followed not long after by price spikes in other commodities (Figure 3.5).[7] This tended to reverse the product price deflationary pressure in the United States and to justify the restoration of the US federal fund rate to more normal levels during 2004–06 (Figure 3.1). At the same time, however, it exacerbated US asset price inflation as oil-exporting countries joined the other transitional economies in building up US dollar-denominated reserve assets.[8] The financial contraction in 2007 was therefore a consequence of more complex forces to which the relative expansion of the Chinese economy was a contributor. Nonetheless, the contraction originated in the United States and spread to varying degrees to other OECD countries and, particularly, to Western Europe. While Figure 3.1 shows that the easing by the Federal Reserve during 2007–08, in response to the credit contraction, reduced long and short government borrowing rates, anecdotal evidence confirms that refinancing rates for private firms increased substantially as risk was repriced (see, for example, Browning and Silver 2008). Our purpose is to examine the direct and

Figure 3.2   **The Mainland China–US real exchange rate since 1995 on producer prices**

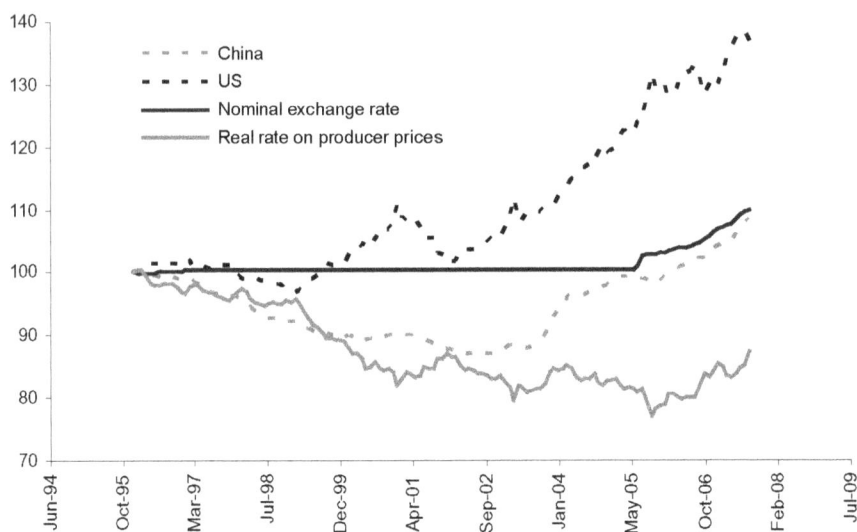

Note: Here the home prices are, for the United States, the Producer Price Index and, for China, the Corporate Goods Price Index. The Chinese index has more coverage of commodities and services than the US one, so this is a less than perfect comparison. The nominal exchange rate is in red, expressed as US$/RMB, so that nominal appreciations are upward movements. The implied real exchange rate is in black, expressed as the value of the Chinese product bundle in terms of the corresponding US bundle. A Chinese real appreciation is therefore an upward movement.
Sources: International Monetary Fund (IMF), 2007. *International Financial Statistics*, April, International Monetary Fund, Washington, DC. Available from www.imfstatistics.org; National Bureau of Statistics (NBS), 2007. *China Statistical Abstract 2006*, China Statistics Press, Beijing.

Table 3.1   **Growth of US labour productivity in the non-farm business sector**

Period	Per cent per annum
1973–95	1.47
1995–2000	2.51
2000–06	2.86

Source: Oliner, S.D., Sichel, D.E. and Stiroh, K.J., 2008. *Explaining a productive decade*, Staff Working Paper 2008-01, Finance and Economics Discussion Series, Federal Reserve Board, Washington, DC.

Figure 3.3   **Rate of US CPI inflation, 1990–2008**

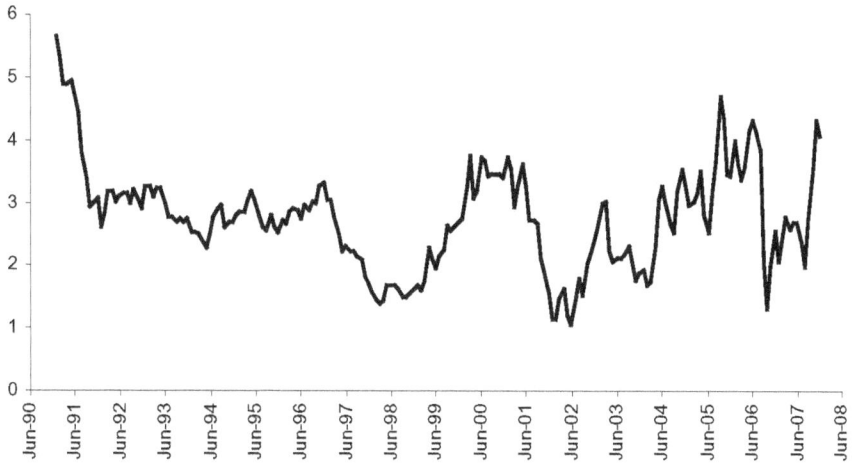

Note: Monthly data percentage change in the CPI during the previous 12 months.
Source: Federal Reserve Board of Governors. Federal Reserve Board of Governors,
Washington DC. Available from wwwfederalreserve.gov.

Figure 3.4   **The average traded price of crude petroleum, 1990–2008**

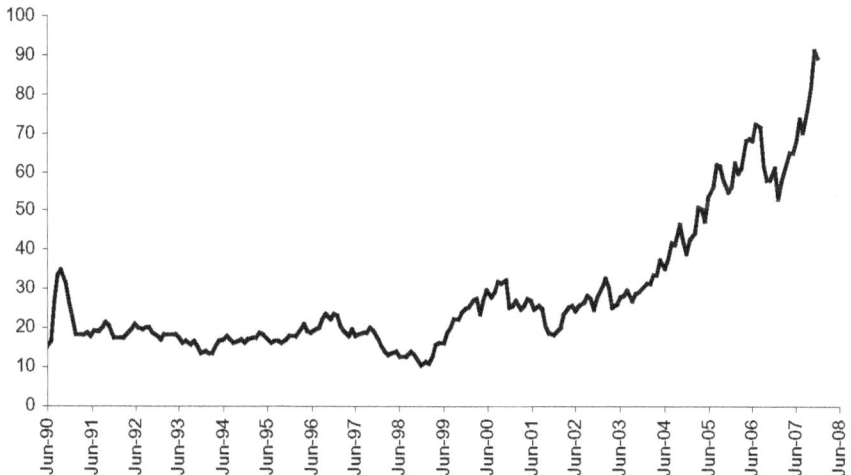

Note: Monthly data, average traded price in US$/barrel.
Source: International Monetary Fund (IMF), 2007. *International Financial Statistics*, April,
International Monetary Fund, Washington, DC. Available from www.imfstatistics.org.

Figure 3.5 **Other commodity price shocks, 1993–2009**

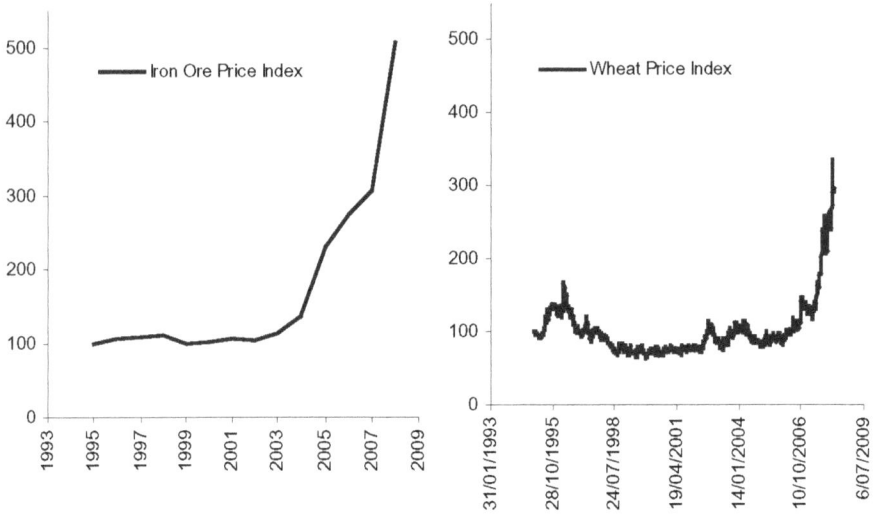

Sources: For wheat, Chicago Board of Trade daily wheat price in US$/bushel, from the Bloomberg Database; iron ore, Hamersley fines, quoted in US cents/dry iron units (dmtu). If the ore shipped is 62 per cent iron (the typical Hamersley grade), the price per tonne of ore is the dmtu price (for 2007, that would be US$0.82) multiplied by 62, which means US$50.84 per tonne, from the IRL Database.

indirect effects of this contraction on China, and thereby on those countries dependent on trade with China. To do this, a dynamic numerical model of the global economy is required.

The model

We use a multi-region, multi-product dynamic simulation model of the world economy, which is an adaptation of the model constructed by Tyers and Shi (2007) and extended for macroeconomic applications by Tyers and Bain (2007) and Tyers and Golley (2008a, 2008b).[9] Only real shocks and their effects are represented.[10] In the version used, the world is subdivided into 14 regions (Table 3.2). Industries are aggregated into three sectors: agriculture (including processed foods), industry (mining, energy and manufacturing) and services (including construction)—the last being little traded in comparison with the other two. Failures of the law of one price for traded goods are represented

36

by product differentiation, so that consumers substitute imperfectly between products from different regions. There are two endogenous sources of simulated economic growth—namely, physical capital accumulation and the transformation of labour from unskilled to skilled. Technical change is introduced in the form of exogenous productivity growth that is sector and factor specific, allowing productivity performance to differ between factors and between tradable and non-tradable sectors.[11]

Regional capital accounts are open so that regional households hold portfolios of assets that are claims over home and foreign capital. Investors in region i have adaptive expectations about real net rates of return, r_i^c, the determinants of which might be summarised simply as

---

Table 3.2 **Regional composition in the global model**

Region	Composition of aggregates
Australia	
North America	Canada, Mexico and the United States
Western Europe	European Union, including Switzerland and Scandinavia but excluding the Czech Republic, Hungary and Poland
Central Europe and the former Soviet Union	Central Europe includes the Czech Republic, Hungary and Poland
Japan	
China	Includes Hong Kong and Taiwan
Indonesia	
Other East Asia	Republic of Korea, Malaysia, the Philippines, Singapore, Thailand and Vietnam
India	
Other South Asia	Bangladesh, Bhutan, Maldives, Nepal, Pakistan and Sri Lanka
South America	Argentina, Bolivia, Brazil, Chile, Colombia, Ecuador, Peru, Venezuela and Uruguay
Middle East and North Africa	Includes Morocco through to the Islamic Republic of Iran
Sub-Saharan Africa	The rest of Africa
Rest of World	Includes the rest of Central America, the rest of Indo-China, the small island states of the Pacific, Atlantic and Indian Oceans and the Mediterranean Sea, Myanmar and Mongolia, New Zealand and the former Yugoslavia

---

Source: The *GTAP 5 Global Database*; Dimaranan, B.V. and McDougall, R.A., 2002. *Global Trade, Assistance and Production: the GTAP 5 Database*, May, Center for Global Trade Analysis, Purdue University, Lafayette.

$$r_i^c = \frac{P_i^Y MP_i^K}{P_i^K} - \delta_i \tag{1}$$

in which P_i^Y is the region's GDP price, P_i^K is the price of capital goods (a separate industry defined in the model) and δ_i is the depreciation rate. Given this rate of return, the determination of investment in each region is complex, but for our purpose it can be characterised simply as follows.[12] It is driven positively by the real net rate of return, r_i^c, and negatively by the rate that must be returned to savers, or the financing cost, r_i. Therefore

$$I_i = I\left(r_i^c, r_i\right) \tag{2}$$

To arrive at r_i, a global interest rate, r^w, is first defined such that $r_i = r^w + \pi_i$, in which π_i is a usually exogenous regional interest premium, which captures the effects of capital controls on the one hand (market segmentation) and differential regional risk on the other. The global rate, r^w, and indirectly r_i, is then derived to clear the global capital market.

$$\sum_i S_i\left(Y_i, r_i\right) = \sum_i I_i\left(r_i^c, r_i\right) \tag{3}$$

in which Y_i is regional income.

Lagged adjustment processes embedded in the $I(r_i^c, r_i)$ ensure that financial capital is not sufficiently mobile internationally to equate r_i^c and r_i in the short term, but that their paths converge in the long term unless exogenous shocks prevent this. General financial reform is represented by a diminution of the interest premium, π_i, which in China's case tends to raise its share of global funds for investment through time.[13] China's average saving rate is high initially, declining through time as its population ages. The baseline simulation therefore maintains a Chinese current account surplus that diminishes after 2010.

A demographic component of the model tracks populations in four age groups, both genders and two skill categories: a total of 16 population groups in each of the 14 regions. The skill subdivision is between production labour (unskilled) and professional labour (skilled).[14] Each age–gender–skill group is represented as a homogeneous sub-population with a group-specific birth and death rate, labour force participation rate and rates of immigration and emigration. Because the non-traded services sector is relatively skill intensive in all regions, trends in skill composition prove to be particularly important for the alignment of real exchange rates. These depend on the rate at which each region's education and social development institutions transform unskilled (production-worker) families into skilled (professional-worker) families. Each year, a particular proportion of the population in production-worker age–gender groups is transferred to professional status. The initial values of these proportions depend on the regions' levels of development, the associated capacities of their education systems and the relative sizes of their production and professional labour forces. Rates of transformation change through time in response to corresponding changes in real per capita income and the skilled wage premium.[15]

The 16 age–gender–skill groups differ in their shares of regional disposable income, consumption preferences, saving rates and labour force participation behaviour. While the consumption–savings choice is parameterised differently between groups, it is dependent for all on group-specific real per capita disposable income and the regional real lending rate. Governments are assumed to balance their budgets while saving and borrowing are undertaken by the private sector. The baseline scenario is a 'business-as-usual' projection of the global economy to 2030, with 1997 as the base year. For validation experiments through 1997–2006, see Tyers and Golley (2008a).

Simulating the North American and Western European slow-down

We compare a baseline business-as-usual simulation to 2030, in which the Chinese economy continues to grow strongly,[16] with one in which a financial contraction retards performance in North America and the European Union. We focus in this section on the characterisation of the downturn in those regions. The analysis is in no way a forecasting exercise; rather, it is to establish a representative pathway for the international economy on which we can superimpose some alternative Chinese policy responses. Recalling that investors in our model are represented as having adaptive expectations, we have not attempted to use it to construct a precise repetition of the events leading up to the US housing bubble and the bubble itself, since the latter arose from ill-formed expectations about future market performance, at least by some investors. Rather, we impose exogenous shocks that combine to represent the real effects of the resulting credit squeeze.

The shocks we use are all transitory, peaking in 2008 with a recovery in the subsequent five years. They apply to the two regions, 'North America' and 'Western Europe', and are weaker in the latter. The first is a rise in the investment interest premiums, π_i, over other regions of the world. This reflects the recent increase in gross returns required by investors in these regions to compensate for perceived increases in risk. The effect of this shock is to raise the financing cost of investment in North America and Western Europe. Second, the productivity of investment in these regions is reduced through shocks to the technologies used in their capital-goods sectors. This is an indirect means of reflecting the recent declines in rates of return on installed capital, r_i^c, in both regions.[17] In effect, this serves to widen the intermediation wedge between marginal investor earnings and financing costs.

Since the pathway to be simulated has only to be 'representative', and since clear data on OECD investment credit costs through 2007 are not yet available, the scale of these shocks is arbitrary. For North America, we raise the investment interest premium by 2 percentage points and capital-goods productivity is reduced by 5 per cent. The corresponding shocks for Western Europe are half the size of those for North America. All shocked variables then return to baseline benchmarks linearly during a recovery period of five years. The effects these shocks have on investment financing rates on the one hand and real rates of return on the other are illustrated in Figures 3.6 and 3.7, which show percentage departures from a baseline simulation in which all regions grow smoothly. Note that the global capital market clearing interest rate, r^w, reduces by 0.9 percentage points because of the contraction in investment demand in North America and Europe. The short-run effect on the domestic financing rate in North America is therefore a rise of 1.1 percentage points, while the corresponding net rise in Western Europe is just 0.1 percentage points.

The wedge between financing costs and the real rate of return is clear from the figures, which also show that the model's investment dynamics lead to some overshooting of rates late in the recovery period for North America and Western Europe.[18] This occurs because the shocks curtail North American and European investment sharply in 2008 but raise investment in other regions, as shown in Figure 3.8. During the recovery, however, investment in North America and Western Europe expands quickly towards their benchmark levels. The 2008 collapse in North American and European investment, however, leaves these regions with capital stocks below baseline levels for many years, the pace of their recoveries notwithstanding. For this reason, real rates of return in these regions rise above baseline levels on the point of recovery and for some years beyond. For Western Europe, the initial shocks are smaller and their effects are muted by the larger North American shocks, which tend to lower financing costs for the rest of the world, including Western Europe.

Figure 3.6   **Simulated effects of the financial contraction on real rates of return and investment financing rates, North America, 2000–2035**

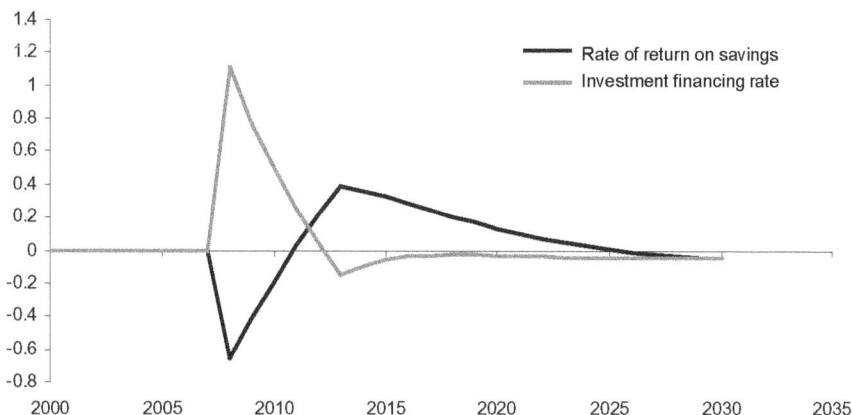

Note: Percentage point departures from the baseline simulation.
Source: Simulations of the model described in the text.

Figure 3.7   **Simulated effects of the financial contraction on real rates of return and investment financing rates, Western Europe, 2000–2035**

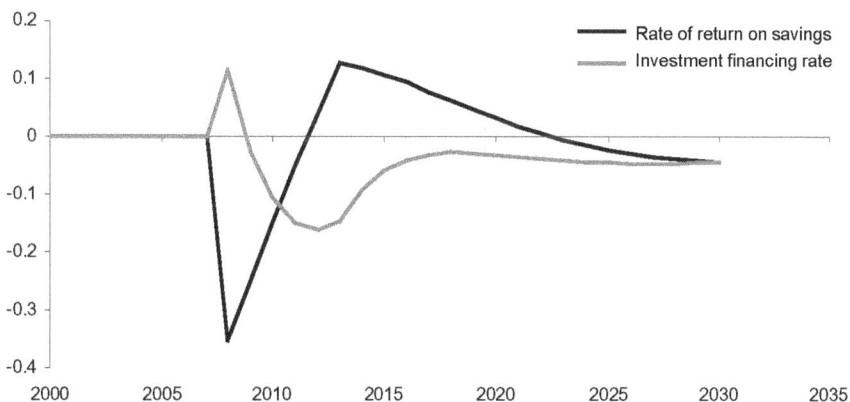

Note: Percentage point departures from the baseline simulation.
Source: Simulations of the model described in the text.

Figure 3.8   **Simulated effects of the financial contraction on investment, 2000–2035**

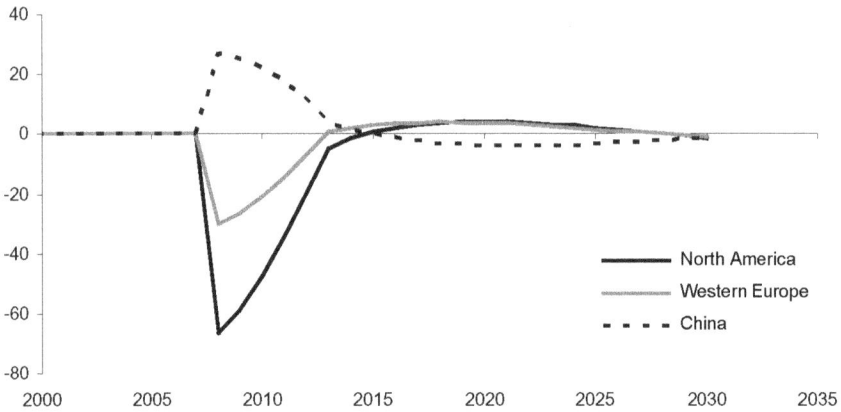

Note: Percentage point departures from the baseline simulation.
Source: Simulations of the model described in the text.

Effects on economic performance

The effects on economic activity in North America, Western Europe and China are indicated in Figure 3.9, which shows percentage departures from baseline GDP levels. The loss of output in North America and Western Europe is quite significant, though it must be clear that the chart measures the extent of their falling behind the baseline. There is no full year of negative growth in either North America or Western Europe, just a slow-down in both.[19] The extent of their falling behind is made larger by the adoption of labour market closures in both shocked regions that maintain the path of real production wages at the baseline level and so cause unemployment. At its peak, 6 per cent of North America's and 2 per cent of Europe's production labour force are rendered unemployed.[20] Due to the flight of investment from North America and Western Europe, indicated in Figure 3.8, there is a surge in investment in China, leading to yet higher growth there in the short run. Since China already invests almost half of its GDP, its capacity to absorb these additional funds could be questioned. For this reason, we explore alternative Chinese scenarios in the next section.

Turning to effects on balances of payment, the North American current account deficit is large by industrial-country standards, although the baseline

Figure 3.9   **Simulated effects of the financial contraction on GDP, 2000–2035**

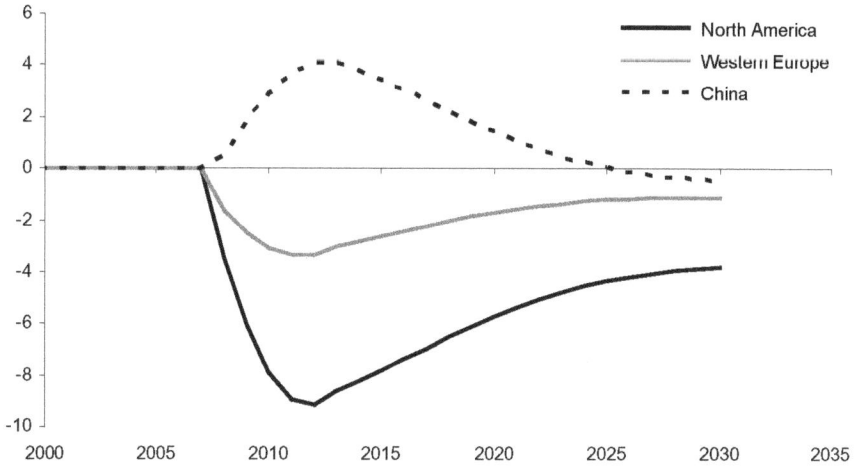

Note: Percentage point departures from the baseline simulation.
Source: Simulations of the model described in the text.

Figure 3.10   **Simulated effects of the financial contraction on regional imports, 2000–2035**

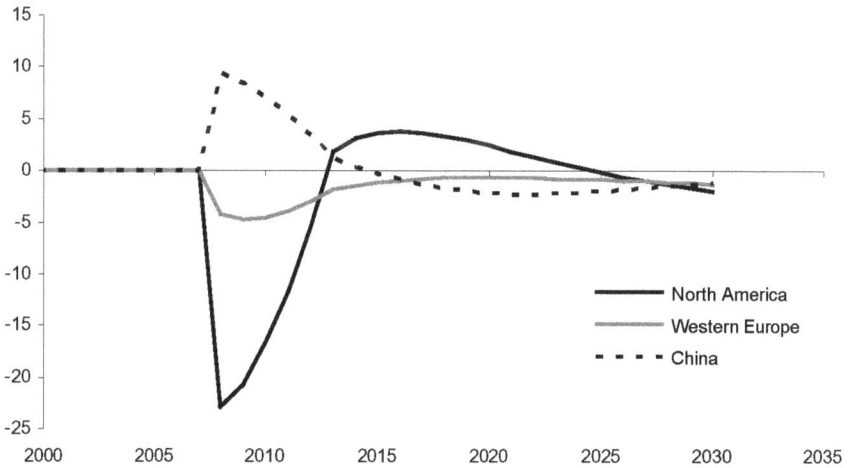

Note: Import volume indices, percentage departures from the baseline.
Source: Simulations of the model described in the text.

scenario has it following a declining trend.[21] Western Europe as a whole exhibits a current account surplus that is smaller in magnitude than the North American deficit, while China's current account surplus is of extraordinary magnitude, exceeding one-tenth of its GDP, driven by its very high saving rate and the resulting surplus of saving over investment.[22] In the baseline projection, this surplus is projected to decline as China's average saving rate falls due to the ageing of its population. After the financial shocks in North America and Western Europe and the resulting flight of investment from those regions, there are very large changes in the balances of payment in these regions. As simulated at least, the US current account deficit is temporarily reversed.

The surge in China's investment after the flight from North America and Western Europe greatly diminishes its current account surplus—however, again temporarily. This change in financial capital movements cushions the effects on China of declining imports in North America and Western Europe. While China's exports take a hit, the closing financial capital imbalance and the associated boost in China's GDP (Figure 3.9) ensures that import growth is more than sustained (Figure 3.10). For China's other trading partners, therefore, and particularly for those supplying it with raw materials and manufacturing components, the loss of markets in North America and Europe is cushioned by a Chinese market that is expanding rather than contracting.[23]

The short-run substitution of investment for exports in contributing to China's GDP requires some difficult structural adjustment in response to relative price changes. The investment generates a surge in domestic demand that raises the prices of Chinese products and services relative to those in North America. It therefore appreciates China's real exchange rate relative to North America.[24] The simulated extent of this is shown in Figure 3.11 to be 10 per cent during 2008. If monetary policy is to emphasise the control of inflation in China, this foreshadows a further 10 per cent appreciation of the renminbi relative to the US dollar in just one year.[25] Otherwise, the rate of inflation must be allowed to accelerate. After four years, however, the path of China's real exchange rate against North America falls below the baseline path. This is because capital accumulation accelerates in China in the early years after the shock and decelerates in North America. Capital costs, and therefore prices, are lower in China in the long run relative to North America. The same pattern is followed by China's real effective exchange rate, except that in the short run China depreciates against some of its other trading partners, so that its short-run real effective appreciation is only small, as shown in Figure 3.12.

Figure 3.11 **Simulated effects of the financial contraction on bilateral real exchange rates relative to North America, 2000–2035**

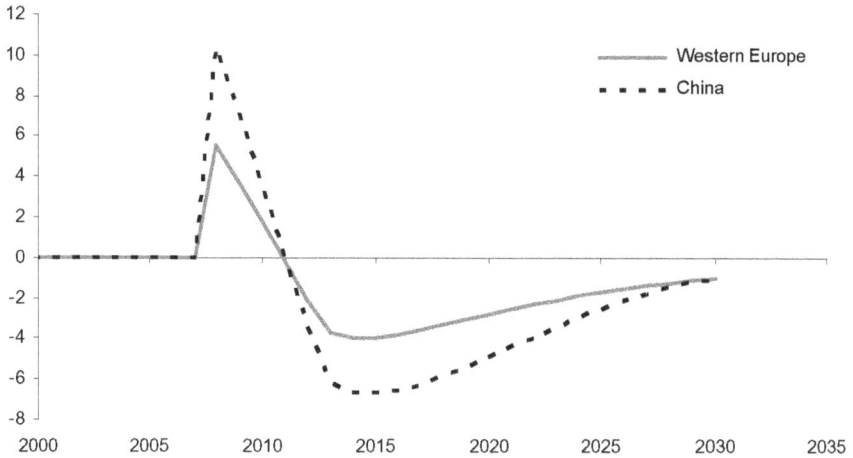

Note: Ratios of the GDP prices of each region with that of North America, percentage departures from the baseline.
Source: Simulations of the model described in the text.

Figure 3.12 **Simulated effects of the financial contraction on real effective exchange rates, 2000–2035**

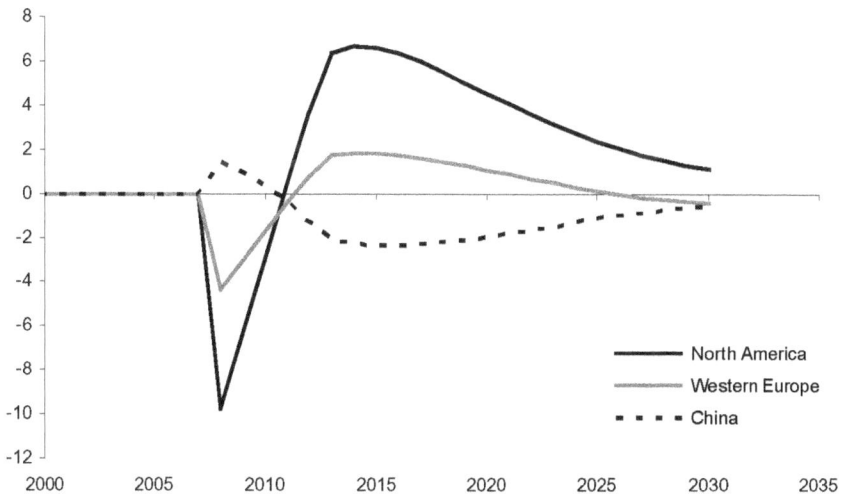

Note: Ratios of the GDP prices of each region with those of all trading partners, weighted by trade shares, percentage departures from the baseline.
Source: Simulations of the model described in the text.

Alternative Chinese policy scenarios

Our standard financial shock scenario might be thought to be optimistic from the perspective of the Chinese economy and its regional trading partners. This is for two reasons. First, it assumes that China is available to absorb additional investment, when it is already investing 45 per cent of its GDP annually and it is arguably already capital heavy.[26] Moreover, the Chinese government might regard the surge of financial inflows as footloose and therefore risky. Second, as seen in the previous section, the surge of funds leaving North America and Western Europe must inevitably drive up China's real exchange rate, causing either faster inflation or more rapid nominal appreciation. Neither of these developments will be palatable to the Chinese government. It might therefore choose a policy response that either retards the inflow of new investment (tighter controls on incoming financial capital) or matches the inflow with increased outflows in the form of reserve accumulation, for which the saving rate would need to be, albeit temporarily, further increased. These alternative policy responses foreshadow two new scenarios to be modelled.

Tighter capital controls that prevent the investment surge

China already maintains effective controls over the inflow and the outflow of financial capital (see Ma and McCauley 2007). Legal inflows are primarily foreign direct investment (FDI), but they include some purchases of domestic assets, including 'B' shares on the Shanghai and Shenzhen exchanges. Illegal inflows have evidently increased in recent years as yields have risen in China relative to the United States and as the renminbi has been allowed to appreciate against the US dollar.[27] Illegal inflows notwithstanding, outflows of financial capital are substantially larger in China than inflows and they have mainly taken the form of official foreign reserve accumulation. The surplus of China's saving over its investment is, by definition, equivalent to the surplus of its exports, generally defined, over its imports. Denominated in foreign currencies, this surplus ends up in the hands of the People's Bank of China (PBC), since outward capital controls do not permit substantial foreign asset holdings by private individuals. In recent years, there has been some relaxation of controls in both directions but the PBC still finds it necessary to acquire foreign reserves in very large volumes each year. The recent surge of illegal inflows has, however, tended to restrain the magnitude of China's net capital account position, appreciating the real exchange rate. The result has been both accelerated inflation and upward flexibility of the renminbi during 2006–07 (Tyers and Bain 2007).

Here we assume that the Chinese government opposes the inflow of additional financial capital on the grounds that it is volatile and therefore risky and that it accelerates inflation.[28] The policy response is to tighten its inward capital controls so as to prevent any surge—maintaining the baseline path of China's capital account flows. Compared with the original (reference) financial shock scenario, this causes the domestic real interest rate to rise by almost 1 percentage point, though it eventually rejoins the baseline path. The simulated consequences of this are summarised for China in Figure 3.13; most important among them are the absence of a significant surge in Chinese investment (5 per cent compared with the 27 per cent indicated in Figure 3.8) and hence a much reduced increase in China's GDP (which peaks at about 1 per cent, compared with the 4 per cent in Figure 3.9) and imports (which peak at 1 per cent but fall below the baseline path thereafter, compared with the peak of 10 per cent in Figure 3.10).

Gross national product falls temporarily, mostly because China holds a substantial stock of foreign assets, the rate of return on which falls, as indicated in Figures 3.6 and 3.7.[29] This fall aside, in this experiment, we deny the Chinese economy the positive aspect of the shock: the increased investment. The prevailing logic would suggest that there would therefore be no compensation for the loss of exports to North America and Western Europe and hence that China's growth rate would fall measurably. The simulation shows, however, little impairment of China's economic performance. Nor is there any significant reduction in China's imports, which support many of its neighbouring economies. The resolution of this puzzle requires a return to the effects on the global capital market. There is a flight of saving from North America and Western Europe. In this experiment, it cannot go to China, so it raises investment in other (mostly developing) regions. In the short run, these investment surges cause real appreciations relative to North America and Western Europe,[30] and, significantly, since China does not participate, real appreciations against China.

Australia, for example—a key supplier of raw materials to China—suffers a short-run real appreciation of 15 per cent against North America and 9 per cent against China. As shown in Figure 3.14, even though China has a real appreciation against North America in the short run, its real effective exchange rate depreciates. This means that China becomes more competitive in other markets as a consequence of the financial shocks and that this is sufficient to allow it to weather the contraction in North American and Western European imports.

Figure 3.13 **Simulated effects of the financial contraction with tighter inward capital controls on Chinese GNP, GDP, investment and imports, 2000–2035**

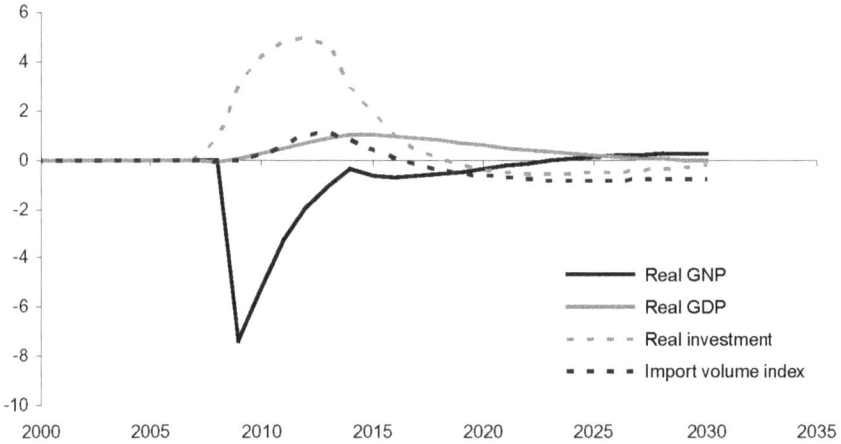

Note: Percentage departures of volume indices from the baseline.
Source: Simulations of the model described in the text.

Figure 3.14 **Simulated effects of the financial contraction with tighter inward capital controls on Chinese real exchange rates, 2000–2035**

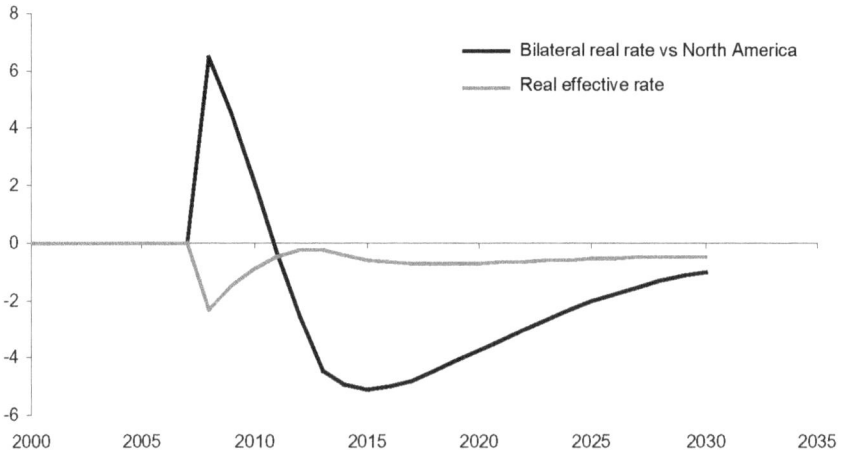

Note: Percentage departures from the baseline.
Source: Simulations of the model described in the text.

Overall, then, apart from a dip in foreign-sourced income, this scenario is neutral from China's standpoint. It maintains China's baseline growth path while insulating against any volatility that would stem from the temporary influx of global saving. Also, from the viewpoint of China's neighbours supplying it with raw materials and components, this scenario is neutral, with no disturbance to the path of China's imports. Its downside risk lies in the comparative tightness that is required in its domestic capital market. Aside from the problems of overcoming the resulting increased incentive for illegal financial inflows, this could place at increased risk debt-financed investments within China and therefore raise the potential for the global financial meltdown to migrate there.

Accelerated reserve accumulation (increased saving)

In this scenario, the Chinese government's inward capital controls are assumed to be ineffective in preventing the investment surge. The government is, however, able to raise the overall saving rate sufficiently to offset any net effect on the balance of payments. One possible mechanism could be through tighter fiscal policy, yielding increased fiscal surpluses that supplement gross saving. As in the past, the thus-expanded surplus of saving over investment would be mopped up by the PBC through the sale of 'sterilisation bonds'. The increased stock of these liabilities would then balance the additional foreign reserves that stem from the corresponding surplus of broadly defined export earnings over import costs, denominated in foreign currency.[31] The particular assumption we make is that the path of China's capital account balance remains exactly as in the baseline scenario. The investment shock is balanced precisely by an increase in total Chinese saving, so that external flows increase in both directions during 2008, netting out at baseline levels.

The results tell us, first, that this scenario would be impossible to achieve in practice if the financial shock is as large as that simulated, since a sudden and prodigious increase in the gross national saving rate—from 50 per cent to 72 per cent—would be required to completely neutralise the investment surge. Nonetheless, some blend of this scenario with the previous one is possible, so we persist with our description of its consequences. The key effects on the Chinese economy, measured as departures from the original baseline scenario, are indicated in Figure 3.15. The increases in investment and in saving in this scenario make the economic implications larger than the capital controls scenario considered previously, in which the principal effects were external, due to real exchange rate realignments. Because the rise in the saving rate robs the economy of consumption expenditure, however, the additional investment does

Figure 3.15 **Simulated effects of the financial contraction with compensating reserve accumulation (temporarily increased saving) on Chinese GNP, GDP, investment and imports, 2000–2035**

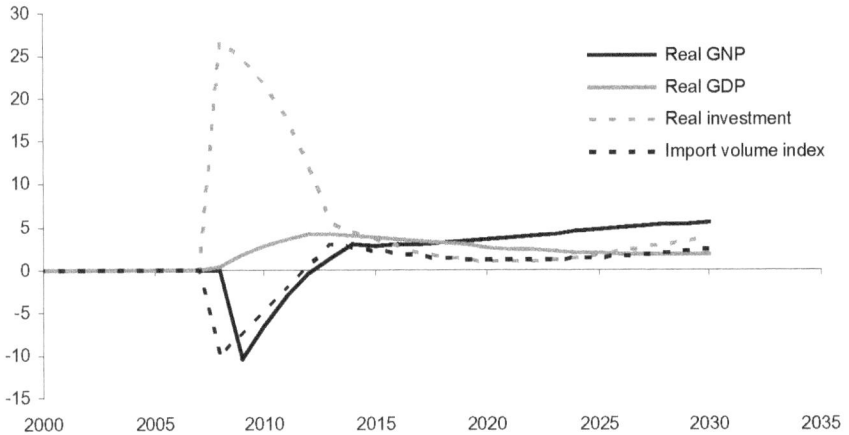

Note: Percentage departures from the baseline.
Source: Simulations of the model described in the text.

Figure 3.16 **Simulated effects of the financial contraction with compensating reserve accumulation (temporarily increased saving) on Chinese real exchange rates, 2000 – 2035**

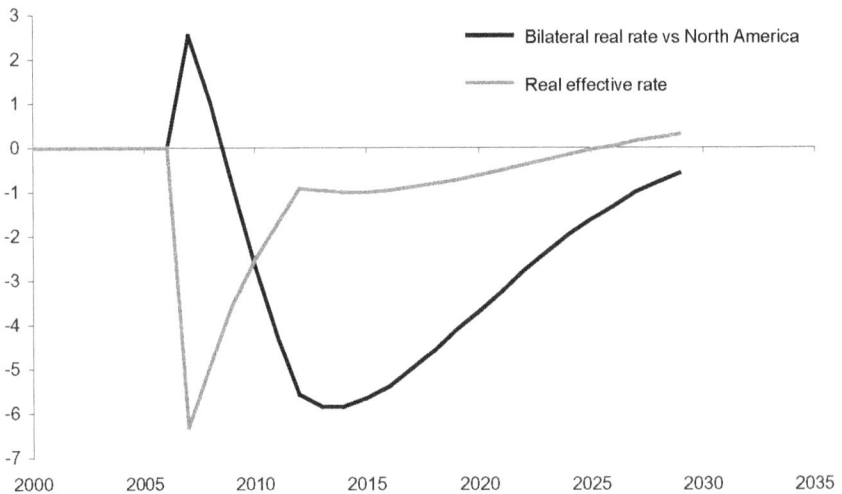

Note: Percentage departures from the baseline.
Source: Simulations of the model described in the text.

not raise GDP any further. The short-run fall in GNP, due to reduced returns on foreign assets, is also of a similar order. What is different about this scenario is that China's exports rise temporarily (by one-fifth, compared with no change under tighter capital controls) and its imports fall, by about one-tenth (again, compared with no change under tighter capital controls).[32]

Again, real exchange rate realignments are decisive here, as suggested by Figure 3.16. Comparing this figure with Figure 3.14, the short-run real appreciation against North America is a mere 2 per cent (compared with almost 7 per cent in the capital controls scenario) and the real effective depreciation is much larger (6 per cent compared with 2 per cent). Imports are very much more expensive in China under this scenario and so decline in the short run. As before, other regions absorb new investment after the flight of saving from the OECD countries and this appreciates their real exchange rates relative to North America. In China's case, however, while the new investment is also accommodated, the real exchange rate appreciates much less because domestic aggregate demand is sapped by the temporary increase in saving and the accelerated accumulation of foreign reserves. Overall, a temporary dip in foreign-sourced income and a substantial contraction in consumption should make this scenario unpalatable to the Chinese government. As for the mainly Asian and Pacific suppliers of China's raw materials and manufactured components, a temporary fall in the size of China's market is offset by increased investment.

Conclusion

Considering that exports make up almost half of China's GDP and most of these are directed to Europe and North America, negative financial shocks in those regions might be expected to retard China's growth. To confirm this quantitatively, shocks that widen the financial intermediation wedge are applied to North America and Western Europe in the context of a dynamic model of the global economy. Contrary to expectation, mitigating factors are also set in train by these shocks that lead to compensating benefits for China that insulate its economy, preserving its comparatively rapid growth path. These mitigating factors take the form of the temporary flight of savings from OECD countries into Chinese investment and real exchange rate realignments.

The mitigating factors are so strong that, as long as China receives a fair share of incremental investment due to the flight of OECD saving, it will be a net beneficiary of financial shocks in North America and Western Europe—at least as measured by its GDP. There are, however, good reasons why the Chinese government might seek to moderate the effects of this investment surge, through tighter inward capital controls, or offset them with increased reserve accumulation. It might, for example, view these additional funds as footloose portfolio capital that could be withdrawn suddenly in the future and therefore increase financial risk at the national level. And whatever form the inflow were to take, it would raise China's domestic aggregate demand and therefore appreciate its real exchange rate, placing pressure on its central bank to either appreciate the renminbi more quickly or allow faster inflation.

The results show that tighter capital controls could eliminate the investment surge but that this would cause other regions' real exchange rates to appreciate relative to China, making exports more competitive and trade diversification easier. If the Chinese government were, instead, to allow the investment boom to take place but to raise the home saving rate so as to offset it with yet faster reserve accumulation, the real depreciation of China relative to its other trading partners would be even larger. Although its import growth would slow temporarily, its GDP would maintain its original growth path, and, while its suppliers of raw materials and manufacturing components would temporarily export less to China, they would also enjoy increased investment sufficient to maintain their own levels of economic activity. The results therefore suggest that, as long as the financial shocks are restricted to North America and Western Europe, China's growth and the imports on which its trading partners rely are unlikely to be hindered significantly.

A key proviso is that the financial shocks do not spread beyond North America and Western Europe. We regard such a spread as unlikely, so do not consider it here. Clearly, if the crisis of financial confidence goes global, financial wealth will diminish in all regions, leading to declines in consumption and employment that could take many years to resolve. The potential for global growth remains considerable, however, and it is difficult to believe that pessimism about the future could become so widespread as to permanently under-price assets essential to that growth.

Finally, while we show that China's continued growth might not depend as closely as had previously been thought on markets for exports in North America and Western Europe, we would do well to remind ourselves on what that growth must depend. The key is continued accumulation and renewal of physical capital. Of course, the transformation of the labour force into skilled workers and professionals is also essential but this occurs in response to wage incentives that depend on capital accumulation. Productivity growth is also important, but this depends on capital accumulation and renewal. So what are the threats to capital accumulation and renewal in China? The main one is a rise in political risk. Should the Chinese government be destabilised, FDI would be repelled and there would be illegal capital flight. Output growth and imports from other regions would slow. The lesson is that continued Chinese growth is not primarily about export markets. It requires that China's government is stable and charts a steady and sensible policy course.

Notes

1   The short-term financial literature on this point is vast: see, for example, Wolf (2007), McKibbin and Stoekel (2007a, 2007b).
2   More so since it appears that these financial shocks have caused a speculative retreat to commodities, which has, at least temporarily, shifted the international terms of trade against China.
3   For evidence supporting the prominence of the components trade in other Asian economies, see Athukorala (2005).
4   It might well be considered to have begun with the US tech boom of the 1990s, however, and its subsequent bust, nonetheless leaving in its wake continued strong US productivity growth, which retarded domestic inflation and made the monetary expansions possible; see Pennings and Tyers (forthcoming) and Oliner et al. (2008).
5   The import surge did not raise US unemployment, although low-skilled workers were less favourably affected (Woo and Xiao 2007).
6   See Oliner et al. (2008). Significantly, they conclude that IT innovations were the strongest contributor to US productivity growth in both periods, but mostly before 2000, when the IT production sectors played key roles. Thereafter, however, while IT continued to be important, the gains came largely from services and were bolstered by one-off industry restructuring, which was unlikely to offer sustained productivity growth in the future.
7   The prices of wheat and iron ore showed extreme behaviour in 2007. Although the growth in demand in the economies in transition was supporting the rising trend in both, the extreme spikes were most likely caused by speculation after falls in OECD equity prices. The consequence was to exacerbate the adverse shift in China's terms of trade.
8   China's current account surplus in that period, and its associated accumulation of US assets, has been the subject of an already large range of literature. The authors' perspectives on this are detailed in Tyers and Bain (2007).
9   The model has its origins in *GTAP-Dynamic*, the standard version of which is a derivative of its comparative static progenitor, *GTAP* (Hertel 1997). The dynamics embodied in the original are described in Ianchovichina and McDougall (2000). Tyers et al. (2008) describe extensions to these dynamics, which emphasise endogenous skill levels.
10   Money is not represented explicitly, necessitating our focus on real effects. Although no focus on China is offered, a similar application using a more complete representation of the global macro-economy is provided by McKibbin and Stoekel (2007).
11   Baseline productivity in the agricultural sectors of developing regions grows more rapidly than that in services. This allows continued shedding of labour by agriculture. In the case of China, Wang and Ding (2006) estimated that there were 40 million surplus workers in China's agricultural sector. While underemployment is not explicit in our model, the assumption of high labour productivity growth in agriculture implies that agriculture is capable of shedding

labour without consequences for its output, as workers are drawn away by urban capital accumulation.

12 See Ianchovichina and McDougall (2000) for details.

13 For further details on the implementation and calibration of the investment interest premium, see Tyers and Golley (2008b).

14 The subdivision between production workers and professionals and para-professionals accords with the International Labour Organisation's occupation-based classification and is consistent with the labour division adopted in the *GTAP Database;* see Liu et al. (1998).

15 China's skill share is projected to rise through time while that in North America remains static. The contrast is due to North America's higher initial skill share, its high rate of unskilled immigration and the higher fertility rate of its low-skilled families.

16 It grows at a rate that is declining through time, mainly because low fertility causes China's labour force to fall after about a decade; see Tyers and Golley (2008b).

17 It acts to raise the region-specific price of capital goods, P_t^K, in Equation 1. Both shocks offer indirect means of representing events in financial markets in the past year. The rises in interest premiums and the declines in expected (and ultimately real) returns are due to weakening optimism about the future, most particularly in the United States, and therefore declining asset prices.

18 Consumption/savings choices by regional collective households are adaptive, responding to changes in real per capita incomes and real returns on saving. We experimented with alternative behavioural assumptions for North America, in one case forcing all the income adjustment on saving and in the other on consumption in that region. The effects of these differences on the global economy were small, so for simplicity we discuss results only from the model's standard specification.

19 This is not the case for gross national product (GNP), which does fall in the short run due to the loss of returns from home and foreign-sourced capital income in North America and Western Europe. In China, in contrast with its GDP, GNP slows (though does not fall absolutely), again due to the loss of foreign-sourced income.

20 The corresponding increments to unemployment rates in both regions are much smaller given their large shares of professional labour, the wages of which remain flexible.

21 The model is not constrained to approach any particular steady state, although all global capital market imbalances do tend to moderate over three decades in the baseline projection.

22 The role of China's very high saving rate in its broad economic behaviour is discussed by Kuijs (2006).

23 This point has already been made by McKibbin and Stoeckel (2007b) and more forcefully by McKibbin (2008).

24 See Tyers et al. (2008) and Tyers and Golley (2008b) for a discussion of China's real exchange rate, its measurement and its determinants.

25 This is a crude interpretation of the results since the simulated appreciation is relative to North America as a whole.

26 That China is arguably already capital-heavy emerges from the discussions in Kuijs and He (2007), Azziz and Cui (2007) and Rosen and Hauser (2007).

27 Illegal inflows enter through legal loopholes, acquisitions in Hong Kong and Macau and via transfer pricing; see Walter and Howie (2006).

28 Or, alternatively, if inflation is to be controlled it requires a politically inexpedient appreciation of the renminbi.

29 As modelled, these assets are held in a global trust that delivers an average global rate of return. Bilateral holdings are not identified. Nonetheless, given the size of the North American and Western European economies, the financial shocks cause a substantial reduction in the rate of return earned.

30 These real appreciations stem from the associated rise in aggregate demand, which tends to raise the prices of home goods more than comparatively elastically supplied foreign goods; see Tyers et al. (2008).

31 During 2007, a portion of the PBC's foreign assets was swapped for renminbi-denominated government debt and placed with China's sovereign wealth fund, the CIC. This was to reduce the currency mismatch on the PBC's balance sheet, a problem that would be exacerbated by a sudden and substantial fiscal surplus; see Tyers and Bain (2007).

32 Imports recover, however, to be larger than the baseline in the long run, once the saving rate is restored to its original declining path.

References

Athukorala, P.-C., 2005. 'Components trade and implications for Asian structural adjustment', in R. Garnaut and L. Song (eds), *The China Boom and its Discontents*, Asia Pacific Press and ANU E Press, The Australian National University, Canberra:215–39.

Azziz, J. and Cui, L., 2007. *Explaining China's low consumption: the neglected role of household income*, International Monetary Fund Working Paper WP07/181, Washington, DC.

Bank for International Settlements (BIS), 2007. *BIS Quarterly Review*, September, Bank for International Settlements, Basel.

Bernanke, B., 2006. Speech to the Chinese Academy of Social Sciences, Beijing, 15 December. Available from www.federalreserve.gov/BoardDocs/Speeches/2006/20061215.

Browning, E.S. and Silver, S., 2008. 'Credit crunch: new hurdles to borrowing', *Wall Street Journal*, 22 January:C1.

Callan, E., 2007. 'Clinton and Obama back China crackdown', *Financial Times*, 5 July. Available from www.FT.com.

Dimaranan, B.V. and McDougall, R.A., 2002. *Global Trade, Assistance and Production: the GTAP 5 Database*, May, Center for Global Trade Analysis, Purdue University, Lafayette.

Edwards, S., 2005. 'Is the US current account deficit sustainable? If not, how costly is adjustment likely to be?', *Brookings Papers on Economic Activity*, 1:211–71.

Eichengreen, B., 2006. 'Global imbalances: the new economy, the dark matter, the savvy investor and the standard analysis', *Journal of Policy Modelling*, 28:645–52.

Federal Reserve Board of Governors. www.federalreserve.gov/, Federal Reserve Board of Governors, Washington, DC.

Hertel, T.W. (ed.), 1997. *Global Trade Analysis Using the GTAP Model*, Cambridge University Press, New York.

Ianchovichina, E. and McDougall, R., 2000. *Theoretical structure of Dynamic GTAP*, GTAP Technical Paper No.17, Purdue University, Lafayette.

International Monetary Fund (IMF), 2007. *International Financial Statistics*, April, International Monetary Fund, Washington, DC. Available from www.imfstatistics.org.

Krugman, P., 2007. 'Will there be a dollar crisis?', *Economic Policy*, July:436–67.

Kuijs, L., 2006. *How will China's saving–investment balance evolve?*, World Bank Policy Research Working Paper 3958, World Bank, Beijing.

Kuijs, L. and He, J., 2007. *Rebalancing China's economy—modelling a policy package*, World Bank China Working Paper No.7, World Bank, Beijing.

Lardy, N., 2006. 'China's interaction with the global economy', in R. Garnaut and L. Song (eds), *The Turning Point in China's Economic Development*, Asia Pacific Press and ANU E Press, The Australian National University, Canberra:75–86.

Liu, J., Van Leeuwen, N., Vo, T.T., Tyers, R. and Hertel, T.W., 1998. *Disaggregating labor payments by skill level in GTAP*, Technical Paper No.11, September, Center for Global Trade Analysis, Department of Agricultural Economics, Purdue University, Lafayette.

Ma, G. and McCauley, R.N., 2007. 'How effective are China's capital controls?', in R. Garnaut and L. Song (eds), *China: linking markets for growth*, Asia Pacific Press and ANU E Press, The Australian National University, Canberra:267–89.

Ma, Y., 2006. *A comparative study of the competitiveness of the domestic and foreign-invested service industries in China*, Centre for Public Policy Studies Working Paper No.176, Lingnan University, Presented at the ACE International Conference, APEC Studies Centre, City University of Hong Kong, 18–20 December.

McKibbin, W.J., 2008. 'Australia has little to worry about in a US downturn', *The Australian*, 3 April:B1.

McKibbin, W.J. and Stoeckel, A., 2007a. *Bursting of the US housing bubble*, Economic Scenarios.Com, No.14, Centre for International Economics, Canberra. Available from www.economicscenarios.com.

——, 2007b. *The potential real effects from the repricing of risk*, Economic Scenarios.Com, No.15, Centre for International Economics, Canberra. Available from www.economics.com.

McKinnon, R.I., 2006. 'China's exchange rate appreciation in the light of the earlier Japanese experience', *Pacific Economic Review*, 11(3):287–98.

National Bureau of Statistics (NBS), 2007. *China Statistical Abstract 2006*, China Statistics Press, Beijing.

Obstfeld, M. and Rogoff, K., 2005. 'Global current account imbalances and exchange rate adjustments', *Brookings Papers on Economic Activity*, 1:67–123.

Oliner, S.D., Sichel, D.E. and Stiroh, K.J., 2008. *Explaining a productive decade*, Staff Working Paper 2008-01, Finance and Economics Discussion Series, Federal Reserve Board, Washington, DC.

Pennings, S. and Tyers, R. (forthcoming). 'Increasing returns, financial capital mobility and real exchange rate dynamics', *The Economic Record*.

Rosen, D.H. and Houser, T., 2007. *China energy: a guide for the perplexed*, China Balance Sheet Paper, May, Center for Strategic and International Studies and Peterson Institute for International Economics, Washington, DC.

Roubini, N. and Setser, B., 2005. Will the Bretton Woods 2 regime unravel soon? The risk of a hard landing in 2005–2006, Presentation at Symposium on the Revived Bretton Woods System: a new paradigm for Asian development?, Federal Reserve Bank of San Francisco and University of California at Berkeley, San Francisco, 4 February.

Tung, C.Y. and Baker, S., 2004. 'RMB revaluation will serve China's self-interest', *China Economic Review*, 15:331–5.

Tyers, R. and Bain, I., 2007. *Appreciating the renminbi*, Working Papers in Trade and Development No.2007/09, Division of Economics, Research School of Pacific and Asian Studies, The Australian National University, Canberra.

Tyers, R., Bu, Y. and Bain, I., 2008. 'China's equilibrium real exchange rate: a counterfactual analysis', *Pacific Economic Review*, 13(1):17–39.

Tyers, R. and Golley, J., 2008a. 'China's real exchange rate puzzle', *Journal of Economic Integration,* 23(3).

——, 2008b. 'China's growth to 2030: the roles of demographic change and financial reform', *Review of Development Economics*.

Tyers, R. and Shi, Q., 2007. 'Global demographic change, policy responses and their economic implications', *The World Economy*, 30(4):537–66.

Walter, C.E. and Howie, F.J.T., 2006. *Privatising China: inside China's stock markets*, John Wiley and Sons (Second Edition), Singapore.

Wang, J. and Ding, S. 2006. 'A re-estimation of China's agricultural surplus labour—the demonstration and modification of three prevalent methods', *Frontiers of Economics in China*, 1(2):171–81.

Wolf, M., 2007. 'Q&A on the debt crisis', *Financial Times*, 4 September, London.

Woo, W.T. and Xiao, G., 2007. 'Facing protectionism generated by trade disputes: China's post WTO blues', in R. Garnaut and L. Song (eds), *China: linking markets for growth*, Asia Pacific Press and ANU E Press, The Australian National University, Canberra:45–70.

Xiao, G. 2006. What is special about China's exchange rate and external imbalance: a structural and institutional perspective, Asian Economic Panel 2007, Brookings-Tsinghua Center and Brookings Institution, Beijing and Washington, DC.

Acknowledgments

Funding for the research described in this chapter came from Australian Research Council Discovery Grant No.DP0557889. Thanks are due to Jane Golley, Warwick McKibbin and Ligang Song for helpful discussions, to participants at seminars at the China Center for Economic Research at Peking University and the Department of Economics at Renmin University of China for useful comments, and to Hsu Pingkun for research assistance.

4

Global production sharing and US–China trade relations

Prema-chandra Athukorala and Nobuaki Yamashita

In the past decade, the widening bilateral trade deficit has been the focal point of US–China economic relations. This is often portrayed as a cause of the overall US current account imbalance. Real public concerns in the United States surrounding the debate about the 'China deficit' are, however, rooted in the perceived economic threat of import competition. In the late 1990s, when imports from China were dominated by traditional labour-intensive manufactures such as clothing and footwear, unskilled workers' employment losses and wage suppression were the prime focus of the debate. More recently, the apparent rising sophistication of imports from China—in particular, the sharp rise in imports of computers and electronic products—has fuelled concerns that the rise of China poses a direct threat to the United States' position as a technology superpower, a concern reminiscent of the economic fears about Japan that pervaded the US policy scene in the 1970s and 1980s.

'Unfair' Chinese import competition is perceived to take a number of forms, including illegal export subsidies, lax enforcement of intellectual property rights, restrictions on imports to and foreign investment in China, and the national currency being kept undervalued through massive intervention in the foreign exchange market (Hufbauer et al. 2006; Mankiw and Swagel 2005; Weisman 2007). These concerns have fuelled calls for new legislation to prevent unfair practices. In February 2005, the US Senate passed the *Byrd Amendment*, a provision that encouraged American companies to file anti-dumping lawsuits by awarding revenue collected from the resultant tariffs to litigating companies. Other China-specific legislation has been proposed in the past two years, including a bill that stipulates declaring exchange rate protection as a form of illegal subsidisation for which US firms can seek compensation. The economic Sinophobia has also begun to spill over to other arenas of US–China relations, including international food-safety standards and US policy postures relating to the entry of Chinese firms in the corporate arena (Shirk 2007:267). The China threat—in particular, the loss of American jobs to China—was a hot issue in the 2000 and 2004 presidential elections, and indications are that it is likely to figure even more prominently in the upcoming (2008) presidential campaign (Easton 2008; Steinbock 2008).

The policy debate about US–China trade relations has so far been based on the conventional notion of horizontal specialisation, in which trade takes place in the form of final goods (goods that are produced from start to finish in a given country). It has largely ignored the continuing process of global production sharing—the break-up of the production process into geographically separated stages[1]—and the resulting trade complementarities between the two countries as dominant players of this new form of international exchange. Global production sharing opens up opportunities for countries to specialise in different slices (different tasks) of the production process depending on their relative cost advantage and other relevant economic fundamentals. Consequently, parts and components are now exchanged across borders at a faster rate than final goods. In this context, decisions about how much to produce and for which market have to be combined with decisions about whether to produce and with what degree of intra-product specialisation. The upshot is that trade-flow analysis based on data coming from a reporting system designed at a time when countries were trading only in final goods naturally distorts values of exports and imports, leading to a falsification of the current account imbalances. The degree of falsification is likely to increase over time as more complex production networks are created with an ever-increasing number of interacting countries (Jones and Kierzkowski 2001a, 2001b). The spread of international production sharing can also diminish the efficacy of exchange rate and tariff policies in influencing trade flows by opening up opportunities for firms to acquire inputs from, and relocate final assembly to, different countries within global production networks, with a view to cushioning their profit margins in the face of such policy changes (Ghosh and Rajan 2007).

Given the current state of data, it is not possible to quantify the effect of international production sharing on bilateral trade imbalances: this would require a major overhaul of the international system of collecting trade data to record domestic value-added content at different stages of production. The COMTRADE database of the United Nations (UN various years) does, however, now provide disaggregated data that permit the separation of parts and components from final goods with a satisfactory coverage of trade in machinery and transport equipment, a commodity class in which most of the global production sharing is concentrated. Data extracted from this source, when combined with the available case study-based evidence of global operations of multinational enterprises, permit us to paint a broad-bush picture of the nature of the continuing process of global production sharing and its implications for US–China trade relations in order to better inform the current policy debate. That is what we aim to do in this chapter. A number of recent studies have alluded to the importance of paying attention to global production sharing in analysing the drivers of the US–China trade deficit (Bergsten et al. 2006; Lardy 2005; Fung et al. 2006; Krugman 2008). To our knowledge, however, this is the first attempt to examine this issue systematically to the extent permitted by the available data.

The remainder of this chapter is presented in four parts. The next section offers an overview of trends and patterns of Chinese trade in order to set the stage for the ensuing analysis. The third section surveys US–China trade patterns with emphasis on emerging patterns in the two countries' involvement in global production networks and their implications for the bilateral trade flows using some fresh datum tabulations separating trade in parts and components and final goods in machinery trade. This is followed by an econometric analysis of the determinants of trade flows. The final section presents concluding remarks.

China's trade performance: an overview

The rise of China as a major trading nation was one of the most momentous developments of the post-World War II era, surpassing even the stunning rise of Germany and Japan. Total merchandise exports from China increased from US$8 billion (about 1 per cent of global exports) in 1978–79, when the process of liberalisation reforms started, to US$1,442 billion (13.4 per cent) in 2005–06.[2] In 2006, China was the second largest exporting nation in the world after Germany and, assuming the current growth rates continue, it will become the largest in about 10 years. During the reform era, until about the mid 1980s, imports followed exports closely with periodic minor trade surpluses or deficits. From then on, exports have persistently outpaced imports, yielding a mild annual surplus averaging about 2 per cent of gross domestic product (GDP), and exceeding 3 per cent only briefly, in 1997–98.

China's phenomenal export expansion has been underpinned by a shift in the commodity composition of exports away from primary products and towards manufacturing (Table 4.1). The share of manufactures in China's total merchandise exports increased from less than 40 per cent in the late 1970s to nearly 80 per cent in the early 1990s, and to 92 per cent in 2005–06. Until about the late 1990s, traditional labour-intensive manufactures—in particular, apparel, footwear, toys and sporting goods—were the prime movers of export expansion. Since then, there has been a notable shift in the export composition away from conventional labour-intensive product lines and towards more sophisticated product lines—in particular, those within the broader Standard International Trade Classification (SITC) category of machinery and transport equipment (SITC 7; henceforth referred to as 'machinery'). Between 1992–93 and 2005–06, the share of miscellaneous manufactures (SITC 8)—a catch-all commodity group encompassing most of the traditional labour-intensive products—declined from 49 per cent to 31 per cent and the share of machinery increased from 17 per cent to 44 per cent.

The expansion of machinery exports has been brought about by China's highly publicised export success in a wide range of 'information and communication technology' (ICT) products (which fall under SITC categories 75, 76 and 77). China's world market share of ICT products recorded a fivefold increase from 5 per cent in 1992–93 to 24.1 per cent in 2005–06. Among them, the share of office machines increased from less than 2 per cent in 1992–93 to more than 28 per cent in 2005–06. Today, China is the world's largest global producer as well as the single largest exporter of personal computers falling in this commodity group. China's world market share of telecommunications and sound-recording equipment (dominated by mobile phones, DVD players and CD players) was 26.2 per cent in 2005–06—up from 7.9 per cent in 1992–93.

Trade data showing this phenomenal structural shift have been used widely—not only in the popular press and policy reports of agencies involved in promoting research and development activities, but in some scholarly writing—to argue that China is rapidly becoming an advanced-technology superpower and the sophistication of its export basket is rapidly approaching the levels of those of most advanced industrial nations (for example, Rodrik 2006; Yusuf et al. 2007). A closer examination of the data, however, suggests that such an inference is fundamentally flawed. In reality, what we observe is the rapid consolidation in China of the final-assembly stages of East Asian-centred global production networks for these products. Ample supplies of relatively cheap and trainable labour and the scales of economy arising from China's vast domestic market (which enables firms to achieve low unit costs)

Table 4.1 China's merchandise exports: composition and world market share, 1992–93 to 2004–05

Product	Composition (%)			World market share (%)		
	1992–93	2000–01	2005–06	1992–93	2000–01	2004–06
Primary products	19.9	11.4	7.6	0.7	0.7	0.8
Manufacturing (5 to 8–68)	79.6	88.4	92.2	4.5	8.0	12.0
Chemicals and related products (5)	5.1	4.9	4.6	1.4	2.2	2.9
Resourced-based manufacturing (6–68)	17.5	15.5	15.9	3.6	6.6	9.5
Machinery and transport equipment (7)	16.1	34.4	46.7	1.9	5.5	11.4
Power-generating machines (71)	0.9	1.2	1.1	1.3	2.7	3.7
Special industrial machinery (72)	0.8	0.8	1.1	0.7	1.5	3.1
Metalworking machinery (73)	0.3	0.3	0.3	1.1	1.7	2.8
General industrial machinery (74)	1.5	2.5	3.4	1.4	3.7	7.4
Information and telecommunication technology products (75, 76, 77)	10.4	26.1	37.0	5.0	9.9	24.1
Office machines (75)	1.6	8.2	14.2	1.7	10.1	28.2
Telecommunications and sound equipment (76)	4.8	8.4	12.6	7.9	13.1	26.2
Electrical machinery (77)	4.1	9.6	10.2	2.8	6.9	13.4
Road vehicles (78)	1.4	2.6	2.8	0.4	0.8	1.3
Other transport equipment (79)	0.7	0.9	0.9	0.3	0.6	1.4
Miscellaneous manufactured articles (8)	40.9	33.6	25.0	13.9	21.6	24.4
Clothing and accessories (84)	19.9	14.1	9.8	19.1	26.3	38.6
Footwear (85)	5.4	3.9	2.4	27.6	42.0	44.2
Baby carriages, toys and games (894)	4.3	3.9	2.6	32.5	54.9	61.5
Total merchandise exports	100	100	100	4.0	6.7	9.6
US$ billion	136	403	1,145			

Note: Standard International Trade Classification (SITC) codes are given in parentheses.
Source: Compiled from United Nations (UN), various years. *Commodity Trade Statistics Database (COMTRADE)*, Statistics Division, United Nations. Available from http://comtrade.un.org/bd/

are contributing factors to China's attractiveness as a global assembly centre. As already noted, China's so-called 'high-tech' exports are concentrated heavily in a single product category: ICT products. The bulk of these products (such as notebook computers, display units, mobile phones and DVD and CD players) are simply 'mass-market commodities' produced in huge quantities and at relatively low unit costs using imported high-tech parts and components; they are not leading-edge technology products.

The share of components in total machinery imports to China increased from 32.5 per cent in 1992–93 to 63.4 per cent in 2004–05, with the import shares of the three ICT products (SITC 75, 76 and 77) recording much faster growth (Table 4.2). In contrast, final goods (total exports minus components) have continued to dominate the export composition. In the past decade, the share of final goods in total machinery exports has remained about 75 per cent, with only minor year-to-year changes. Given the fact that the production of parts and components is generally more capital and technology intensive than final assembly, these figures clearly suggest that China's export success has so far been underpinned largely by its comparative advantage in international production arising from its labour abundance. When components are netted out, more than 80 per cent of total Chinese manufacturing exports can still be treated as labour-intensive products.

The bulk of assembly activities in China are carried out by affiliates of multinational enterprises ('foreign-invested enterprises', FIEs) from imported components within their global production networks (Dean and Tam 2005; Naughton 2007; Sung 2007). The share of FIEs in total exports from China

---

Table 4.2 **Share of parts and components in China's manufacturing trade, 1992–93, 2000–01 and 2005–06**

Product	Exports			Imports		
	1992–93	2000–01	2005–06	1992–93	2000–01	2005–06
Total manufacturing	17.93	32.56	40.77	6.86	14.98	18.39
Machinery and transport equipment (7)	36.51	58.98	67.65	33.95	38.50	36.32
ICT products (75, 76, 77)	59.90	72.70	79.20	32.53	37.91	34.83
Miscellaneous manufacturing (8)	20.66	15.49	10.89	1.47	1.69	2.35

Note: Standard International Trade Classification (SITC) codes are given in parentheses.
Source: Compiled from United Nations (UN), various years. *Commodity Trade Statistics Database (COMTRADE)*, Statistics Division, United Nations, New York. Available from http://comtrade.un.org/bd.

increased from less than 2 per cent in 1980 to more than 58 per cent by 2005. They accounted for 88 per cent of total information-technology products exported from China in 2005. The FIEs are mostly wholly foreign owned, and their activities in China are concentrated overwhelmingly in the final-assembly stage of production, which is the most labour-intensive layer of a production process spread across many countries. Basic research and product design and capital and human capital-intensive stages of the production process are carried out in the home countries of the multinational enterprises or in other Asian countries that are in a more advanced stage of industrial development than China. Affiliates of US multinationals account directly for only a small share of total exports by FIEs (about 10 per cent),[4] with affiliated Taiwanese, Hong Kong and Korean firms accounting for the lion's share (of more than 80 per cent). US multinationals seem, however, to play a major role in parts and components supplies for all export-oriented assembly firms from their production bases in China and other countries—in particular, those located in Southeast Asia.[5]

US–China trade patterns

Bilateral trade between the United States and China has grown persistently since the early 1980s, with the rate of growth accelerating from about the mid 1990s and again after China's accession to the World Trade Organization (WTO) in 2001 (Figure 4.1). The value of US imports from China rose from US$16 billion in 1990 to US$307 billion in 2006. Since 2003, China has been the second largest source of US imports, after Canada but ahead of Mexico and Japan. US exports to China have also grown persistently during this period, but from a low base and at a slower rate. Total exports in 2006 amounted to a mere US$55 billion, up from US$5 billion in 1990. Bilateral economic ties between the two countries have therefore been characterised by a steadily growing trade imbalance: the trade deficit increased from US$11 billion to US$205 billion in 2006—the largest deficit the United States has ever had with any country. The bilateral trade deficit as a percentage of US GDP increased from 0.2 per cent in 1990 to 0.9 per cent in 2000, and then to 1.9 per cent in 2007. The deficit with China has been the United States' single largest bilateral trade deficit since 1999. As part of its WTO accession, China substantially reduced barriers to import trade, becoming the fastest growing market for US exports. China's WTO accession also gave foreign companies confidence to move their assembly plants within global production networks to China. As a result, China's exports kept growing.

Figure 4.2 illustrates the US–China trade deficit in the context of the United States' growing overall trade deficit. At the same time that the US deficit with

Figure 4.1   **US–China trade, 1990–2006**

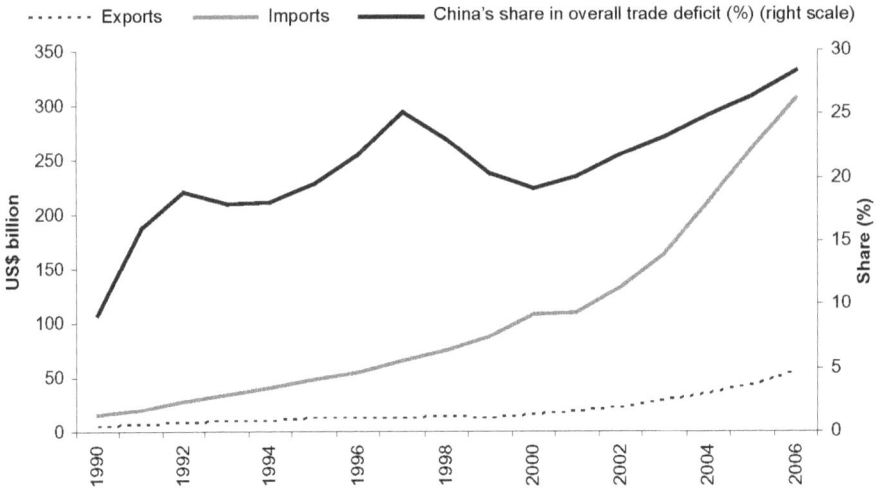

Figure 4.2   **US trade deficit: China's share in comparative perspective, 1990–2006**

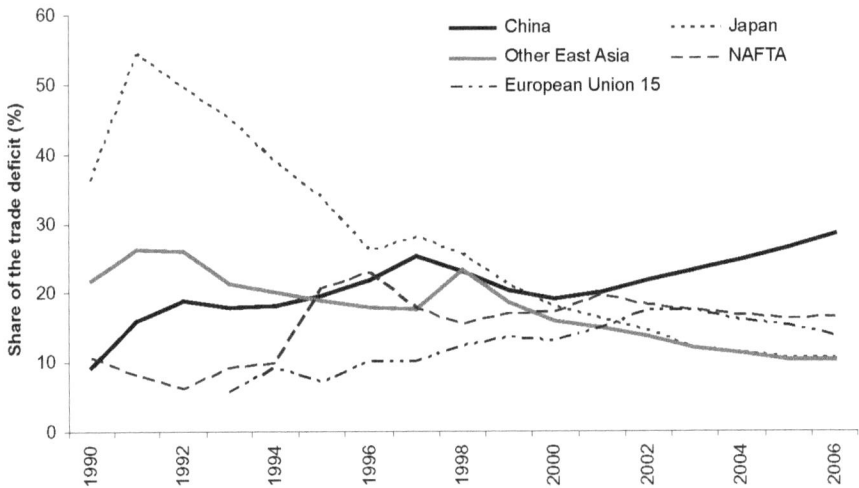

* the United States had a small trade surplus with the European Union during 1990–93
NAFTA = North American Free Trade Agreement
Source: Based on data compiled from United Nations (UN), various years. *Commodity Trade Statistics Database (COMTRADE)*, Statistics Division, United Nations, New York. Available from http://comtrade.un.org/bd.

China increased, the overall US deficit with all other countries has expanded. In 2006, the US–China bilateral deficit amounted to 28 per cent of the total US trade deficit—in other words, almost three-quarters of the total US trade deficit with the rest of the world. This comparison suggests that current US policy concerns about Chinese trade have been driven primarily by the perceived threat of import competition from China rather than by the broader economic issue of a widening overall trade imbalance.

Figure 4.2 also shows that, from about 1999, the widening US–China deficit has been significantly counterbalanced by a sharp decline in the relative importance of US bilateral trade deficits with Japan and other East Asian countries. Between 1999 and 2006, the increase in China's share in the total US trade deficit from 20.4 per cent to 28.4 per cent was accompanied by a decline in the respective figure for Japan—from 21.1 per cent to 10.5 per cent. The share of the other East Asian countries also declined—from 16 per cent to 10.3 per cent—between these two years. These contrasting patterns point to the fragility of any analysis of trade imbalances based only on 'reported' bilateral trade figures in an era of global production sharing—a point to be investigated using disaggregated trade data in the next section.

Figure 4.3 tells the Chinese side of the story.[6] The widening China–US trade surplus in the past 10 years has been accompanied by widening bilateral deficits with Japan and the other East Asian countries. From 2004 to 2006, the combined deficit with Japan and the other East Asian countries amounted to 85 per cent of the China–US trade surplus. In fact, China has had an overall trade deficit with the world, excluding the United States, for several years now. There is evidence that China's widening trade deficits with its regional trading partners are closely associated with China's increasingly important role as the main centre of final assembly within regional production networks. At the beginning of the reform era, export-oriented firms in Hong Kong, Taiwan and Korea that were involved in the production of conventional labour-intensive products such as clothing, footwear, toys and travel goods relocated their production to China. This was followed, from about the early 1990s, by rapid relocation to China of the final-assembly stages of 'high-tech' industries—in particular, ICT industries from these countries and also from Japan, the United States and other industrial countries. During this period, the major member countries of the Association of Southeast Asian Nations (ASEAN; Singapore, Thailand, Malaysia and the Philippines) also began to participate in the production networks in a big way as suppliers of parts and components for final-assembly activities in China (Athukorala 2008). This massive restructuring of production processes within global production networks naturally set the stage for a shift

Figure 4.3　**China's bilateral trade balance, 1992–2006** (US$ billion)

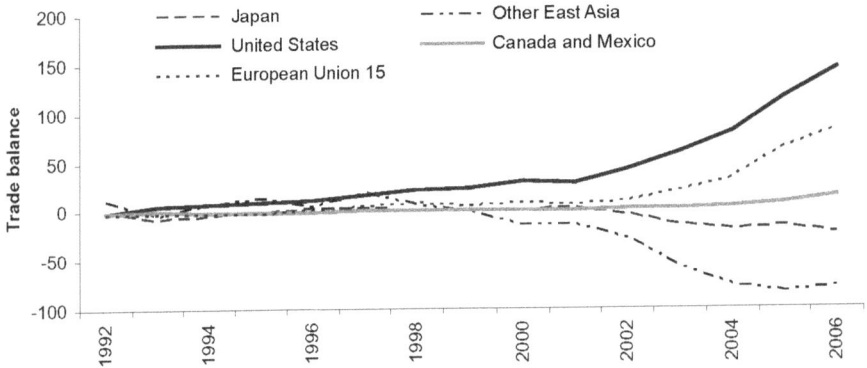

Source: Based on data compiled from United Nations (UN), various years. *Commodity Trade Statistics Database (COMTRADE)*, Statistics Division, United Nations, New York. Available from http://comtrade.un.org/bd.

in China's growing bilateral trade surplus with the United States (and some other industrial countries), which provided markets for final (assembled) goods in its regional trading partners in the form of narrowing trade deficits and/or growing trade surpluses.

The discussion so far in this section suggests that the US–China trade imbalance could be largely a structural phenomenon, quite distinct from the overall trade imbalance of the United States, and related largely to the continuing process of global production sharing and the pivotal role played by China in this new form of international exchange. We now turn to disaggregated analysis of US–China trade flows in order to broaden our understanding of the phenomenon.

In the reform era, until about the early 1990s, traditional conventional labour-intensive manufactured goods (miscellaneous manufactures) dominated Chinese exports to the US, reflecting China's general pattern of export specialisation at the time. Since then there has been a palpable shift in the commodity composition of imports away from these products and towards machinery and transport equipment—in particular, ICT products. Between 1995–96 and 2005–06, the share of miscellaneous manufactures in total imports from China declined from 58.8 per cent to 38.5 per cent, accompanied by an increase in the share of machinery from 26.3 per cent to 44.1 per cent (Table 4.3). The share of ICT products increased from 22.4 per cent to 37.6 per cent, and this commodity group contributed to more than 40 per cent of the total

Table 4.3 **Geographic profile of US trade** (per cent)

Partner country/ country groups	Primary products		Manufactured goods								Total trade	
			Total		Machinery (SITC 7)		ICT products		Misc. manufactures (SITC 8)			
	1995–96	2005–06	1995–96	2005–06	1995–96	2005–06	1995–96	2005–06	1995–96	2005–06	1995–06	2005–06
a) IMPORTS												
China and Hong Kong	1.4	1.7	9.5	21.6	4.5	18.2	8.1	33.4	27.4	39.7	7.8	16.0
China	1.3	1.7	7.9	21.0	3.7	18.0	6.5	33.0	22.6	37.8	6.5	15.5
East Asia	6.2	4.2	37.1	24.1	47.1	32.1	63.5	36.6	26.8	17.0	30.7	18.7
Japan	0.8	0.5	19.2	10.8	26.2	16.1	24.9	9.2	8.3	4.3	15.4	8.1
Korea	0.3	0.7	3.7	3.3	4.5	4.6	7.5	5.3	3.0	1.0	3.0	2.6
Taiwan	0.3	0.2	4.8	2.8	4.9	3.2	8.3	5.1	5.6	2.3	3.9	2.1
ASEAN	4.7	2.7	9.4	7.3	11.5	8.3	22.8	16.9	9.9	9.5	8.4	6.0
NAFTA	37.1	33.3	25.2	23.6	28.5	28.3	17.2	19.5	13.5	13.7	28.0	26.6
Mexico	8.9	9.4	8.5	10.4	10.2	14.0	10.9	15.7	7.1	7.8	8.6	10.2
European Union 15	10.3	9.9	19.4	19.8	17.2	17.3	9.3	7.2	15.3	13.1	17.9	17.4
World	100	100	100	100	100	100	100	100	100	100	100	100
b) EXPORTS												
China and Hong Kong	5.3	10.8	4.1	6.2	4.0	6.6	4.7	9.5	3.3	5.4	4.3	6.8
China	3.1	9.7	1.8	4.3	1.8	4.7	1.1	5.6	1.2	3.1	2.0	5.0
East Asia	35.1	19.4	23.6	16.6	24.9	18.1	31.0	24.8	23.5	17.8	25.2	16.8
Japan	20.3	9.4	9.1	5.3	8.6	4.8	10.2	5.0	12.4	8.2	10.9	5.9
Korea	6.2	3.7	4.0	3.1	4.4	3.3	4.3	4.3	3.0	2.9	4.3	3.1
Taiwan	4.0	2.3	3.0	2.4	3.1	2.6	3.6	3.4	2.1	2.5	3.1	2.3
ASEAN	4.6	4.0	7.5	5.8	8.8	7.4	13.0	12.2	5.9	4.1	6.8	5.5
NAFTA	21.0	34.6	32.1	36.4	32.5	36.5	27.9	33.3	28.3	31.3	30.0	35.8
Mexico	7.3	14.0	8.8	13.0	8.3	12.8	10.3	17.1	9.9	10.7	8.5	13.1
European Union 15	18.0	14.2	21.2	21.1	21.0	18.7	22.7	17.3	22.6	26.3	20.9	20.1
World	100	100	100	100	100	100	100	100	100	100	100	100

Source: Compiled from United Nations (UN), various years. *Commodity Trade Statistics Database (COMTRADE)*, Statistics Division, United Nations, New York. Available from http://comtrade.un.org/bd.

Table 4.4   Commodity composition of US trade by partner country

a) IMPORTS

	Year	China and Hong Kong		East Asia					NAFTA		European Union 15	World
		Total	China	Total	Japan	Korea	Taiwan	ASEAN	Total	Mexico		
Primary products	1995–96	3.2	3.5	3.7	1.0	2.0	1.6	10.3	24.3	18.8	10.5	18.3
	2005–06	2.7	2.7	5.7	1.7	7.0	2.9	11.5	32.0	23.4	14.6	25.6
Manufactures	1995–96	95.4	95.7	94.7	97.4	96.5	97.0	88.0	70.3	77.1	84.7	78.3
	2005–06	96.0	96.1	91.6	95.4	91.1	93.8	85.8	63.1	72.5	80.9	71.0
Machinery and equipment	1995–96	26.4	26.3	70.3	77.9	67.8	58.5	62.7	46.5	54.4	44.0	45.8
	2005–06	43.3	44.1	65.2	75.7	68.6	57.4	52.4	40.3	52.0	37.7	38.0
ICT products	1995–96	23.0	22.4	46.2	36.0	55.3	48.2	60.7	13.7	28.3	11.6	22.3
	2005–06	37.0	37.6	34.6	20.2	36.7	43.5	49.9	13.0	27.3	7.3	17.7
Miscellaneous manufactures	1995–96	58.8	58.5	14.6	9.0	16.8	24.2	19.8	8.1	13.8	14.4	16.8
	2005–06	38.5	37.7	14.1	8.2	6.3	16.9	24.4	7.9	11.8	11.6	15.5
Total	1995–96	100	100	100	100	100	100	100	100	100	100	100
	2005–06	100	100	100	100	100	100	100	100	100	100	100

b) EXPORTS

Primary products	1995–96	21.6	27.6	24.6	32.8	25.5	22.4	11.9	12.4	15.1	15.3	17.7
	2005–06	24.4	29.8	17.6	24.1	18.3	15.1	11.3	14.7	16.3	10.8	15.3
Manufactures	1995–96	75.2	71.1	73.2	65.4	72.2	74.5	85.6	83.8	80.8	79.7	78.3
	2005–06	73.8	68.7	79.8	72.8	79.7	82.9	86.0	81.9	80.0	85.1	80.8
Machinery	1995–96	45.6	43.8	48.1	38.3	49.2	48.8	62.6	52.9	47.1	49.1	48.7
	2005–06	46.8	45.4	51.4	38.4	51.6	53.1	64.7	48.7	46.6	44.6	47.8
ICT products	1995–96	23.9	12.6	27.1	20.5	21.8	25.2	41.8	20.5	26.4	24.0	22.0
	2005–06	26.3	20.8	27.5	15.7	25.8	27.1	41.5	17.4	24.4	16.1	18.7
Miscellaneous manufactures	1995–96	9.5	7.7	11.7	14.2	8.8	8.5	10.9	11.8	14.4	13.5	12.5
	2005–06	9.0	7.1	12.0	15.7	10.7	12.3	8.5	9.9	9.2	14.8	11.3
Total	1995–96	100	100	100	100	100	100	100	100	100	100	100
	2005–06	100	100	100	100	100	100	100	100	100	100	100

Source: Compiled from United Nations (UN), various years. *Commodity Trade Statistics Database (COMTRADE)*, Statistics Division, United Nations., New York Available from http://comtrade.un.org/bd.

increment in imports from China between these two periods. China's share in total US ICT-product imports increased from 6.5 per cent in 1995–96 to 33 per cent in 2005–06 (Table 4.4). This was underpinned by a sharp decline in the combined share of the other East Asian countries—from 63.5 per cent to 36.6 per cent. China's share in US ICT imports in 2005–06 was almost two times that of Mexico (15.7 per cent).

To gain further insights into the growing importance of overseas assembly as a source of imports for the United States and the pivotal role played by China in this international division of labour, we disaggregated data for machinery trade into parts and components and final goods (reported trade: parts and components). The results of this exercise are presented in Figure 4.4 and Table 4.5. Table 4.5 gives data for the share of parts and components in total US imports and exports of machinery and the subcategory of ICT products therein. Data for ICT products disaggregated into final goods and parts and components are plotted in Figure 4.4.

The share of components in US machinery exports is generally much higher across all partner countries compared with that of imports (Table 4.5). Moreover, on the import side, the shares have recorded a notable decline across all import-trading partners. This decline is much sharper for the ICT products subcategory within the broader category of machinery and transport equipment. These patterns are generally consistent with the United States' comparative advantage in skill and capital-intensive activities in production processes within global production networks in vertically integrated industries. Within this broader context, one can observe two peculiarities relating to China's role in international production sharing in relation to its trade with the United States.

First, the share of parts and components in US exports to other East Asian countries (in particular, to the countries in ASEAN) is much higher compared with that of exports to China. This pattern is consistent with the case study-based findings that US firms located in East Asian countries undertake further processing/assembly of parts and components originally designed and/or produced in the United States as part of their engagement in China-centred regional production networks. United States-based multinational enterprises have a long history of engagement in parts and components assembly and testing in Southeast Asia, dating back to the setting up of processing plants by National Semiconductors and Hewlett Packard in Singapore in the early 1970s. Many more US firms entered this arena and their production locations expanded to Thailand, Malaysia and the Philippines in subsequent years and, more recently, to Vietnam. At the formative stage, the activities of US multinational enterprise affiliates involved assembly/testing of simple components and the re-

Figure 4.4 **US trade in ICT goods disaggregated into parts and components and final goods, 1990-2006** (per cent)

Imports
Figure 4.4a **Final Goods**

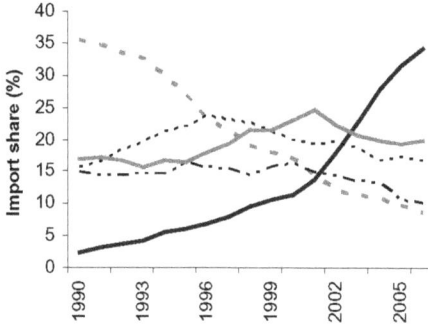

Exports
Figure 4.4b **Final Goods**

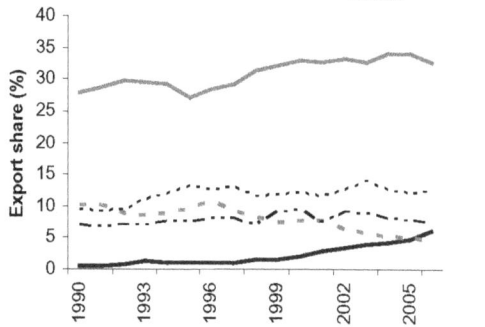

China
Japan
Korea and Taiwan
ASEAN

Figure 4.4c **Parts and components**

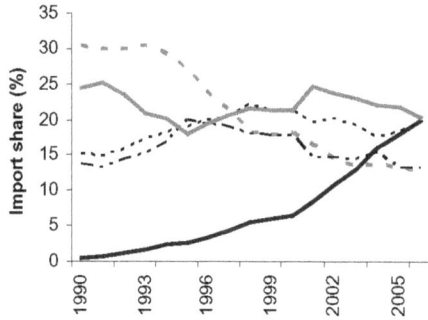

Figure 4.4d **Parts and components**

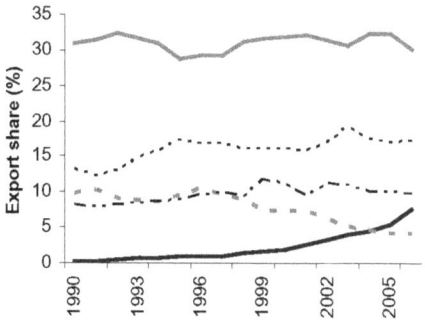

Figure 4.4e **Total (reported) imports**

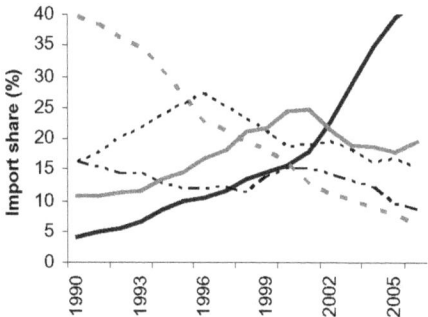

Figure 4.4f **Total (reported) exports**

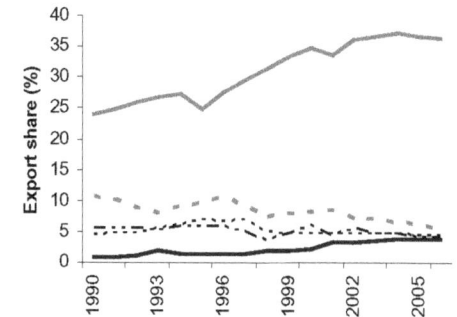

importing of the assembled components to the United States to be incorporated in the final products. Over time, more and more sophisticated stages of the production process were relocated in Southeast Asia, resulting in multiple border crossings of parts and components before they were incorporated in final production not only in the United States, but in other countries covered

Table 4.5   **Share of parts and components in US machinery trade** (per cent)

	Imports Share		Exports	
	1995–96	2005–06	1995–96	2005–06
(a) Machinery and transport equipment*				
China and Hong Kong	32.1	24.4	45.6	56.0
China	25.0	24.2	36.1	50.8
East Asia	45.6	36.8	57.5	62.1
Japan	42.2	33.3	51.1	49.4
Korea	60.3	31.0	51.2	58.2
Taiwan	54.9	52.6	55.4	58.4
ASEAN	43.6	40.9	67.7	73.2
NAFTA	35.7	34.6	58.8	52.7
Mexico	42.7	37.7	68.9	61.9
European Union 15	43.7	38.9	54.3	52.4
World	42.1	34.9	54.4	52.4
ICT products				
China and Hong Kong	31.9	20.9	59.2	72.7
China	23.5	20.7	51.2	72.8
East Asia	51.8	44.6	71.3	77.4
Japan	51.8	51.3	60.7	53.6
Korea	70.4	38.6	64.4	78.3
Taiwan	57.6	52.9	78.6	81.1
ASEAN	43.5	40.4	79.8	85.7
NAFTA	55.6	39.0	63.2	57.3
Mexico	50.5	36.2	70.4	65.9
European Union 15	54.9	48.9	54.9	51.1
World	51.2	36.1	60.9	61.0

* including ICT products
Source: Compiled from United Nations (UN), various years. *Commodity Trade Statistics Database (COMTRADE)*, Statistics Division, United Nations. Available from http://comtrade. un.org/bd/

by the US multinationals' networks (through their own affiliated firms and arm's-length trade relations) (Lipsey 1998; Athukorala 2007). US multinational affiliates in Southeast Asia have, in fact, expanded assembly activities in the region since the emergence of China as the global assembly centre for ICT products and other machinery, and these firms supply parts and components to their own affiliates and other firms involved in assembly operations in China (Athukorala 2007).

Second, the share of parts and components in US imports from China is remarkably low compared with the figures for the other East Asian countries, as well as with the global average. In years for which data are reported, parts and components account for about 20 per cent of total ICT imports to the United States—that is, final goods accounted for nearly four-fifths of total imports. Consequently, the increasing trend of China's penetration in the US ICT markets is much sharper (Figure 4.4) in terms of data for final goods than for figures based on the standard (gross) trade data. Third—and related to the two previous points—two-way trade in parts and components seems to account for a much larger share of trade between the United States and other East Asian countries (in particular, ASEAN countries) compared with trade with China. It seems that China's comparative advantage in global production sharing is still concentrated disproportionately in final assembly.

Determinants of trade flows

It is clear from the discussion so far that Chain's emergence as an important player in global production networks is an important structural factor behind the widening trade deficit between the United States and China. We now turn to a more formal examination of the determinants of US trade, distinguishing between imports and exports, and focusing on the behaviour of the trade flows of final goods and parts and components within machinery exports. The purpose is to examine whether trade with China has a specific effect on the overall international trade patterns of the United States beyond what can be expected in terms of the standard determinants of bilateral trade flows. The analytical tool used for this purpose is the gravity equation, which has become a standard tool for analysing bilateral trade flows. For the purpose of our analysis, we augmented the basic gravity model in a number of ways to yield the following specification

$$Ln\ TRD_{i,j} = \alpha + \beta_1 lnGDP_i + \beta_2\ lnGDP_j + \beta_3\ lnPGDP_i + \beta_4\ lnPGDP_j$$
$$+\ \beta_5 lnDST_{i,j} + \beta_6 ADJ_{i,j} + \beta_7 lnRULC_{i,j} + \beta_8\ lnRER_{i,j} + \beta_9 DCH$$
$$+\ \beta_{10} DJP + \beta_{11} DTW + \beta_{12} DAS + \gamma T + \varepsilon_{ij} \tag{1}$$

in which *i* and *j* refer to the reporting country (the United States) and the partner country, and *Ln* denotes natural logarithms. The variables are listed and defined below, with the postulated sign of the regression coefficient for the explanatory variables in brackets.

TRD	trade (imports [*MP*] or exports [*EX*]) between *i* and *j*
GDP	real GDP (+)
PGDP	real GDP per capita (+)
DST	the distance between the economic centres of *i* and *j* (–)
ADJ	a binary variable assuming the value of 1 if *i* and *j* share a common land border and 0 otherwise (+)
RULC	relative unit labour costs of manufacturing between *j* and *i* (*EX* +; *MP* –)
RER	an index of bilateral real exchange rates, which measure the international competitiveness between *j* and *i* (*EX* +; *MP* –)
DCH	intercept dummy variable for China (+ or –)
DJP	intercept dummy variable for Japan (+ or –)
DTW	intercept dummy variable for Taiwan and Korea (+ or –)
DAS	intercept dummy variable for the five major member countries of ASEAN (Indonesia, Malaysia, Philippines, Singapore and Thailand)
T	a set of time dummy variables to capture year-specific 'fixed' effects
α	a constant term
ε	a stochastic error term, representing the omitted other influences on bilateral trade.

The first four explanatory variables (*GDP*, *GDPP*, *DST* and *ADJ*) are the standard gravity-model arguments that do not require further discussion. Among the remaining variables, the relative unit labour cost (*RULC*, relative manufacturing wages adjusted for labour productivity) is presumably a major factor impacting on the global spread of fragmentation-based specialisation (Jones and Kierzkowski 2001a, 2001b). In a context in which capital and components have become increasingly mobile, the relative cost of production naturally becomes an important consideration in cross-border production. The inclusion of the real exchange rate, *RER*, which captures the international competitiveness of traded-goods production, is based on a similar reasoning. Another important determinant of trade flows suggested by the theory of production fragmentation is the cost of 'service links' connecting 'production blocks' in different countries. There is no unique measure of the cost of service links; however, in our model, distance (*DST*), adjacency (*ADJ*) and per capita income (*PGDP*) capture certain aspects such as costs. Technological advances during the post-World War II era have certainly contributed to a remarkable reduction in international communication costs. There is, however, evidence that geographical 'distance' is still a key factor in determining international transport costs—in particular, shipping costs—and delivery time (Evans and Harrigan 2003). Timely delivery can in fact be a more important influence on vertical trade than final trade because of multiple border crossings involved in the value-adding chain. The common border dummy (*ADJ*) captures possible additional advantages of proximity that are not captured by the standard distance measure (the greater cycle distance between capital cities). The inclusion of *PGDP* as an explanatory variable allows for the fact that more industrialised countries have better ports and communication systems and other trade-related infrastructure than developing countries as well as better institutional arrangements for contract enforcement that facilitate trade by reducing the cost of maintaining 'service links'.

The China dummy (*DCH*) is expected to capture the 'China effect' over and above the other variables. Dummy variables are also included for Japan (*DJP*), Taiwan and Korea (*DTW*), ASEAN (*ASN*) and Mexico (*DMX*), guided by the empirical regularities in trade patterns observed in the previous section. We observed that China's rapid export expansion in standard labour-intensive manufactures and ICT products has been in direct competition with these countries. It is therefore important to control for any unobserved fixed effects relating to these countries for precise estimation of the 'China effect'. Finally, the time-specific fixed effects (*T*) are included to control for general technological change and other time-varying factors.

The model was estimated using annual data for machinery trade in the period 1992–2005 for all countries, each of which accounted for 0.1 per cent or more of total manufacturing trade in 2000–01. There were 41 US trading partner countries, which satisfied this criterion. Of these, Hong Kong was combined with China because of its peculiar trade links with the latter.[8] Our data set therefore relates to 40 countries. Data for bilateral exports are compiled from the importers' records (CIF) of the UN COMTRADE database (UN various years). The data were disaggregated into components and final products following the procedures detailed in Athukorala (2006). The data sources for other variables and methods of variable construction are explained in the appendix.

We use the random-effect estimator as our preferred estimation technique. The alternative fixed-effect estimator is not appropriate because our model contains a number of time-invariant variables. Note that our panel data set relates to bilateral trade (imports and exports) for the United States (not bilateral trade flows of all countries under study). This means that the reporting country's (the United States) GDP and PGDP have only 'within variation' in the data panel. It is not possible therefore to retain one or both of these variables and time dummies in the same regression because of multiple co-linearity. After undertaking experimental estimations to see the senility of the results to alternative specifications (with income variables only and time dummies only while keeping all other variables the same), we opted for the version with time dummies. It turned out to be superior to the alternative in terms of the overall fit, and in the economic plausibility and statistical significance of the coefficient estimates of the other variables. This specification choice means that the estimated coefficients of time dummies capture the reporting country's (the United States) income effect and other time-specific factors impacting on trade flows. Relating to the latter, the most noteworthy developments are China's accession to the WTO in 2001 and the subsequent tariff reductions, and the abolition of Multifibre Arrangement (MFA) with effect from 2005. The common border dummy (*BRD*) could not be retained in the final estimation because of its high correlation with the distance variable. This is not surprising given the United States' high intensity of trade with its two neighbours, Mexico and Canada. We also tested additional dummy variables for Canada and North American Free Trade Agreement (NAFTA) membership (in place of the Mexico dummy) in experimental runs. Both variables turned out to be statistically insignificant over and above the other variables, and they had no significant effect on the size/statistical significance of the other coefficients in the regressions. The regression results are reported in Table 4.6.

The results for the distance variable (*DST*) provide strong support for the hypothesis that transportation and other distance-related costs are important determinants of trade flows. For total manufacturing, there is evidence of a symmetrical effect of distance on trade: the distance coefficient of import and export equations is remarkably similar in magnitude (1.18 and 1.12, respectively). Interestingly, at the disaggregated level, the distance coefficient for components and final goods of machinery imports are much larger than the coefficients of other manufacturing and total manufacturing.[9] This difference is consistent with the hypothesis that vertical specialisation, given the multiple border crossings involved in the production process, is much more sensitive to transport costs. The distance coefficients on exports of parts and components and final goods are smaller in magnitude than the respective coefficients on the import side. This asymmetry in the distance effect is consistent with the increased concentration over time of US machinery exports, in particular ICT exports, in 'high value-to-weight' segments of the production process within global production networks—a process that seems to have helped US exporting firms to overcome trade barriers associated with distance. The distance coefficient of machinery parts and components exports is, however, much larger (1.26) than that of final machinery (0.91), presumably because speedy delivery is a relatively more important determinant of success in components exports trade than in trade in final goods. This inference is consistent with the recent tendency (observed in the previous section) of US multinationals to expand components assembly and testing in Southeast Asia for supply to the final assembler in China (and other countries in the region). The coefficient of per capita GDP is highly significant and similar in magnitude (about 0.3) in the four export equations. This perhaps reflects the heavy industrial-country bias in the geographic patterns of US exports. On the import side, the coefficient is significant only in the two machinery equations. Moreover, the coefficient of the final goods equation is almost twice that of parts and components. It seems that inter-country differences in the stage of economic development are important to imports to the United States taking place within global production networks, in particular final goods.

The coefficient of the relative unit labour cost variable (RULC) is statistically significant with the expected (negative) sign only in the equation for final machinery imports. It suggests that, other things being equal, a 1 percentage-point difference among exporting countries is associated with 0.35 per cent difference in growth of exports of this product category to the US market. This unique result is consistent with the important role played by relatively low unit labour costs in the rapid penetration of ICT products and other assembled goods from China in the US market.

Turning to results for the real exchange rate (RER), on the import side, its coefficient is barely significant with the unexpected (positive) sign in the equation for final machinery and is not different from zero in the other three equations. On the export side, the coefficient carries the expected (positive) sign in all four equations and it fails to achieve significance only in the machinery parts and components equation. The coefficients are, however, rather small—less than 0.1 in all cases. Overall, there is no evidence here to suggest that the exchange rate plays a significant role in determining the United States' widening trade gap.[10] These results are generally consistent with the available evidence that global production sharing considerably weakens the link between the degree of exchange and trade performance, particularly when it comes to the components trade (Gron and Swenson 1996; Swenson 2000).

Table 4.6   Determinants of US manufacturing imports and exports: regression results, 1992–2005[1]

Explanatory variables[2]	Total manufacturing		Machinery and transport equipment				Other manufacturing[3]	
			Parts and components		Final goods			
a) Imports								
Ln GDP exporter	0.87***	(3.95)	0.98***	(3.99)	0.79***	(3.24)	0.82***	(4.12)
Ln PGDP exporter	0.37*	(1.56)	0.37*	(1.76)	0.610**	(3.05)	0.24	(0.80)
Ln distance (DST)	-1.18***	(3.19)	-1.65***	(4.56)	-1.651***	(3.44)	-0.912***	(2.56)
Ln relative unit labour cost (RULC)	-0.01	(1.01)	+0.02	(0.13)	-0.37***	(2.98)	+0.05	(0.24)
Ln real exchange rate (RER)	0.01	(0.62)	-0.00	(0.28)	0.04*	(1.71)	-0.00	(0.08)
China dummy (DCH)	2.83***	(3.96)	2.16***	(3.58)	4.25***	(5.34)	2.62***	(3.76)
Japan dummy (DJP)	1.32**	(2.31)	1.42**	(2.57)	2.13***	(3.12)	1.23*	(1.89)
ASEAN dummy (DAS)	2.83***	(5.31)	3.61***	(5.98)	4.17***	(6.89)	1.85***	(2.99)
Korea and Taiwan dummy (DKT)	1.53***	(5.58)	2.24***	(8.66)	2.60***	(8.12)	0.819***	(3.14)
Mexico dummy (DMX)	1.065*	(1.63)	1.03*	(1.89)	1.33**	(1.82)	0.961	(1.08)
Constant	1.459	(0.21)	0.918	(0.12)	3.614	(0.43)	1.099	(0.14)
Observations	467		467		467		467	
R² overall	0.67		0.68		0.69		0.60	
R² within	0.79		0.57		0.63		0.72	
R² between	0.67		0.70		0.68		0.61	
RMSE[3]	0.18		0.33		0.29		0.21	
b) Exports								
Ln GDP importer	0.88***	(7.97)	0.907***	(5.81)	0.81***	(7.06)	0.775***	(4.87)
Ln PGDP importer	0.30**	(2.05)	0.37**	(2.06)	0.37***	(3.71)	0.341**	(2.13)
Ln distance (DST)	-1.12***	(2.96)	-1.261***	(3.15)	-0.910**	(2.38)	-1.079***	(2.63)
Ln relative unit labour cost (RULC)	-0.02	(0.42)	-0.01	(0.11)	+0.02	(0.20)	-0.05	(1.10)
Ln real exchange rate (RER)	0.03*	(1.65)	0.01	(0.29)	0.072***	(3.27)	0.03**	(2.28)
China dummy (DCH)	1.03**	(2.35)	1.03**	(1.97)	1.42***	(4.14)	1.26**	(2.37)

Japan (DJP)	-0.51*	(1.78)	-0.58*	(1.86)	-0.49	(1.02)	-0.53**	(2.07)
ASEAN dummy (DAS)	2.15***	(6.13)	2.87***	(5.73)	1.705***	(5.13)	1.665***	(5.20)
Korea and Taiwan dummy (DKT)	1.10***	(6.11)	1.52***	(7.74)	0.896***	(4.86)	0.99***	(5.21)
Mexico dummy (DMX)	1.025*	(1.81)	1.25*		0.65	(0.89)	0.634	(1.17)
Constant	1.609	(0.37)	0.05	(0.04)	-0.25	(0.06)	2.73	(0.57)
Observations	464		464		464		464	
R^2 overall	0.67		0.62		0.69		0.65	
R^2 within	0.61		0.49		0.28		0.67	
R^2 between	0.69		0.66		0.73		0.65	
RMSE[4]	0.18		0.28		0.29		0.15	

*** 1 per cent statistical significance (based on the standard t-test)
** 5 per cent statistical significance (based on the standard t-test)
* 10 per cent statistical significance (based on the standard t-test)

[1] Estimated by applying the random-effect estimator to annual data for bilateral trade for 41 countries in the period 1992–2003. The standard errors (SEs) of the regression coefficients have been derived using the Huber-White consistent variance–covariance ('sandwich') estimator.

[2] for variable definitions and details on variable construction, see the appendix

[3] total manufacturing (SITC 5 through 8–68) less machinery and transport equipment (SITC 7)

[4] root mean square error

Note: Results for the time dummies are not reported.

The coefficient of the China dummy (*DCH*) is positive and statistically significant in all equations.[11] It is much larger in the import equations, indicating that, after controlling for the standard determinants of trade flows, exports from China have penetrated the United States at a rate much higher rate (on average, 16 times) than those from other countries. The coefficient of *DCH* in the final machinery export equation is strikingly large (4.25) and is almost twice that in the equation for parts and components (and total manufacturing). This result is consistent with the dominant 'assembly bias' in the emerging patterns of China's export specialisation, which we observed in the previous section. The differences in magnitude among the coefficients of *DCH*, *DAS*, *DKT* and *DMX* in each of the four import equations are also consistent with the observed differences in relative export performance. The much larger coefficient of the ASEAN dummy in the component equation (3.61) is particularly noteworthy. As discussed, the explanation seems to lie in economic history: the early choice of the region by multinational companies as a location for components assembly and testing in their global production networks.

On the export side, there is no evidence to suggest that US firms perform relatively poorly in exporting to China compared with exports to other countries. The coefficient of *DCH* is greater than unity and is statistically significant in all cases, suggesting that, once other determinants are controlled for, on average, exports to China have grown almost three times faster than exports to other destinations. The results for the dummy variables also do not reveal any notable difference in the rates of expansion of exports to the United States from China and Mexico when other relevant variables are controlled for—in particular, the specific adjacency/distance advantage of Mexico. A comparison of the results for China and ASEAN corroborate our earlier observation of the growing complementarity among these countries in their trade links with the United States within global production networks.

Concluding remarks

The evidence harnessed in this chapter supports the view that, in a context in which international fragmentation of production is becoming the symbol of economic globalisation, the standard trade-flow analysis leads to misleading inferences about the sophistication of China's emerging export patterns. Although China has displayed a rapid increase in exports of high-tech products in recent years, the real value added in China is generally not in high-tech activities. When components are netted out, it becomes clear that China is specialising in labour-intensive niches within otherwise skill-intensive sectors. China's general patterns of trade are much in line with its underlying comparative advantage in labour-intensive production.

In the contemporary international economy, in which global production sharing is expanding rapidly, the real story behind the US–China trade gap is much more complicated than what is revealed by the standard trade-flow analysis undertaken with data coming from a data-reporting system developed at a time when countries were trading predominantly (if not solely) in final goods. The widely held view that China's rapid market penetration of the US economy is driven by unfair trade practices needs to be re-examined in light of the fact that the two economies are deeply interconnected and interdependent within global production networks. The growing trade deficit between the two countries has been underpinned by China's emergence as the main point of final assembly in Asian production networks based on its ample supply of labour and moves by US firms to supply high-end parts and components from their Asian bases. In sum, the deficit is to a large extent a structural deficit driven by the process of global production sharing. It is akin to the substantial structural surplus in the oil-exporting countries (based on their specific resource endowment), which the rest of the world has become accustomed to living with.

Given the current state of China's factor-market conditions (as surveyed in a number of recent studies, including Cooper 2006; Meng and Bai 2007; Naughton 2007), one can speculate that China's trade patterns are unlikely to change dramatically in the short to medium term. China still has about half of its labour force employed in agriculture, where its productivity is, on average, barely one-eighth of that in industry and about one-quarter that in the service sector. Agriculture still accounts for more than 45 per cent of total employment in the country even though agriculture's share in GDP is only 13 per cent. GDP per worker in the economy as a whole is three times the value added per worker in agriculture. The country still remains very rural, with an urbanisation rate of about 40 per cent of the total population—much lower than the 'normal' level of 60 per cent consistent with China's income level. These features, coupled with the high skilled–unskilled wage differential (which, according to some estimates, has risen from 1.3 to 2.1 in the past decade), suggest that China still has much potential for moving unskilled workers out of agriculture and into manufacturing and other productive urban-sector activities. For this to happen, the global trading environment needs to remain accommodative and Chinese policies need to be receptive to gains from specialisation on the basis of comparative advantage.

Given the current state of data, in this chapter, we have focused solely on US–China trade in goods. The inferences therefore need to be qualified for the fact that the difference between merchandise trade and services trade has become increasingly blurred because of the continuing process of production fragmentation. US firms that have shifted components production/assembly and final assembly activities and that manage 'service links' involved in the global production networks from their home bases undertake knowledge-based or information technology-enabled services (Brown and Linden 2005). In other words, as part of the continuing process of global production sharing, the related services—particularly knowledge-based or information technology-enabled services that are beyond the traditional notion of internationally traded services, such as transportation, travel and tourism—have become increasingly tradable. There is evidence that exports of these new production-related services (and the related employment opportunities) have significantly expanded in recent years (Mann 2006). The surplus in US services trade (which has persisted since the late 1970s) has expanded rapidly in recent years, reaching US$75 billion in 2006. The largest subcategory—the export growth of which has far outpaced growth in all other services—in the services account is 'other private services' trade, which captures many of the information technology-related services, management and consultancy services and business, professional and technical services, all of which are central to the process of global production sharing.[12] An analysis that overlooks these exports could overstate the magnitude of the US–China trade imbalance, presumably by a wide margin.

Notes

1   In the recent literature on international trade, an array of alternative terms has been used to describe this phenomenon, including 'international production fragmentation', 'vertical specialisation', 'slicing the value chain' and 'outsourcing'.

2   The data reported in this chapter, unless otherwise stated, come from the UN COMTRADE database. Throughout the chapter, inter-temporal comparison calculations are made for the two-year averages relating to the end points of the period under study so as to reduce the impact of year-to-year fluctuations of trade flows.

3   According to available estimates, 70 per cent or more of assembled products are sold domestically (Bergsten et al. 2006:90).

4   Exports by US multinational affiliates in China to the United States accounted for only 6 per cent of total US imports from China in 2004 (Bosworth and Collins 2008:Table 5)

5   For instance, the typical notebook computer made in a Taiwanese-owned factory in China has processing chips made by Intel in Malaysia, an operating system made by Microsoft, a CD display screen sourced from Taiwan or Korea and hard-disk drives sourced from Japan. Domestic value added (the cost of labour, components sourced within China and the profit earned by foreign-owned companies in China) is only one-third of the value of output (Dean and Tam 2005).

6   It is important to note that Chinese estimates of the US trade deficit have always been lower than the United States' own figures because of the different ways the United States treats products traded to and from China that pass through Hong Kong. According to estimates by Fung et al. (2006), the official US data tend to overstate the real deficit by about 17 per cent, while the degree of underestimation involved in the Chinese official estimate is as high as 33

per cent. This discrepancy does not, however, seem to create a serious problem in examining overall trends in the trade gap.

7　For details on this decomposition procedure, see Athukorala (2006).

8　We also treated Hong Kong as a separate country in experimental runs and found that results were insensitive to this alternative specification.

9　The differences are statistically significant at the 1 per cent level or better.

10　In experimental regression runs, we also interacted *RER* with *CHD* and failed to detect any specific Chinese effect on the link between *RER* and trade flows.

11　Note that, as the model was estimated in natural logarithms, the percentage equivalent for any dummy coefficient is [(dummy coefficient) – 1]* 100).

12　From 1995 to 2005, US exports of 'other services' grew 143 per cent, compared with 44 per cent growth in all other services, and accounted for 90 per cent of the overall US services trade surplus in 2005—up from 38 per cent in 1995 (CEA 2007).

References

Athukorala, P.-C., 2006. 'Product fragmentation and trade patterns in East Asia', *Asian Economic Papers*, 4(3):1–28.

——, 2007. *Multinational Enterprises in Asian Development*, Edward Elgar, Cheltenham, UK.

——, 2008 (forthcoming). 'Singapore and ASEAN in the new regional division of labour', *Singapore Economic Review,* 53(3).

Bergsten, C.F., Gill, B., Lardy, N.R. and Mitchell, D., 2006. *China: the balance sheet*, Public Affairs, New York.

Bosworth, B. and Collins, S., 2008. 'United States–China trade: where are the exports?', *Journal of Chinese Economics and Business Studies*, 6(1):1–21.

Brown, C. and Linden, G., 2005. 'Offshoring in the semiconductor industry: a historical perspective', in L. Brainard and S.M. Collins (eds), *The Brookings Trade Forum 2005: offshoring white-collar work: the issues and implications*, Brookings Institution Press, Washington, DC:270–333.

CEPII, various years. *Institutional Profiles Database*, CEPII. Available from www. cepii.fr/anglaisgraph/bdd/form_instit/login.asp

Council of Economic Advisors (CEA), 2007. *Economic Report of the President 2007*, US Government Printing Office, Washington, DC.

Cooper, R.N., 2006. *How integrated are Chinese and Indian labor into the world economy*, Background paper for A.L. Wintersand and S. Yusuf, 2007, *Dancing with Giants: China, India, and the global economy*, World Bank, Washington, DC. Available from http:/econ.worldbank.org/dancingwithgiants.

Dean, J. and Tam, P.-W., 2005. 'The lap top trail', *The Wall Street Journal*, 9 June.

Easton, N., 2008. 'Make the world go away', *Fortune*, 157(2):104–6.

Evans, C. and Harrigan, J., 2003. *Distance, time and specialization*, NBER Working Paper 9729, National Bureau of Economic Research, Cambridge, Mass.

Fung, K.C., Lawrence, J.L. and Xiong, Y., 2006. *Adjusted estimates of United States–China bilateral trade balances: an update*, Stanford Center for International Development Working Paper No.278, Stanford University.

Gosh, A. and Rajan, R.S., 2007. 'A survey of exchange rate pass-through in Asia', *Asian-Pacific Economic Literature*, 21(2):13–28.

Gron, A. and Swenson, D.L., 1996. 'Incomplete exchange-rate pass-through and imperfect competition: the effect of local production', *American Economic Review*, 86(2):71–6.

Hufbauer, G.C., Won, Y. and Sheth, K., 2006. *US–China Trade Disputes: rising tide, rising stakes*, Institute for International Economics, Washington, DC.

Jones, R.W. and Kierzkowski, H., 2001a. 'A framework for fragmentation', in S. Arndt and H. Kierzkowski (eds), *Fragmentation: new production patterns in the world economy*, Oxford University Press, New York:17–34.

——, 2001b. 'Globalization and the consequences of international fragmentation', in R. Dornbusch, G. Calvo and M. Obstfeld (eds), *Money, Factor Mobility and Trade: the festschrift in honor of Robert A. Mundell*, MIT Press, Cambridge, Mass.:365–81.

Krugman, P., 2008. Trade and wages reconsidered, Paper presented at the 2008 spring meeting of the Brookings Panel on Economic Activity. Available from http://www.princeton.edu/~pkrugman-bpea-draft.pdf

Lardy, N.R., 2005. 'China: the great new economic challenge', in C.F. Bergsten (ed.), *The United States and the World Economy: foreign policy for the next decade*, Institute for International Economics, Washington, DC:121–41.

Lipsey, R.E., 1998. 'Trade and production networks of US MNEs and exports by their Asian affiliates', in J. Dunning (ed.), *Globalization, Trade and Foreign Direct Investment*, Elsevier Science Press, Amsterdam.

Mankiw, N.G. and Swagel, P.L., 2005. 'Antidumping: the third rail of trade policy', *Foreign Affairs*, 84(4):107–12.

Mann, C.L., 2006. *Accelerating the Globalization of America: the role for information technology*, Institute for International Economics, Washington, DC.

Meng, X. and Bai, N., 2007. 'How much have the wages of unskilled workers in China increased? Data from seven factories in Guangdong', in R. Garnaut and L. Song (eds), *China: linking markets for growth*, Asia Pacific Press and ANU E Press, The Australian National University, Canberra:151–75.

Naughton, B., 2007. *China's Economy: transition and growth*, MIT Press, Cambridge, Mass.

Rodrik, D., 2006. 'What's so special about China's exports?', *China and the World Economy*, 2(1):153–72.

Shirk, S.L., 2007. *China: fragile superpower*, Oxford University Press, New York.

Soloaga, I. and L. A. Winters., 2001. ' Regionalism in the Nineties: What Effect on Trade?", *North American Journal of Economics and Finance*, 12 (1): 1-15.

Steinbock, D., 2008. 'US presidential election 2008: policy implications for US–China trade and investment', *China and the World Economy*, 16(3):40–56.

Sung, Y.-W., 2007. 'Made in China: from world sweatshop to a global manufacturing centre?', *Asian Economic Papers*, 6(3):43–72.

Swenson, D.L., 2000. 'Firm outsourcing decisions: evidence from US foreign trade zones', *Economic Inquiry*, 38(2):175–89.

United Nations (UN), various years. *Commodity Trade Statistics Database (COMTRADE)*, Statistics Division, United Nations. Available from http://comtrade.un.org/bd.

US Department of Commerce (USDC), various years (a). 'Income and employment by industry', *National Economic Accounts*, Section 6, Bureau of Economics. Available from http://www.bea.gov/bea/dn/nipaweb/SelectTable.asp?Selected=N#S6.

——, various years (b). *Survey of US Direct Investment Abroad*, Bureau of Economics. Available from http://www.bea.doc.gov/bea/uguide.htm#_1_23.

Weisman, S., 2007. 'Fourth senate seek plenty for China', *New York Times*, 14 June. Available from http://www.nytimes.com/2007/06/14/business/world

World Bank, various years. *World Development Indicators*, World Bank, Washington, DC.

Yusuf, S., Nabeshima, K. and Perkins, D., 2007. 'China and India reshape global industrial geography', in A.L. Winters and S. Yusuf (eds), *Dancing with Giants: China, India, and the global economy*, World Bank, Washington, DC. Available from http:/econ.worldbank.org/dancingwithgiants.

Table A4.1 **Data set used in regression analysis: definition of variables, source and variable construction, and the country coverage**

Variable	Definition	Data source/variable construction
EXP	Value of US bilateral trade in US$ measured at constant (2000) price	Trade data (in current US$) compiled from importer records of UN (various years) (http://www.bls.gov/ppi/home.htm), deflated by the manufacturing sub-index of the US producer price index
GDP, PGDP	Real GDP and real per capita GDP (at 1995 price)	World Bank (various years)
DIST	Weighted distance measure of the French Institute for Research on the International Economy (CEPII), which measures the bilateral great-circle distance between major cities of each country	CEPII (various years)
ADJ	A binary dummy variable, which takes the value of 1 for countries that share a common land border and 0 otherwise	CEPII (various years)
RULC	The ratio of unit labour costs (ULC) in country j and country i, where ULC is measured as the ratio of the average manufacturing wage to manufacturing value added per worker—both measured in US$. In construct, an increase (a decrease) in RULC indicates a deterioration (an improvement) in country j's cost competitiveness relative to i (the United States, in this case)	Annual manufacturing wages data for the United States (USDC various years [a]) All other countries: (USDC various years [b])
RER	Real exchange rate: Equation A1 $$RER_{ij} = NER * \frac{P_j^W}{P_i^D}$$ in which NER is the nominal bilateral exchange rate index (US$ price of foreign currency), PW is the price level of country j measured by the producer price index and PD is the domestic price index of country i measured by the GDP deflator. By construct, an increase (decrease) in RER_{ij} indicates a deterioration (an improvement) in country j's competitiveness in traded-goods production vis-à-vis i (the United States, in this case)	Constructed using data obtained from World Bank (various years) Following Soloaga and Winters (2001), mean-adjusted RER is used in the model. This variable specification assumes that countries are in exchange-rate equilibrium at the mean.

Table A4.2 **Country coverage**

Argentina	Finland	Malaysia	Slovakia
Australia	France	Mexico	Slovenia
Austria	Germany	Netherlands	South Africa
Belgium	Hungary	Norway	Spain
Brazil	India	Philippines	Sweden
Canada	Indonesia	Poland	Switzerland
China and Hong Kong	Ireland	Portugal	Taiwan
Costa Rica	Israel	Republic of Korea	Thailand
Czech Republic	Italy	Russian Federation	Turkey
Denmark	Japan	Singapore	United Kingdom

5

Rebalancing equity and efficiency for sustained growth

Justin Yifu Lin

Income distribution is currently one of the most conspicuous problems in China. Since the reforms of 1978, China has maintained rapid economic growth, with an average gross domestic product (GDP) growth rate of 9.7 per cent per annum and a yearly foreign trade growth rate of 17.4 per cent. Growth was particularly strong between 2003 and 2007, during which time China maintained a high GDP growth rate of more than 10 per cent and a foreign trade growth rate of 28.5 per cent per annum. As a country in rapid transition, however, many social and economic problems have developed as the reform deepens. For example, through the end of the 1980s and into the early 1990s, the reform of state-owned enterprises (SOEs) was one of the key talking points in the Chinese economy. It was acknowledged publicly that one-third of SOEs were in deficit, another one-third were known to be in deficit (but this fact was not acknowledged), while only the remaining third were turning a profit. The profitability of SOEs is no longer the main problem in China, as many of the small SOEs have been privatised and the larger SOEs are earning healthy profits. The key issue now is how to improve the competitiveness of SOEs in the global market. A further example of the transition in China's economy is its financial system, which was once very fragile, and which carried a high ratio of non-performing loans. This issue was particularly serious in the four state-owned banks. After several years of reform, however, the ratio of non-performing loans has decreased significantly in the four state-owned banks, and three of them have introduced strategic investors and are listed publicly. Moreover, the stock-market has experienced a full cycle, with a bull market emerging in 2006 and 2007.

While the old problems were solved, new ones have emerged. With the disappearance of the original chief concerns about the Chinese economy, those concerns that were traditionally of only secondary importance are now centre stage. First, investment is growing too quickly, leading to insufficient consumption. The growth rate of investment in fixed assets was 27.7 per cent in 2003, 26.6 per cent in 2004, 25.7 per cent in 2005 and 24 per cent in 2006. More importantly, such high investment growth rates were reached with a very low inflation rate. In the national income account, too much investment leads to a level of consumption that is insufficient to sustain economic growth. In the past five years, China's investment grew faster than its comparatively high increases in consumption of 10 per cent per annum, leading to a sharp accumulation of production capacity. In this case, consumption becomes insufficient when matched with this increase in productive capacity. China needs to export more in order to deal with its excessive production capacity. This is why China has experienced a high growth rate in exports and a rapid accumulation in its foreign trade surplus during recent years. This is a new problem for China.

Second, too many capital and resource-intensive industries have been developed. China's growth is highly dependent on resources. According to Ma Kai (2007), chair of the National Development and Reform Commission of China, in 2006, China contributed 5.5 per cent of global GDP, but it used 15 per cent of the world's energy, 30 per cent of the world's steel and 54 per cent of the world's concrete. Unemployment is another problem for developing capital-intensive industries. Since 1980, although the Chinese growth rate has been very high, the ability of the Chinese economy to create employment has been continually decreasing. The main reason is that China has developed too much capital-intensive industry rather than building up those industries that are labour intensive.

Third, investment has been concentrated mainly in just a few industries. This situation was especially serious in 2003 and 2004. I call this the 'wave phenomena' (Lin 2007a), highlighting that investment tends to wash into certain industries like waves, often leaving the industry with an excess capacity to produce. Take the Chinese steel industry as an example: in 2007, the country's productive ability reached 450 million tonnes, whereas it was only 190 million tonnes in 2002. While China needed to import a great deal of steel in the past, now it is essential for China to export it.

Moreover, there is excess liquidity in the Chinese economy. Large banks usually hold excess reserves. This reduces the effectiveness of monetary policy. Chinese experts usually argue about whether interest management or gross management should be the monetary tool. In some cases, interest management is not as effective as gross management—for instance, when changing the reserve ratio. Because of the existence of excess reserves, raising the reserve ratio sometimes has little influence on credits.

Furthermore, a bubble plagues the real estate market. It is reasonable policy to make real estate the pillar of industry in China. On one hand, surplus labour in rural areas will move into urban areas as a continuation of urbanisation, which will lift the demand for housing; on the other hand, as income levels grow, larger, more expensive houses are demanded by those already living in cities. The demand for real estate is therefore, and will continue to be, high. Property developers in China tend, however, to contradict the normal experience of house construction patterns in industrialised economies. According to this experience, small houses are generally in demand at the first stage of development. When income levels rise, people begin to upgrade the size of their houses and, as a result, there is higher demand for larger properties. The current trend in the Chinese real estate market, however, is for the widespread construction of large houses with insufficient numbers of smaller abodes under construction. This trend must be reversed.

Apart from these, many new social problems have emerged. Among them, income disparity is one of the most notable (Lin and Chen 2007a, 2007b, 2008; Lin 1999, 2005; Lin and Liu 2003). In 1978, the ratio of urban disposable income to rural net income was 2.6:1; this ratio grew to 3.3:1 in 2006. China now has one of the biggest income gaps between urban and rural areas in the world. Simultaneously, a large group of low income-earning citizens has emerged in the larger cities because of widespread unemployment (Lin and Liu 2003). At present, China's Gini coefficient exceeds 0.45, which means it has passed the 'safe line' for social stability, according to many experts. There are, however, many other problems emerging in China, such as incomplete services in health care and the educational system, the prices of which are too high compared with income levels.

Living conditions have improved greatly thanks to the high economic growth rate during the past years. Basic shortages of food and clothing have been solved, even in provincial areas. Social discontent is, however, spreading, largely because of the social problems mentioned above. As the old adage suggests, 'shortage is not a problem, but inequality really matters'. When the income gap becomes more significant, many kinds of problems and contradictions arise. Many people even think that conditions in the past were better than those today. In the past, although they were very poor, people felt content because incomes were more equal; now, however, even though people are generally richer than previously, inequality is causing widespread discontent.

There have been many appeals to cure the problem of income disparity, and redistribution is usually suggested as the tool. In a discussion about personal income tax in 2005 and 2006, for example, many people suggested that China should raise its tax rate for the rich as a way of subsidising low-income groups in the form of transfer payments. The highest personal income tax rate in China has, however, already reached 45 per cent—one of the highest tax rates in the world. Considering this, I would suggest that the main problem with China's tax system is with enforcement, not the rate itself. Critics of this viewpoint express an important social sentiment—that is, many people are unhappy because they believe that some people in China are too rich. This is not, however, the case. Compared with other countries, the rich in China are not rich at all; therefore, we still need to activate the rich to create more social wealth—at least in the present. It is our task to ensure that in the course of development, the income of the poor grows faster than that of the rich, but it should not be accomplished by redistribution.

Under these conditions, it is very difficult to build a harmonious society. Based on the characteristics of the current stage of economic and social development, the Communist Party of China formulated a policy of 'advancing with the times', and put forward a 'scientific concept of development' at its sixteenth National Congress, the contents of which included two key points. On one hand, development should be the ultimate policy goal; on the other hand, in the course of development, we must try to achieve 'five areas of coordination': coordination between urban and rural areas, between regions, between society and the economy, between humans and nature, and between domestic and international markets. At the seventeenth National Congress in 2007, the party put forward a policy of building a harmonious society (Hu 2007), in order to resolve livelihood issues, including education and health-care difficulties, the gap between urban and rural areas, social security and so on. The government's working reports from 2007 and 2008 have put forward many concrete policy measures to solve these livelihood issues.

Comparative advantage: realising equality and efficiency simultaneously

How does one carry out the scientific concept of development and simultan-eously construct a harmonious society in China? The current guidelines for policymaking dictate that primary distribution emphasises efficiency, and redistribution solves equality. This statement suggests that since China aims to emphasise individual livelihoods and social harmony, the Chinese government should play a more active role in realising equality through distribution.

However, this course of action will be unproductive, if not counterproductive, if we fail to realise equality and efficiency simultaneously in primary distribution and, instead, resort only to redistribution. Equality and efficiency should be realised simultaneously in primary distribution, and redistribution, acting as a complementary tool, should solve the residual problems left by primary distribution. It is worth emphasising that in certain circumstances, redistribution can be helpful. In the current system, however, primary distribution emphasises efficiency and redistribution solves equality, which means simply that we should not take the equality problem into consideration in the process of primary distribution. This, it seems, is counterproductive.

Can China really achieve its aim of realising equality and efficiency simultaneously in primary distribution?

Such a policy has at least two meanings. First and foremost, China must maintain a high economic growth rate in order to ensure sustainable growth of average personal incomes. After 29 years of rapid growth, the average personal income in China reached US$2,400 in 2006. This is still very low, reflecting 4–5 per cent of the average personal income in the United States. This suggests that China should continue to emphasise efficiency to speed up economic growth. The second meaning of China's policy is that the country must ensure that incomes for the poor grow faster than those for the rich in the course of economic development, in order to realise equality and efficiency simultaneously.

The question remains, is there a path to the realisation of equality and efficiency through initial distribution? The answer is yes. The imperative is to develop according to China's comparative advantage. If China can choose the industries, products and technologies that are suitable to its comparative advantage, it will be able to improve the distribution conditions and realise the 'five areas of coordination' put forward in its scientific development concept, while at the same time maintaining a high economic growth rate (Lin et al. 1999; Lin 2004a, 2004b, 2004c, 2007b).

The discussion about comparative advantage has lasted for more than 20 years, with some arguing that China should adopt a *competitive* advantage as opposed to the traditional *comparative* advantage. This is a mistaken view of the concepts. In fact, competitive advantage is based on the notion of comparative advantage. Failing to follow comparative advantage means that competitive advantage cannot be achieved. Mike Porter (1990), a professor at the Kennedy School at Harvard University in the United States, first put forward the concept of competitive advantage. He argued that four factors determined competitive advantage: first, whether the country had relatively lower prices of a type of resource compared with the international price—that is, the country had a resource price advantage. In this case, the country should utilise its price advantage and develop industries intensively using the competitive resource instead of other resources. Second, did the developing industry have a large enough domestic market? Where a new industry can rely on high domestic demand, it will have a much better chance of being internationally competitive. Third, does the new industry enjoy a good clustering effect in the home country, which can improve its efficiency? Finally, can the new industry enjoy a competitive domestic market, instead of one that is monopolistic?

Among the four determining factors mentioned by Porter (1990), there are really only two independent factors: the scale of domestic markets and factor prices. The scale of the domestic market is largely fixed, being determined by population size and income levels. In the domain of factor prices, Porter (1990) argues that we must take full advantage of the resources with price advantage, where the price advantage really equates to comparative advantage. The remaining two factors are determined inherently by the price advantage. First, the industrial clustering effect cannot be achieved unless the industry has a comparative advantage. For example, the electronic-processing industry cannot be clustered in the United States as it is in Dong Wan, China, because this industry is highly labour intensive and wages in the United States are very high. Similarly, capital-intensive industries cannot be clustered in these countries, because interest rates in developing countries are high due to the shortage of capital. Second, only industries with comparative advantage can form a competitive domestic market. Without comparative advantage, government protection and subsidies are needed in order to build the industry—therefore, competitive markets cannot develop. All this leads to the conclusion that the competitive advantage theory, in nature, refers only to two basic factors: comparative advantage and the scale of the domestic market.

This leads to the question of which of the two factors is more important? Doubtless, it is good to have a large domestic market; however, if a country follows its comparative advantage, its products can be sold on the international market. In this case, the scale of the domestic market matters little. For example, Finland, whose population is only a little more than five million, has companies that sell predominantly to the international market instead of to the relatively small domestic market (examples include Nokia). Therefore, we can see that competitive advantage is based on comparative advantage, and economies with a competitive advantage can grow rapidly and achieve 'high efficiency'.

Generally speaking, China's market advantage remains its abundance of low-paid workers. Accordingly, many labour-intensive industries can be developed easily, including manufacturing, services and the segments of capital-intensive industries that are relatively labour intensive. Further job opportunities can therefore still be created. On the other hand, competitive power and market share can be increased if China develops according to its comparative advantage, and more profits can be realised. With a surplus, the country can speed up capital accumulation and the capital can be deepened rapidly because the growth of the labour force is relatively steady because of China's restrictions on population growth. The country will change its factor endowment gradually from being a labour-abundant country to one that is capital abundant. This transitional process is also typical in the development history of industrialised countries. When capital becomes abundant and labour becomes relatively scarce, the price system begins to function and wages grow, while interest rates decline. This results in depreciation of the assets of the rich (those with capital) and appreciation of the assets of the poor (low wage earners). As a result, income distribution begins to even out. For example, after World War II, some countries, including Japan, Taiwan, Korea and Singapore, continually improved their income distribution through rapid economic development (and they have since been dubbed the 'East Asian miracles'). Many empirical studies that demonstrate this effect (see, for example, Lin, Cai and Zi 1999b).

In the discussion about comparative advantage, some commentators worry that China might have difficulty developing if its focus remains on labour-intensive industries whereas industrialised countries develop capital-intensive industries. This is, however, a misunderstanding of the situation. Development is achieved in stages. Leap-frogging a stage of development could result in a complete failure to develop; in contrast, moving step by step can lead to rapid development. Industries in industrialised countries are mostly capital and technology intensive, while industries in developing countries are mostly labour and resource intensive. Different industrial structures are, however, the result of natural endowments, not the causes. Industrialised countries are relatively rich in capital but short of labour, so the price of capital is low but the price of labour is high. In order to maintain competitive power, industrialised countries have no choice but to substitute labour for capital by developing capital and technologically intensive industries. The United States has protected its labour-intensive industries such as textiles. If, however, China exports more in these industries, it will hurt the industrial-country markets. Why? Because wages in the United States are much higher than in China, which leads to relatively high production costs in labour-intensive industries in the United States. In contrast, wages in China are relatively low and the industries with competitive power are therefore the labour-intensive ones. It is said that Chinese industries currently lack competitive power; this is not the case. If Chinese industries lack competitive power, why do other countries always raise 'dumping' complaints against China? The industries with comparative advantage are always competitive.

Since competitive advantage is based on comparative advantage, for China to catch up, it is important that it upgrades its endowment structure more quickly than its competitors. Generally, there are three components of factor endowment: capital, labour and natural resources. Because natural resources are relatively fixed, upgrading the factor endowment structure means that China must increase per capita capital rapidly. China can catch up with the industrialised countries over time if its per capita capital increases faster than in industrialised countries. If developing countries continue to grow according to their comparative advantage, their factor endowment structure can be upgraded rapidly and they will, as a result, reach the same level of development as industrialised countries. The logic employed here is simple, but powerful. Output can be used for either current consumption or accumulation of capital. The ratio of output used for accumulation is determined largely by the rate of return on capital. High rates of return on capital lead to high ratios of accumulation. At the early stage of development, the rate of return on capital is very high in developing countries because of the general shortage of capital, leading to a high capital accumulation rate.

A further advantage specific to developing countries is that they are less technologically advanced than industrialised countries (Lin 2003). On the surface, this seems to be a clear disadvantage; however, several gains are afforded from not being at the forefront of technological advancement. Using this advantage, developing countries can upgrade their industries continuously and the return rate on capital does not therefore decrease (or decreases only very slowly). Under these conditions, the capital accumulation rate in developing countries can remain continually higher than the rate in industrialised countries. The industries with the biggest potential for technological 'catch-up' are those that use existing technologies and produce products with established demand, and this means that developing counties can introduce new technologies, absorb them and re-innovate at very low cost. As a result, technology can be updated frequently, leading to a slow decrease or even stable rate of return on capital. The capital accumulation rate will also remain high in developing countries. In contrast, technology in industrialised countries is positioned at the frontier of the global technology chain, so invention and innovation drive development with high costs and high probability of failure. Although industrialised countries hold most of the world's patents, their industries and technologies update more slowly than in developing countries because the latter are less technologically advanced and can take advantage of the industrialised world's technology. This is why industrialised countries typically have a lower rate of return on capital and a lower propensity to invest compared with developing countries. After 10–20 years, or at most two generations, the average levels of capital per capita in developing countries can catch up with those in industrialised countries. By making full use of comparative advantage and the technological gap, for example, East Asian countries successfully shortened the gap or even caught up with industrialised countries within only one generation during the 1960s and 1970s.

Several years ago, I put forward a concept I called 'run fast with small steps'. This means that a developing country needs to keep each step of its industrial and technological upgrading fairly small, but it must also ensure that the frequency of upgrades is high. The goal, in China at least, should be to catch up with industrialised countries within two to three decades, or even within one generation. Most people see high-technology industries only in industrialised countries and low-technology industries only in developing countries. The level of industry and technology certainly influences national power and the income levels of a country and, as a result, this leaves some hoping that the process of development can be achieved through shortcuts. This catch-up strategy, however, which aims to develop industries and technology as quickly as possible, cuts across the grain of comparative advantage.

Industrialised and developing countries have distinct comparative advantages and therefore industries with comparative advantage in industrialised countries might not be suitable for developing countries, such as China. If developing countries adopt a catch-up strategy—that is, they set up capital-intensive industries across the board—these industries will lack competitive power in the international market. This leads to inefficient resource allocation across the whole economy, which in turn leads to low efficiency in economic development and a range of other problems, including over-concentration of capital and reductions in levels of employment. Moreover, a large number of low-income wage earners are shut out of the labour market and cannot share the fruits of economic development, which leads to an increase in explicit and implicit unemployment. Worse still, since industries under comparative advantage-defying strategies have no competitive power in an open and competitive market, they can survive only if protection and subsidies shield the industries from competition. Governments invest in capital-intensive industries in centrally planned economies, but in market economies, only the rich can do this.

This begs the question, where are the subsidies coming from? They will come, directly or indirectly (through taxes, subsidy policies and so on), only from people who do not invest in the capital-intensive industries—namely, the poor. Providing subsidies to the rich by extracting from the poor will lead to a worsening of income and wealth distribution. Without the adoption of a permanent-residency registration policy, as used in some countries, the rural unemployed swarm into big cities and form slums because they cannot enter the formal employment market. Of course, developing countries can set up several high-level industries, large companies and brands, as can be found in industrialised countries, but the profits of these industries, by and large, come from government protection and subsidisation, which is a kind of wealth transfer rather than real value creation by the firms. The labour force would be better served by working in labour-intensive industries that have comparative advantage, but they cannot do this because there is not enough capital to work with. Since wealth transfer and the uncompetitive capital-intensive industries create nothing, the remaining capital that can be funnelled into labour-intensive industries is meagre. Ironically, it is these industries that have the potential to create the highest levels of wealth for the developing country's economy. Furthermore, this slows increases in the factor endowment as well as economic growth.

Developing countries that adopt this inappropriate catch-up strategy share the following common economic path: in the beginning, the economy grows rapidly through investment, which is protected and subsidised by the government. As time passes and investment slows, economic growth slows and the country begins to seek foreign capital, which is followed by a short period of rapid growth. The result is that the country is left with industries that cannot create wealth but which have drawn on large foreign loans that must be repaid. If they fail to repay these debts, financial and social crises are inevitable. Moreover, the potential exists for the so-called 'bad market economy' to rear its ugly head. Industries developed under this catch-up strategy need protection and subsidisation from the government; in market economies, however, people who invest in these industries prefer to seek more protection and subsidisation from the government rather than enhancing productive efficiency. This leads to a large number of rent-seeking activities in these economies.

After World War II, Latin America adopted the 'import substitution' strategy, developing various capital-intensive industries, which amounted essentially to a catch-up strategy. This approach to development resulted in a lack of employment, unequal distribution of public finance and the emergence of slums in big cities. Under the resulting conditions of social disharmony and mass unemployment, Latin American governments moved towards 'democratic politics', in which they put forward appealing, populist social welfare policies. This led to large fiscal deficits and financial crises. These issues have continued to plague the development of Latin American countries to this day.

The current distribution system in China emphasises efficiency in primary distribution; however, China does not define efficiency by competitive market power without government protection and subsidisation. Firms that benefit from protection and subsidisation enjoy high profit levels. Essentially, however, these profits are a kind of wealth transfer that will inevitably lead to social instability. If China seeks to solve its equity problem through redistribution, it will risk slumping like Latin America. Therefore, I do not agree with the theory that 'primary distribution emphasises efficiency, and redistribution solves equality'. Rather, I believe that equity and efficiency problems should be solved through the primary distribution of wealth and that redistribution should only complement this.

Equity and efficiency can be realised simultaneously by adopting a comparative advantage-following strategy for development. 'Comparative advantage' is a term used by economists, but which of China's industries are *really* in accordance with its comparative advantage? China's factor endowment structure is changing and improving. Labour-intensive and capital-intensive industries are relative concepts and there are large factor differences between the regions of China: coastal regions such as Shanghai and Shenzhen; inland regions such as Anhui, Jiangxi, Hunan and Hubei; and western regions such as Xinjiang, Ningxia and Gansu. How can development best utilise comparative advantage, taking these vastly differing experiences into account? From the viewpoint of an entrepreneur, profits are more important than the concept of comparative advantage. Profits are determined by the prices of products and factors. Developing according to comparative advantage requires a perfect price system rather than the development of industries according to comparative advantage. The price system must sufficiently and flexibly reflect the relative scarcity of every factor in the factor endowment structure. If a factor is relatively rich, its price is relatively low and vice versa. If the velocity of a factor's accumulation is relatively large, implying that it changes from being relatively scarce to relatively abundant, its price will decrease from relatively high to relatively low, and vice versa. If such a price system is established, entrepreneurs will use relatively cheap factors to minimise costs and maximise profits in competitive markets. Consequently, setting up a sufficiently competitive market system is vital. Development following comparative advantage leads to a high velocity of capital accumulation. Through this, an upgrade of products and technologies leads to an upgrade of industries from labour intensive to relatively capital and technology intensive. In this development process, income distribution will be in labour's favour.

The aim of the reforms in China is to set up a perfect socialist market economy. Because China has developed according to comparative advantage, it has been maintaining fast growth for more than two decades since the beginning of the reforms and the era of openness. In the era of the planned economy, the Chinese government ignored comparative advantage and developed various capital and technology-intensive industries. To set up these industries, the government depressed factor prices artificially, which led to a shortfall of capital. The government then allocated scarce capital to favoured industries through administrative measures. China adopted a gradual process of reform from 1978 onwards. The government now provides some protection and subsidisation for industries that have no comparative advantage while, at the same time, it places some constraints on those industries that have comparative advantage and that enjoyed preferential pricing in the past. Consequently, economic stability is obtained while the economy secures rapid growth. At present, China is the world's third-largest trading country and exports predominantly labour-intensive goods; however, as the economy grows and capital accumulates, the capital and technological composition of export products is increasing. This is the necessary stage of development that follows on from comparative advantage.

The gradual process of reform also creates some issues. Thus far, China is yet to set up a sufficiently competitive market system. The government continues to intervene in the allocation of certain resources. This is a hangover from the era of protection and subsidisation provided to industries without comparative advantage. The distortions leading to these issues are discussed below.

Distortions of financial structures

Before the reform period began, there were no banks or stock-markets in China. After the reform, the government began to develop and perfect the financial system; however, the main task of the financial system during that period was to serve large firms. The process of appropriating funds from government was replaced by bank loans in 1983, meaning that firms could no longer appropriate national public finance but instead had to secure loans from banks at low interest rates. The low-interest loans played the role of subsidisation. To meet the large firms' needs, the government kept depressing interest rates. Meanwhile, the government set up large state-owned banks to provide subsidies for the large SOEs. The current financial system in China, which has inherited aspects of the previous system, comprises mainly the four large state-owned banks. The amount of RMB held by these four banks as a proportion of total RMB used in the financial system is 75 per cent. These four banks provide loans to the large firms. In addition, the stock-market has begun to develop. Of course, the firms that had the capacity to enter the stock-market were the larger firms in the market. Labour-intensive firms utilising comparative advantage are mainly small and middle-sized. Under this highly concentrated financial system, small firms cannot access finance from banks through loans, which restricts their development, and this in turn reduces employment opportunities. All of these factors lead to the following four wage problems.

First, a large number of middle-sized and small firms in labour-intensive industries, such as manufacturing and services, cannot receive financial support. According to the results of an industry survey in 2004, there were 39.2 million individual firms in China. Today, however, the number has fallen to 25.8 million. The drop is due partly to the fact that some small and medium-sized firms experience normal closures, mergers and transfers. A large proportion of these firms, however, disappear because they cannot access enough finance. Second, almost all rural households cannot obtain financial support. As a result, the problem highlighted above is equally prevalent in rural areas. Developing modern agriculture requires financial support. Setting up a plastic canopy costs 10–20,000 yuan and a modern henhouse costs more than 100,000 yuan. Peasants do not have these sums of money and have great difficulty obtaining loans from banks. Therefore, industries that should be able to profit from comparative advantage cannot access the necessary financial support. Meanwhile, capital is over-abundant, which means that the allocation of capital is inefficient. Third, employment creation is relatively insufficient. On one hand, there are high rates of unemployment in rural areas; on the other, a large number of excess rural labourers cannot transfer out. This is the main reason for the growth in income inequality within cities, and between cities and rural areas. Fourth, capital is over-intensive. The firms able to borrow money from banks are generally the larger, former state-owned enterprises and large private businesses that emerged after the reforms began. These firms can borrow money at a fairly low cost (the current interest rate for loans in China is 6 per cent, which is lower than in industrialised countries). These firms will naturally invest capital into capital-intensive industries, which is also the main cause of the over-intensification of capital. Since sufficient employment cannot be realised and those with excess capital (the rich) invest with subsidies, income distribution suffers. Given this, since the marginal propensity of consumption (MPC) of the rich is low while the MPC of the poor is high, the economy will suffer from insufficient consumption due to poor income distribution.

Many social phenomena are related to income distribution inequality. If the banks' capital is over-concentrated, it is easy to see a 'wave phenomenon' in investment because in industrial economies, it is not easy to judge which industries will be profitable. As with the case of the Internet in the 1990s, everyone believed it would be profitable and capital flowed into all types of Internet businesses. This led to overblown prices and the bubble finally burst, with dramatic effects. This phenomenon occurs every one or two decades by accident in industrialised countries. Developing countries can, however, avoid these shocks to the system by learning from the experiences of industrialised economies, and using them as a guide. This means that large amounts of money are injected into particular industries and investment tends to concentrate in that industry. This continuing investment builds on itself within the industry, thus forming the wave phenomenon. Distortions in the financial structure can be a big problem and, if not adjusted, the problems of income distribution might also remain unresolved.

Distortion of prices of resources

In China's formerly planned economy, the prices of capital and resources were depressed artificially. Once the reform process began, China set about adjusting these price distortions; however, only some distortions have been dealt with. During the 1990s, fees and taxes on resources, when taken together, accounted for 1.8 per cent of resource prices. This has led to two problems. First, the deregulation of resource prices, particularly for coal, petroleum, copper and iron ore, resulted in convergence with international price levels. These prices are therefore much higher than before. As the resource sector increased in value, the fees and taxes as a proportion of resource prices fell to 0.5 per cent. The second issue it that while mining firms were once state-owned, in the mid 1980s the sector was opened up to private and foreign-owned companies. As a result of high resource prices and relatively low taxes, the resource sector has become highly profitable. While the value of the exploitation of a resource can be upwards of 10 billion yuan, the cost of securing these rights from Chinese authorities is only tens of million yuan, which equates to about 1 per cent. Consequently, those who receive these rights can reap substantial profits. Several concerns arise from the disparity between the high private earnings of resource companies and the low incomes derived from them. First, income distribution is further skewed in the direction of inequality. Second, as local governments decide on resource exploitation rights, social values are damaged and adverse rent-seeking activities are common. Finally, corrupt practices and corner-cutting in the mining industry tend to lead to poor practices and frequent accidents.

The administrative monopoly

Market monopolisation poses some serious problems for the Chinese economy. Monopolisation tends to be exercised mainly by SOEs in sectors such as electricity services, telecommunications and, in some cases, education. Revenue derived from monopolistic activities is retained by firms and not passed on to the national treasury. This has a distortional effect and further contributes to income disparities. Where primary distribution does not solve the equity problem, the Chinese government lacks the incentive or ability to undertake redistribution to close income gaps. This has come about because reform is incomplete and because development strategies that defy comparative advantage have been followed. If China is able to improve its reform strategy, equity and efficiency can be realised through primary distribution. The remaining issues, such as incomplete markets, asymmetrical information, economies of scale and social security, can be solved through redistribution.

Specific measures

What are some of the ways in which China can achieve equity and efficiency simultaneously through primary distribution? First, reform of the market economy needs to be deepened and, most significantly, the financial structure must be improved. Market economies achieve productivity through a combination of capital and labour. China's most competitive industries tend to be labour intensive, generally in manufacturing or services. The proportion of the tertiary, or services, industry as a part of China's GDP decreased from 40 per cent in 2006 to 39.5 per cent in 2007, due mainly to insufficient capital support to the services industry. Moreover, the large number of rural households that cannot obtain financial support and thus modernise their agricultural practices contributes to the slowing of the tertiary sector.

From the experience of industrialised countries, it is clear that banks tend to begin small and, as development takes hold, increase in size. In the early stages of economic development, labour-intensive industries dominate the market and, as a result, small and medium-sized banks service small and medium-sized firms. As development progresses and capital in the economy increases, so does the scale of firms, as banks and stock-markets emerge. Chinese reform has, however, been predominantly top down, meaning that the State has set up large-scale SOEs, which are served by state-owned banks and stock-markets. The trouble is that there are no small or medium-sized financial institutions to service the loans of smaller enterprises. To overcome this problem, China must develop its small and medium-sized financial sector to encourage the growth of small firms. At present, the Chinese government has begun to allow the establishment of small rural banks, which it is hoped will meet the needs of rural households. This represents a healthy step forward for the Chinese economy (Lin et al. 2006).

There are, however, issues arising from the design of this policy. First, the threshold conditions for setting up a small bank are too low—set at just RMB500,000. Moreover, one of the key articles of regulation states that any new bank must jointly operate with an existing commercial bank which must hold at least 20 per cent stock shares of the new bank. This is difficult to realise in practice. It is certainly prudent to ensure that new banks are well backed; however, it is still difficult for rural households to access loans because of the continued reluctance of the commercial lenders to provide rural loans. The regulations demand that commercial banks buy 20 per cent of the shares in the new bank, which means that these larger banks take on a considerable financial responsibility in the performance of the smaller, rural-focused banks. This erodes any incentive to lend to rural households. As a result of the legislation, the goal of setting up a small banking sector is compromised. For the sector to function properly, two things must change. First, the entry threshold must be raised by between RMB10 and 50 million. This could be coupled with a requirement that each loan be backed by greater collateral in the form of assets. Second, strict supervision should be imposed on the sector.

The second specific measure that China should undertake to achieve efficiency and equity through primary distribution is to impose a significant tax on resources. As mentioned above, the taxes and fees on resources amount to only 1.8 per cent. In the United States, however, such royalties account for 12 per cent of resource prices and tax on petroleum drilled from US waters is set at 16 per cent. In addition, quantity imposition of tax must be replaced by an ad valorem system: when the price of a product rises above a certain level, a high rate of profit tax should be imposed to keep profits reasonable. There are, however, difficulties in designing an appropriate policy to achieve this because the former state-owned mining companies bear a heavy social burden, propped up by government subsidies provided through lower taxes and fees. In contrast, there is a range of private firms that have no social burden. The most effective solution to this imbalance is for the Chinese government to shed the social responsibilities from state-owned enterprises and impose resource taxes according to the law of the market economy.

Finally, the Chinese government should focus on cancelling administrative monopolies and should begin strengthening supervision of monopolistic industries into which competition cannot be introduced. Once competition is introduced into monopolistic industries, prices will go down and profits will be lower. In those industries in which competition is not suitable, such as the electricity industry, the government must strengthen its supervision of prices, costs and revenue allocation.

If these three aspects can be achieved successfully, China will be able to realise equity and efficiency simultaneously in primary distribution. Income differences between urban and rural regions will decline once a complete market system is fully in place. The main reason for the superabundance of liquidity is that small firms are unable to secure loans and large firms have a great deal of unutilised capital due to restrictions on investment. Furthermore, as the number of middle-sized and small banks in China increases, the issue of superabundant liquidity will decline, income distribution will become more equitable, the marginal propensity of consumption will increase and incentives for investment will decrease.

Functions of the government under a development strategy that pursues comparative advantage

If China follows a strategy of comparative advantage, industries with comparative advantage can produce for export while the products of industries without comparative advantage can be imported. This will balance international and domestic markets. Under such a strategy, protection of industry will no longer be required because Chinese firms will be competitive in world markets. This will also reduce rent-seeking activities and reduce adverse effects on social values. The government will still have an important role to play. The Chinese government should focus on the activities that are generally undertaken by market economy governments, as well as attending to certain unique functions that are essential for developing-economy governments.

First, the Chinese government should take responsibility for education services. If the economy develops through a policy of exploiting comparative advantage, the velocity of upgrades in the industrial and technical structures will be fairly high. This has been the case for certain East Asian economies that have transformed from poor agricultural economies to modernised economies within one or two generations. This has been coupled with an increase in per capita incomes, which now rival those in industrialised economies. To meet the needs of rapid upgrades to the industrial and technical structures, China must produce greater numbers of skilled workers and administrators who can adapt to changes in the market. To achieve this, it is essential that the government invests more in education.

Second, greater attention needs to be directed towards improving the quality of Chinese institutions. The relative backwardness of a developing country is not represented only by its level of economic development; it is judged against the level of development of its institutions, particularly its legal and financial institutions. Improvements in institutional quality have explicit positive externalities. This implies that firms will apply sub-optimal quantities of resources to this objective; therefore, to achieve economic development, the government should seek to improve institutional structures and frameworks step by step.

Third, China must develop a social security strategy of sorts. In every economy, there are those who represent a net cost to the economy, such as the aged, children and those individuals without families to support them. Given that these groups cannot always enter the workforce, their survival depends on government support. Through the process of rapid industrialisation, people are more prone to change jobs. During periods of transition from one form of employment to another, those without work will also depend on government support to survive, and this might come in the form of unemployment benefits.

In addition, market failures occur as a result of asymmetrical information, which requires proper government intervention. For example, one might ask the question: why does banking require supervision? The answer is that there is a problem of information asymmetry between the supplier and the purchaser of capital. Without supervision of the system, exploitation is inevitable. Looking further afield, externality issues in areas such as environmental protection will also require government supervision.

Finally, China's government must focus on macroeconomic issues because of the volatility in domestic and overseas markets. This is a significant responsibility for government generally. The phenomenon of 'waves' of investment was discussed above—this happens occasionally in industrialised economies and frequently in developing economies. An important role for government could be to provide signals on the level of investment in particular industries, thus guiding further inflows of capital. This would reduce the severity of over or under-investment. Further areas of government guidance in the economy should be the setting of industry standards, supervision of sectors, such as banking, and setting thresholds for investment.

Conclusions

It is problematic to suggest that primary distribution emphasises efficiency, and redistribution solves equality, partly because there are problems with how we understand efficiency. It is unreasonable to define efficiency as China's ability to compete in industries in which industrialised countries have an established market share. Furthermore, trying to achieve this type of efficiency fails to recognise that China's interests are best served by pursuing industries that make use of comparative advantage. Failing to follow comparative advantage principles will lead to insufficient employment and the widening of income gaps. Meanwhile, if China attempts to solve its equity problem through redistribution, it runs the risk of plunging into the 'Latin America trap'. Consequently, China should realise equity and efficiency simultaneously in primary distribution to promote rapid and healthy economic development. The focus should also be on bridging the income gap within cities, and between cities and rural areas, by lifting employment rates. In doing so, China will be able to avoid the development pattern that concentrates on resources, avert damage to the natural environment and realise the 'five balanced aspects', including unification of markets at home and abroad, and creating a harmonious society.

References

Hu, J., 2007. Hold the great banner of socialism with Chinese characteristics to strive for new victories in building a well-off society, Speech given to the seventeenth National People's Congress of the Communist Party of China, Beijing, 15 October.

Kai, M., 2007. Changing the style of economic growth, realizing rapid and healthy growth, Speech given to Annual Senior Forum on Chinese Development, Beijing, 18 March.

Lin, J.Y. and Chen, B., 2007a. *Development strategy, financial repression and inequality*, CCER Working Paper, China Center for Economic Research, Beijing.

——, 2007b. *Priority to develop heavy industries, urbanization and income disparity between urban and rural areas*, CCER Working Paper, China Center for Economic Research, Beijing.

——, 2008. *Development strategy, technology choice and inequality*, CCER Working Paper, China Center for Economic Research, Beijing.

Lin, J.Y. and Liu, M., 2003. 'Convergence and income distribution in Chinese economic growth', *World Economy*, 26(8):3–13.

Lin, J.Y. and Liu, P., 2003. 'Development strategy, equality and efficiency', *China Economic Quarterly*, 2(2):479–501.

Lin, J.Y., 1999. 'The result of Chinese economic reform and the disparity between different areas', *Economic Development Review*, 1–2:7–22.

——, 2003. 'Backward advantage or backward disadvantage: a discussion with Yang Xiaokai', *China Economic Quarterly*, 2(4):989–1,004.

——, 2004a. *Development Strategy and Economic Development*, Peking University Press, Beijing.

——, 2004b. *Development Strategy and Economic Reform*, Peking University Press, Beijing.

——, 2004c. *Viability, Economic Development and Transition*, Peking University Press, Beijing.

——, 2005. *Development strategies and regional income disparities in China*, CCER Working Paper, China Center for Economic Research, Beijing.

——, 2007a. 'Wave phenomena and the restructure of macroeconomics in developing countries', *Economic Research*, January.

——, 2007b. Development and transition: idea, strategy, and viability, Marshall Lectures of Cambridge University, 31 October–1 November.

Lin, J.Y., Cai, F. and Li, Z., 1999a. *The Chinese Miracle: development strategy and economic reform*, Shanghai Sanlianc Bookstore and Shanghai People's Publishing House, Shanghai.

——, 1999b. 'Development strategy and comparative advantage: re-explanation of the East Asian miracle', *Chinese Social Science*, 5:8–10.

Lin, J.Y., Sun, X. and Jiang, Y., 2006. *A primary study on the appropriate financial structure theory in the course of economic development*, CCER Working Paper, 6 June, China Center for Economic Research, Beijing.

Porter, M.E., 1990. *The Competitive Advantage of Nations, Free Press*, New York.

6

Rural–urban migrants
A driving force for growth

Xiaodong Gong, Sherry Tao Kong, Shi Li and Xin Meng

A spontaneous transfer of rural labour from agriculture to secondary and tertiary industries accompanies economic growth in most countries. It is often believed that this movement is pushed partly by the increase of surplus labour in the agricultural sector and, more importantly, pulled by higher returns to labour in urban manufacturing and service industries. Through this process, modern sectors with high productivity expand and engage more labour to move into modern sectors, leading to an improvement in overall productivity, which in turn gives an impetus to economic growth.

The world has witnessed China's extraordinary pace of economic change in the past three decades. Underlying China's reforms are the economic incentives and possibilities made available by a wide range of institutional changes that stimulated rapid economic growth. Since the 1990s, a most extraordinary phenomenon in China is the large-scale demographic movement induced by the relaxation of regulations that restricted labour mobility since the early years of the communist regime. While the exact size of the migrant population is debatable, official estimates suggest the number of rural migrant workers in 2005 was about 126 million (NBS 2006)—amounting to the largest peace-time movement of people in world history.[1]

This massive reallocation of rural labour to modern urban sectors is, at a micro level, in response to the rapid rise in demand for labour in cities where growth has been concentrated and because the rewards for labour are significantly higher in urban than in rural areas. At a macro level, the reallocation coincides with the restructuring process of the economy from an essentially agrarian society to a predominantly industrial economy. At the beginning of the reform period, the agricultural sector occupied more than 70 per cent of the total labour force and accounted for 28.2 per cent of gross domestic product (GDP). In stark contrast, secondary and tertiary industries engaged more than half of the total labour and contributed to nearly 95 per cent of GDP in 2006 (NBS 2007). During the economic reform, the large population of low-cost rural workers in labour-intensive manufacturing industries contributed substantially to export-led growth. Figure 6.1 shows that the increase in migrant labour is correlated with the growth of GDP per capita income. Figure 6.2 demonstrates that the size of rural–urban migration is positively associated with the increased contribution to GDP of the secondary and tertiary sectors (NBS 1998–2003).[2]

Further investigation of the role of migration reveals that as an integral part of the reform experience, rural–urban migration has served as an important fuel for China's economic growth.

The established economic growth literature has reached a census that economic growth is determined largely by the accumulation of a number of proximate sources—that is, physical and human capital, labour and the enhancement in total factor productivity (TFP). Typically, a growth-accounting method is used to provide a first cut of an economy's immediate sources of growth. Notwithstanding the continuing debate since Young's (1995) seminal paper about the primary source of China's growth—factor accumulation versus productivity growth—the increase in productivity has played a crucial role in China's rapid development.[3]

Studies looking at the sources of China's economic growth have documented that more than one-third of this growth can be attributed to productivity increases (Islam and Dai 2007; Chow 2002; Fan et al. 1999; Woo 1997; Wang and Yao 2001; Borensztein and Ostry 1996; Hu and Khan 1997).[4]

In particular, after incorporating human capital, which was often absent from the early studies, Wang and Yao (2001) found that the contribution of productivity growth during the reform period (1978–99) was about 24 per cent. Using a dual approach, Islam and Dai (2007) estimated an annual TFP growth rate of 2.26 per cent for the period 1978–2002.

For analytical purposes, TFP growth can be decomposed further into at least: 1) a shift of productive factors—that is, labour and capital from a lower productivity sector to a higher productivity sector; and 2) productivity improvement within each sector owing to technical efficiency gains, technological progress and improvements in institutional quality, incentive structures and so on. It is primarily through the first channel that rural–urban migration contributes to improvements in productivity. Modern economic theories elaborating the

Figure 6.1   **Migration and economic structural changes, 1997–2005**

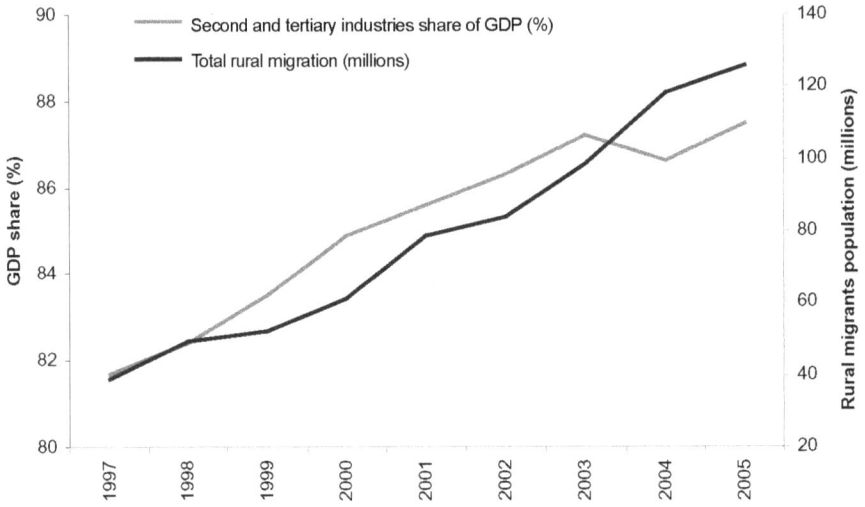

Figure 6.2   **Migration and GDP per capita growth, 1997–2005**

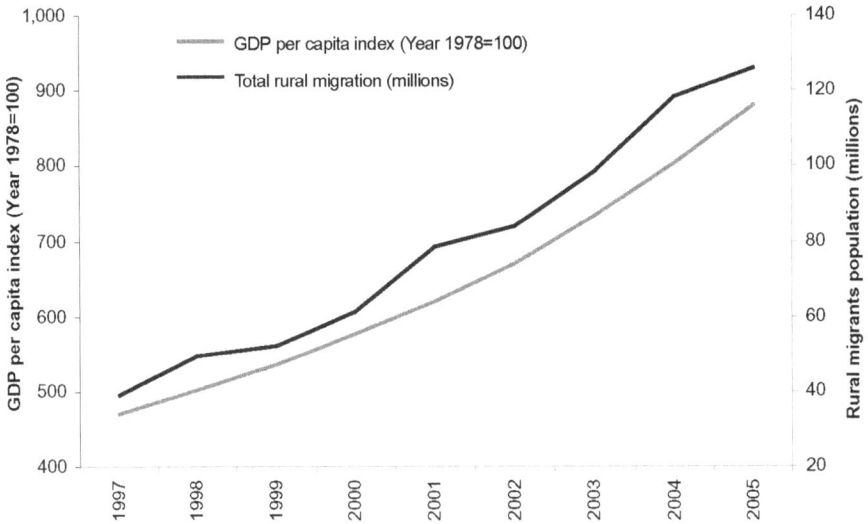

linkage between migration and economic growth can be traced back to the two-sector Lewis–Fei–Ranis model of development, in which expansion of the high-productivity urban sector and the process of rural–urban labour transfer were considered (Lewis 1954; Fei and Ranis 1961). In the context of productivity, this labour transfer moves towards equalising marginal returns to labour input across sectors, and therefore provides a direct means to enhance allocated efficiency and hence productivity gain.

In the case of China, despite the lack of consensus about the precise productivity growth figure, it is recognised widely that much of the TFP growth has been driven by the reallocation of rural surplus labour from agriculture to secondary and tertiary industries, where labour productivity is significantly higher (Young 2003; Woo 1997; Fan et al. 1999; Zhang and Tan 2007). Empirically, Fan et al. (1999) develop an analytical framework to investigate the sources of China's growth, taking into account the structural change featured in China's reform period. They find that labour productivity differs considerably across various sectors. For example, average labour productivity in 1978 was 903 (indicated in 1978 constant yuan per worker), while in the agricultural sector it was 362 yuan, and in urban industries and service sectors it was 3,335 yuan and 1,737 yuan, respectively.[5]

By 1995, average labour productivity had improved by an annual growth rate of 6.54 per cent to 2,648 yuan per worker. Labour productivity in agriculture, urban industries and service sectors reached 1,026, 5,601 and 5,177 yuan per worker, respectively. Moreover, Fan et al. (1999) conclude that a significant improvement in the overall allocative efficiency gain for the reform period (1978–95) is due primarily to rural reforms and shifts of surplus labour to non-farm activities—particularly to urban sectors.[6]

In addition, indirect channels of productivity enhancement through internal migration can be achieved through agricultural productivity gain due to increased endowment per unit of rural labour and the utilisation of remittances. A reduction in agricultural labour has the potential to increase the land/labour and capital/labour ratios, which in turn might have a positive impact on agricultural productivity. Meanwhile, the large amount of income generated by migrants sent back to rural communities has a substantial impact on rural development. The Consultative Group to Assist the Poor (CGAP), a consortium of 31 public and private development agencies housed within the World Bank, documented that in 2005 nearly US$30 billion worth of remittances were sent back to rural homes (CGAP 2006). A number of studies have investigated the impact of remittances and returning migrants on rural development (Taylor et al. 2003; Giles 2006; Du et al. 2005; Rozelle et al. 1999; Zhao 1999a, 1999b,

2002). Taylor et al. (2003) provide evidence that remittances help loosen capital constraints and stimulate crop production. Murphy (1999) and Zhao (2002) examined household investments and new productive projects developed by returned migrants in the rural economy. More generally, the productive investment made possible by remittances, new information and knowledge in employment and economic activities as well as further education or human capital investment are all indirect yet important channels through which internal migration can positively affect productivity growth and hence economic performance as a whole. Given the enormous scale of internal migration and its profound impacts, it is undoubtedly one of the most remarkable events associated with China's rapid economic growth in the past 30 years.

What do we (not) know about migrants?

Despite its massive scale and important role in the process of economic growth, internal migration in China has been discussed in a relatively small body of the economics literature. A number of studies have looked at issues such as migration decisions (Zhao 1997, 1999a, 1999b, 2001; Hare 1999; Zhu 2002), the impact of migration on rural incomes and productivity (Rozelle et al. 1999; Taylor et al. 2003; Li 2001), the labour market performance of migrants (Meng 2001; Meng and Zhang 2001) and migrants' assimilation into urban society (Ma and Xiang 1998; Zhu 2001).

Using household-level data, a number of authors (for example, Zhao 1997, 1999a, 1999b, 2001; Hare 1999; Zhu 2002) analyse the probability of migration. In general, they find that young, unmarried male workers are the most likely to migrate, and education does not play a significant role in migration decisions. Hare (1999) finds that the length of migration spells is determined primarily by the demands from home villages and the level of wages in cities; male workers in families with larger areas of land per capita or who earn less in cities tend to have shorter spells. Zhu (2002) employs data collected in urban areas of Hubei Province and investigates the impact of the rural–urban wage gap on migration decisions. Using a structural probability approach, the author finds that the income gap is important and has a non-linear effect on the probability of migration. Although these studies provide some information about the motivations for migration, a thorough investigation of migration choices requires comparison of the incomes and welfare of migrants in destination cities with what they would have achieved had they not migrated.[7]

This comparison is, however, not always feasible as existing surveys are often so limited in terms of the information collected that they cannot provide the data necessary for comparison.

A second issue that has been addressed in the existing literature is the impact of migration on the incomes of migrant families and rural productivity. While Li (2001) and some other studies mentioned earlier find that remittances have a significantly positive effect on rural household income, the conclusion on this issue is less than unanimous. Using data from a survey of 787 farm households in Hebei and Liaoning Provinces, Rozelle et al. (1999) and Taylor et al. (2003) estimate a recursive model of yields, migration and remittances, and find a direct negative net effect of migration on household cropping income, but not on crop yields. Moreover, remittances compensate partially for this lost-labour effect, as hypothesised by the 'new economics of labour migration' (NELM), by loosening constraints on production in the imperfect market environment. In conclusion, Taylor et al. (2003) find that participating in migration increases household per capita income, for those left behind, by between 16 per cent and 43 per cent. These results potentially offer support for the general recognition of migration as an important path out of poverty. A recent study (Du et al. 2005) estimates that, controlling for household and area characteristics, having a migrant worker in the family increases household per capita income by a margin of 8.5–13.1 per cent. Empirically estimating the overall poverty-reduction effects of migration is, however, a challenging task, as these effects can take place through direct and indirect channels. Investigation into how migration affects poverty needs to be based on information of changes in income and any productive investment induced by migration, which includes the contribution to physical and human capital accumulation. Data limitations once again often become the biggest obstacle for further in-depth research in this area.

In terms of the performance of migrants in the urban labour market, existing studies highlight discrimination against migrants who are employed predominantly in low-income and 'three-Ds' (dirty, dangerous and demeaning) jobs, with longer working hours and much lower earnings than their urban counterparts (Zhao 1999; Meng and Zhang 2001). Using data from Shanghai, Meng (2001) studies employment choices and finds that individuals with more human capital choose to be self-employed rather than being employed by others. As important as it is for these studies to capture the situation at one particular point, the dynamics of migrants' labour-market performance can reveal the evolution of their economic status and shed light on important questions of changes in migrants' economic conditions and income mobility. Research in this area is, however, virtually non-existent due to the unavailability of longitudinal migration data.

Labour-market outcomes constitute only a partial view of what happens (or does not happen) as a result of migration. The impact of migration as a socioeconomic process is extensive. Understanding migration requires going beyond analysis focused on the income effects *per se*. For instance, questions about the extent to which the health and general well-being of migrants are affected and how they integrate into the urban society are important but under-researched topics. A summary of the initial contributions made by a number of descriptive studies finds that most migrants do not have health insurance (Xiang 2005) and they experience deteriorating health conditions (Zhang 2005; Zheng 2005). The majority lives in crowded conditions and they have no private sanitation facilities (Shen 2002; Guo et al. 2005; Qi and He 2005). Their residential clusters are often based on common origins (Ma and Xiang 1998), which effectively segregate them from the local urban population. Additionally, there is prejudice (Zhu 2001) and discontent between the local urban population and the population of migrants (Li 1995). Apart from these basic descriptions, however, little is known about the extent of health effects of migration or the processes of and channels for rural migrants' adjustment, assimilation and integration into cities.

The substantial and far-reaching impacts of migration also manifest in the lives of migrants' children. While nearly 15 million children follow their migrating parents to cities, more than 20 million children are left behind in home villages (State Council Research Group 2006). What impact does the city have on those children who migrate with their parents? Do they receive better health and education as beneficiaries of the more developed urban infrastructure? Or, is the outcome the opposite due to the higher costs of living and the unfair policies in place in cities? Likewise, are those children left behind better off in terms of nutrition and schooling due to the extra income provided by remittances? Or, does the lack of parental care take its toll on these children's development? Questions such as these are important not only for the understanding of the current situation, but to shape the future society. Unfortunately, except for a few studies looking at school enrolment and educational outcomes (Lu 2005; Giles and de Brauw 2005), little progress has been made in understanding the profound and multifaceted consequences of migration on children.

An overall assessment of the existing literature reveals that our knowledge about China's migration is piecemeal and largely inadequate. Until now, little has been known about the overall impact of migration on migrants themselves, on their children—in the sending and receiving communities—and how these impacts evolve over time. Much of the limitation in the existing research is caused primarily by the scarcity of data. Researchers have long been plagued by a lack of well-designed longitudinal survey data for migrants and their appro-priate comparison groups. The RUMIC Project was established to respond to this urgent need. The cornerstone of the project is a large-scale household survey, consisting of 5,000 migrant households in cities, 5,000 urban resident households and 8,000 rural households with and without migrant workers. The survey is intended to cover the 18,000 households for a five-year period (2008–12). The rest of this chapter provides a brief introduction to the survey before outlining some preliminary findings from the pilot and first-wave data.

RUMIC: sampling strategy and census results

The RUMIC survey is conducted in 15 cities across nine provinces or metropolitan areas: Shanghai, Guangdong, Jiangsu, Zhejiang, Anhui, Hunbei, Sichuan, Chongqing and Henan.[8] Whereas the first four locations are the largest migration destinations, the remaining five are among the largest migration sending areas.[9]

China's National Bureau of Statistics (NBS) conducts the rural and urban households survey using existing random samples within its annual income and expenditure survey framework. There is, however, no existing sampling frame for migrant workers in cities, and the most fundamental challenges of developing an unbiased sampling frame involve randomly sampling the migrant population with little reliable advance information of its distribution. The reason for the lack of background information is due primarily to the largely incomplete official residential registration of migrants in cities. Most existing migrant surveys use administrative records of residential addresses as the basis for sampling; we believe, however, that at the very least a large number of migrant workers who live at their workplaces, such as factory dormitories and construction sites, are left out, and hence such a sampling framework is inescapably biased.

To address this issue, a unique sampling strategy has been developed and is employed in the RUMIC surveys. Essentially, the sampling frame is based on information collected in a census of migrant workers at their workplaces. More specifically, the census is conducted in a number of randomly selected city grids within the defined city's boundary. A sampling frame is formed using the aggregate number of migrants. Based on this, a random sample of migrant workers is selected and interviewed by enumerators for survey data collection. Using Shanghai as an example, within the identified city boundary, the city is divided into more than 2,000 equal-sized grids (0.5 km by 0.5 km),[10] of which 50 are drawn randomly for census operation. Within each of the 50 grids, a census of workplaces (rather than residential addresses) is undertaken. During this process, information such as the number of migrant workers at every workplace, including formal and informal sectors, and basic descriptive information (size and industry categories of the businesses) is recorded. The total number of

117

migrant workers in Shanghai can be aggregated from these census data. As a final step of sampling, a simple random sample of 500 migrants is selected for questionnaire-based face-to-face interview. Following this approach, the RUMIC surveys are able to rectify the sample biases of most existing urban-based surveys in which residential addresses are used as a basis for sampling, and thereby can include most migrant workers within the defined city boundary.[11]

Table 6.1 presents some basic statistics for the migrant population from the census conducted in the 15 survey cities in December 2007. The first column shows the number of grids in which a census was conducted, while column two indicates the total number of migrants listed in these grids. Column three presents the migrant population density per census grid. Among the 15 cities, Donguan in Guangdong Province has the most densely populated migrant group (2,614 migrants per 0.25 sq km), followed by Shenzhen (2,295), Wuxi (1,691) and Guangzhou (1,171). Based on the per-census grid average migrant number (column three) and the total number of grids within the defined city boundary, the total number of migrants in each city is aggregated (column five). It appears that Dongguan, Shanghai, Guangzhou, Hangzhou and Shenzhen are the top-five migration destination cities, which seems to coincide with anecdotal evidence. Based on our census results, the total number of migrants in the 15 survey cities amounts to 12 million, which is slightly less than 10 per cent of the official figure of 126 million for 2005 (NBS 2006).

Additionally, six broad categories are aggregated from a 72-industry classification based on census information of the industry categories of businesses (Table A6.3 provides a detailed description of the 72 industries and the aggregation method). The industrial distribution of the migrants presented in Table 6.2 indicates that among the 541,792 migrants listed in the RUMIC census: 8.4 per cent work in the construction industry; 18.3 per cent in manufacturing; 2 per cent in education and government organisations; 4.7 per cent in employment, real estate and other types of agencies, and consulting companies; 34 per cent in services, including public services, security, transportation, communication, hotels, restaurants, repairs and recycling; and the remaining 32.7 per cent work in wholesale and retail trade. It is worth noting that industry distributions vary considerably across cities. For example, 41 per cent of migrants work in the manufacturing industry in Wuxi, while more than half (55 per cent) of the migrants in Dongguan work in wholesale/retail trade.

For comparison purposes, the bottom four rows of Table 6.2 present the results for four previous migrant surveys, among which the first three are city-based and the last is a rural household-based survey. The Institute of Economics (IE) at the Chinese Academy of Social Sciences (CASS) conducted two of the urban-based migrant household surveys (hereafter referred to as 'IE surveys'). The IE surveys covered 13 cities[12] with 780 households (1,785 individuals) in 1999 (hereafter referred to as 'IE 1999') and 12 provinces (27 cities) with a total of 2,000 households (5,327 individuals) in 2002 (hereafter referred to as 'IE 2002'). On average, the IE surveyed less than 100 households in each city. Another urban-migrants survey for comparison is the *China Urban Labour Survey* (CULS) (IPLE 2001) conducted by the Institute of Population and Labour Economics (IPLE) within CASS in 2001. It was undertaken in five large cities (Shanghai, Wuhan, Shenyang, Fuzhou and Xian) with 340 migrant households and 2,365 individuals. These three surveys were based on a random sample of residents. As noted earlier, this sampling approach effectively missed all the migrants who lived at their workplaces from the sampling frame.[13]

The last row of Table 6.2 shows official information published by the NBS (hereafter referred to as 'NBS 2004') drawn primarily from the 2004 national rural household survey (Sheng and Peng 2005).[14]

Relative to the three urban-based migrant surveys, the RUMIC census results indicate a much higher proportion of migrants working in manufacturing and construction sectors, and a lower proportion working in the wholesale/retail trade sector. These differences are likely caused by the exclusion of migrant workers living at business premises in the sampling frame of IE surveys and the CULS. There are other possible factors that might contribute to the differences, such as the locality differences between the RUMIC census and the other surveys, the sample size differences and timing differences. In contrast, the NBS results indicate a much higher proportion of migrants in the manufacturing sector (greater than 30 per cent) and a much lower proportion in wholesale/retail trade (4.6 per cent). There are several potential explanations for such a large discrepancy: first, while the NBS survey was conducted in sending areas with information reported by household members who had remained behind, the RUMIC census was conducted in the destination cities and information was collected by enumerators. The completely different data-collection procedures employed could lead to different degrees of measurement errors. Second, the coverage of these surveys differs. Whereas the NBS undertook a national survey covering all 31 provinces, the RUMIC census covered only 15 cities. Third, as far as an individual city is concerned, the RUMIC census does not necessarily include all the districts that are considered to be administratively part of that city. In order to keep the census within a manageable scale, the census excludes some districts that are far from the city centre. It is possible that some manufacturing enterprises are located in these areas, and the exclusion of these enterprises could potentially explain the large differences between the RUMIC and NBS surveys with regards to the proportion of migrants in the manufacturing sector. Nevertheless, within the defined city boundary, the RUMIC census results are well positioned to represent the industry distribution of employed migrants. Another, more general explanation for the differences in the surveys' results is that the RUMIC census was conducted much later than the other surveys.

What have we learned from the survey?

As demonstrated in Table 6.2, after rectifying the sampling biases, the RUMIC survey presents a somewhat different portrait of migrant workers than the existing urban-based surveys. Furthermore, departing from the conventional focus on migration determinants and labour market discrimination, the RUMIC survey has collected much information concerning the broad welfare of migrants (their jobs, income, health and mental health status), their children's education and health, and the extent to which migrants assimilate into the city communities. These data allow researchers to explore a number of important aspects of migration that have so far been under-researched because of data limitation.

Between May and July 2007, a large-scale (1,000 households) pilot survey was undertaken in Shanghai, Wuxi, Guangzhou and Shenzhen. The first wave of the formal survey began in March 2008 and is currently under way. The following discussion is based on the pilot data and the first batch of data for approximately 2,000 migrant households from seven cities (Shanghai, Nanjing, Wuxi, Hangzhou, Ningbo, Wuhan and Chengdu) from the first wave of the survey.

Basic family and individual characteristics of migrants

Table 6.3 presents some basic family and individual characteristics of migrants. The RUMIC surveys (Table 6.3, columns one to three)[15] show that about 40 per cent of migrant households are not single-person households. While about 50–70 per cent of the migrants are married, one-third of these live separately from their spouses. Further, close to 60 per cent of households have children, but up to three-quarters of these have left their children behind in the countryside. This general description of migrant-household structure coincides with anecdotal evidence that underscores the breakdown of migrant workers' normal family relationships as a result of the physical distance between family members.[16]

This impression does not, however, concur with the results obtained from the previous urban-based migration surveys. For example, the IE surveys (columns four and five) indicate a much higher proportion of households with married

Table 6.1 Census: migrant numbers by city

	No. of census grids (1)	Census total migrants (2)	No. of migrants per census grid (3)=(2)/(1)	Total no. of grids (4)	Implied total no. of migrants (5)=(4)*(3)	Total no. of migrants to be sampled (6)	Weight (7)=(6)/(5)
Guangzhou	40	46,822	1,171	1,048	1,226,736	400	0.000326068
Shenzhen	30	68,851	2,295	474	1,087,846	300	0.000275774
Dongguan	30	78,432	2,614	886	2,316,358	300	0.000129514
Shanghai	50	40,293	806	2,050	1,652,013	500	0.000302661
Nanjing	40	30,352	759	893	677,608	400	0.000590311
Wuxi	20	33,823	1,691	677	1,144,909	200	0.000174686
Hangzhou	40	45,540	1,139	512	582,912	400	0.00068621
Ningbo	20	18,637	932	249	232,031	200	0.000861955
Wuhan	40	39,060	977	1,004	980,406	400	0.000407994
Chongqing	40	39,792	995	640	636,672	400	0.000628267
Zhengzhou	35	28,604	817	655	535,303	350	0.000653835
Chengdu	40	36,145	904	590	533,139	400	0.000750274
Hefei	35	21,152	604	618	373,484	350	0.000937122
Louyang	20	8,600	430	296	127,280	200	0.001571339
Bangbu	20	5,689	284	121	34,418	200	0.005810837

Table 6.2   **Industry distribution of migrant workers in 15 cities (census results)**

City	Construction	Manufacturing	Education and government organisations	Various types of agencies	Services	Wholesale and retail trade	Total no. of migrants
Chengdu	10.69	3.08	0.57	6.37	45.83	33.45	36,145
Hangzhou	7.06	17.64	2.48	2.69	36.98	33.15	45,540
Nanjing	16.86	15.86	1.19	10.21	38.16	17.73	30,352
Ninbuo	26.06	8.99	5.07	10.14	28.21	21.55	18,637
Shanghai	8.35	10.85	1.50	11.88	41.29	26.13	40,293
Wuhan	5.17	28.90	3.08	2.88	36.49	23.47	39,060
Wuxi	9.40	40.82	2.00	6.03	22.01	19.75	33,823
Bangbu	7.91	13.78	0.30	1.44	30.32	46.24	5,689
Chongqing	9.22	4.68	1.24	0.82	48.76	35.28	39,792
Dongguan	1.24	26.29	3.23	1.72	12.55	54.97	78,432
Guangzhou	1.30	23.23	2.88	6.07	40.09	26.43	46,822
Hefei	10.10	5.35	1.29	8.94	36.33	37.99	21,152
Luoyang	8.33	12.92	2.23	0.52	38.85	37.15	8,600
Shenzhen	7.21	21.81	0.99	2.77	36.32	30.91	68,851
Zhengzhou	21.48	8.21	1.27	1.60	34.49	32.96	28,604
Total	8.36	18.25	2.04	4.68	34.00	32.68	541,792
IE (1999)	4.31	12.68	3.75	0.48	36.60	42.19	1,254
IE (2002)	5.49	9.80	2.91	0.70	34.08	47.02	3,407
IPLE (2001)	10.25	7.30	3.00	0.91	38.68	39.86	2,205
NBS (2004)	24.70	30.30	..	..	40.40	4.60	..

migrants (87–89 per cent in IE surveys versus 51–68 per cent in the RUMIC) and a much smaller incidence of family separations (9–22 per cent)—at least 50 per cent lower than that observed in the RUMIC surveys. Similarly, the CULS survey (column six) indicates a much higher proportion of migrants who are married and/or have children, and a lower proportion of migrants with children left behind in the countryside. Different sampling strategies might hold the key for understanding these discrepancies. It is not unreasonable to suggest that migrant workers living at their workplace are more likely to be single, or living separately from their spouse. As these migrants are systematically excluded from the IE and CULS surveys, an under-representation of the singles and those who live apart from their spouses is a plausible conjecture.

The bottom half of Table 6.3 reports individual characteristics of migrants. The RUMIC and previous surveys show that, on average, about 55 per cent of the migrants are males, and their average age is 28–29 years, although the CULS has a higher proportion of males and their average age is older, which is related to the fact that the CULS surveys only those aged 16 and above. Notwithstanding the similar basic demographic characteristics at the mean, some important differences are revealed between the RUMIC and IE surveys when the age structure is examined. Figure 6.3 shows the age distribution by gender for the four different samples. In contrast with the RUMIC samples, in which the mode of migrant age is just more than 20 years, in the IE surveys it is about 35 years. Moreover, more than 85 per cent of migrants in the RUMIC sample are participants in the labour force—almost 20 percentage points higher than in the IE surveys. In terms of the duration from first migrating to the city to the time of the survey, the RUMIC survey records an average of seven to eight years—about one year longer than that suggested by the IE survey. Perhaps due to a younger sample, the average years of schooling in the RUMIC surveys is about 1.5 years more than in the IE surveys. Indeed, Figure 6.4 shows early birth years are associated with a shorter average number of schooling years in all samples. Even within a same age cohort, however, average years of schooling in the RUMIC survey are still longer than those of the IE survey samples—again, indicating a possible sampling bias in the previous urban-based migrant surveys.

Job and job-related welfare

A crucial aspect of migration studies concerns the characteristics and conditions of migrants' employment. Table 6.3 provides a summary of migrants' job characteristics. The RUMIC survey data show that the unemployment rate of migrant workers is as low as 1–2 per cent. About 20–23 per cent of the sample

Table 6.3 Family and individual characteristics

	Pilot survey (4 cities)	RUMIC First wave (7 cities)	First wave (7 cities) weighted	Other urban-based migrant surveys (IE 1999)	(IE 2002)	(IPLE 2001)
Number of households	1,002	2,055	2,055	800	2,000	2,463
Number of individuals	1,720	3,349	3,349	1,785	5,318	2,463
Percentage of households with two or more people in the city	44.41	38.69	39.10	86.89	89.44	60.86
Percentage of households with married rural migrants	50.80	54.60	53.71	82.04	91.44	65.26
Households with rural migrant couple living separately	31.04	32.62	30.51	21.79	8.65	35.95
Percentage of households with children	58.27	38.19	39.63	n.a.	n.a.	60.82
Households with children left behind	76.99	56.29	52.31	n.a.	n.a.	47.01
Number of months lived in host city in the past 12 months	10.73	11.11	11.23	10.82	11.35	n.a.
Male (percentage)	54.64	55.18	53.59	52.66	52.39	60.58
Age of total sample	28.78	29.29	29.13	27.97	28.36	30.35
Age of labour force	30.47	31.48	31.48	32.86	34.58	30.36
Years of schooling (aged >=16)	9.05	9.11	9.12	7.31	7.82	8.17
Percentage in the labour force (aged 16–65, employed + unemp. + family workers)	87.88	86.53	86.30	65.83	66.51	91.56
Average years since first migration (aged >=16)	6.95	8.07	8.03	5.82	6.73	5.92
Number of cities lived in	4.50	2.22	2.02	n.a.	n.a.	n.a.

n.a. not applicable

workers are self-employed while the rest work in the wage/salary sector. This description is in stark contrast with what is revealed by the IE and CULS surveys, in which about 50–65 per cent of sample workers are self-employed. In terms of occupational distribution, 70 per cent of migrants in the RUMIC surveys are engaged in production, service or wholesale/retail trade, whereas about half of the migrants in the IE surveys are business owners in the private sector.[17]

That migrants work incredibly long hours is a long-established fact (for example, Du et al. 2006). The RUMIC survey confirms that, on average, migrants work 60–63 hours a week. Self-employed migrants work 10 hours longer than the average, which translates into 10–11 hours daily, seven days a week.

Migrant workers earn, on average, about 1,500 yuan a month, which is double the amount observed from the IE surveys and 60 per cent higher than that found in the CULS (Table 6.5). Their hourly earnings are about 6 yuan, which is also more than double that indicated in the previous surveys. Whether such a comparison of results directly implies improvement in migrants' economic conditions is questionable. Fortunately, additional information is collected in the RUMIC surveys for the year respondents started their first job in the city and the wages they received in the first month. A series of averages for the first month's wages for the first job is constructed for the past 15 years using these two instruments. This information can serve as a crude but useful indicator of the change in earnings of unskilled migrants. Compared with the earlier surveys, this measure standardises the sampling method and the structure of the questionnaire. This datum series is presented in Figure 6.5, showing a small but gradual increase in starting salaries for migrants. For the period 1999–2006, the increase was 1.1 per cent per annum, whereas between 2002 and 2006 the increase was 0.1 per cent per annum. Such slow growth in nominal terms implies that the real earnings of a newcomer to rural-to-urban migration have fallen over time.

Earnings alone sometimes do not represent the full income of migrant workers. Work-related welfare often includes subsidies or the provision of a meal, accommodation, social insurance and so on. Much of the information in this respect has, however, never been collected in previous surveys. Table 6.5 offers some basic information revealed from the RUMIC surveys. It shows that about 65–75 per cent of migrants receive meal subsidies, with an estimated value of about 200–230 yuan a month, and 50–65 per cent receive accommodation subsidies with an estimated value of 120–200 yuan a month. Relative to the CULS, the RUMIC surveys identify that the proportion receiving a subsidy and the amount of subsidy seem to have increased. Although the government has issued requirements for compulsory social insurance participation since the

Figure 6.3 Age distribution of migrants from different samples

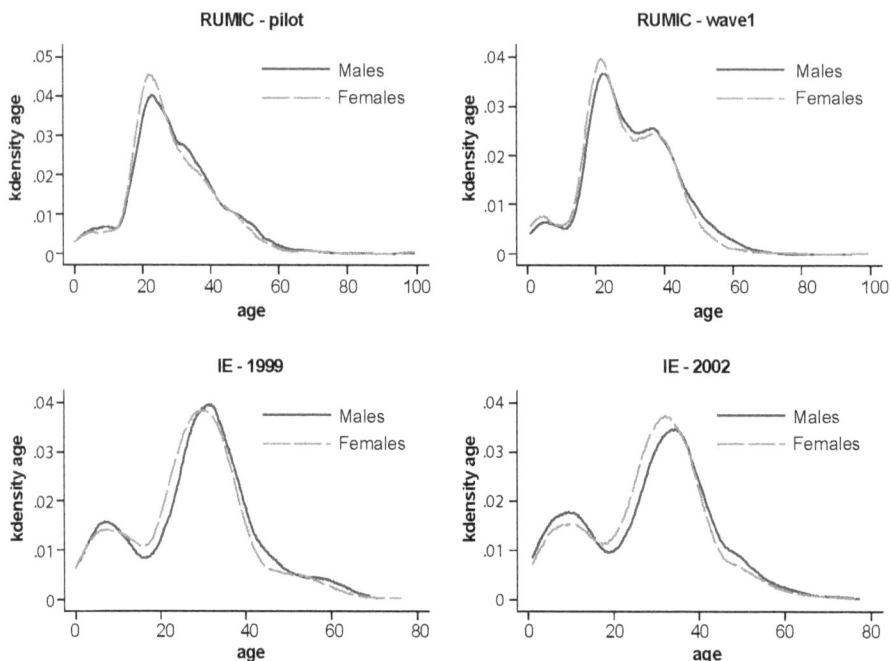

Figure 6.4 Years of schooling by birth year, 1950 – 1990

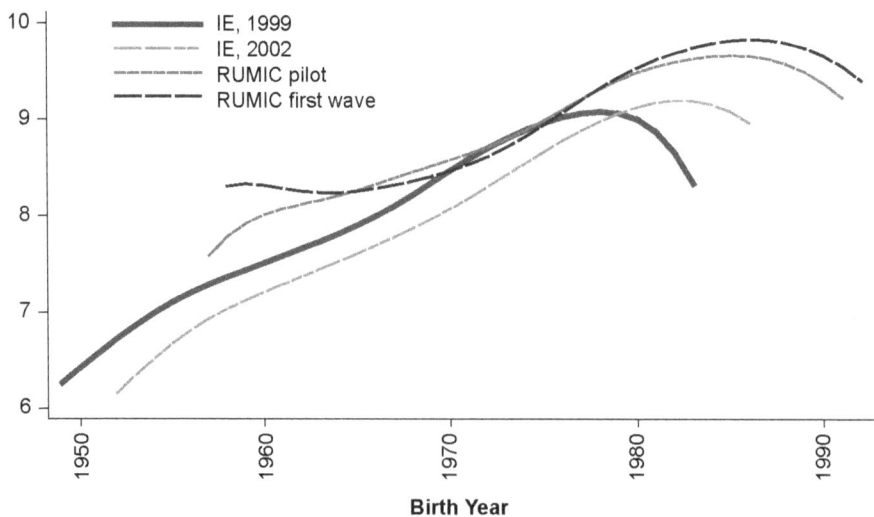

Table 6.4　Job characteristics

	RUMIC			Other urban-based migrant surveys		
	Pilot survey (4 cities)	First wave (7 cities)	First wave (7 cities) weighted	(IE 1999)	(IE 2002)	(IPLE 2001)
Number of people in the labour force	1,444	2,898	2,898	1,175	3,543	2,255
Age of labour force (aged 16–65, employed + unemp.)	30.47	31.35	31.34	32.74	34.63	30.39
Unemployment rate (%)	1.72	1.07	0.88	2.64	2.79	1.29
Work status: Self-employed	19.02	23.38	23.48	64.16	65.55	51.62
Wage-earner	80.98	76.62	76.52	35.84	34.45	48.38
Occupation:						
1. Owner of private sector business	4.82	3.05	3.24	42.19	52.53	n.a.
2. Professional	1.26	0.60	0.54	4.57	4.43	n.a.
3. Clerk or official	9.14	6.10	5.59	4.06	2.43	n.a.
4. Worker in wholesale/retail trade	15.99	22.14	19.80	7.94	7.28	n.a.
5. Worker in service sector	25.67	33.01	29.56	25.88	19.73	n.a.
6. Production worker	28.27	20.02	26.10	6.56	6.99	n.a.
7. Other	14.87	15.08	15.16	8.8	6.63	n.a.
Hours worked per week	59.42	62.75	60.98	66.1	69.66	71.10
Self-employed	75.41	78.18	77.31	69.62	73.77	77.21
Wage-earner	56.03	58.02	55.95	59.93	61.53	64.56
Workplace size (people)	181	154	172	n.a.	n.a.	89

n.a. not applicable

early 2000s, only a small number of migrants participate in any of four social insurance programs. For example, the share of migrants with work-related injury insurance or unemployment insurance is less than 25 per cent. Nevertheless, the share of migrants participating in insurance programs has increased compared with the IE survey data.[18]

Income and poverty dynamics, living conditions and the broader well-being of migrants

It has long been known that rural–urban migrants under the 'guest-worker' system have very low 'discount rates'. As they regard themselves as working in cities 'temporarily', with their future ultimately in their rural home villages, they work extremely hard, live a very modest life in cities and try to save as much as possible for the future (for example, Du et al. 2006). Existing surveys have not been able to provide adequate information for researchers to examine the well-being of migrant workers. The RUMIC survey, in contrast, pays special attention to issues related to the well-being of migrants. A number of indicators of migrants' living standards, their health and mental health conditions are summarised below.

Table 6.6 presents information on income, expenditure, poverty and living conditions based on the RUMIC survey data. It shows that average per capita monthly income is about 1,700 yuan for the pilot survey and 1,400 yuan for the first wave of the survey. The difference might reflect only differentials across cities covered by the two surveys. Indeed, if Shanghai and Wuxi, the only two cities covered by both surveys, are compared, per capita income in the pilot survey is 1,418 yuan and in the first wave it is 1,668 yuan—translating into an increase in nominal income of about 17.6 per cent over two years. In the absence of consumer price index (CPI) data, the change in official minimum standards of living lines (the 'Dibao line'; see Table A6.2) is used as a proxy to estimate the proportional change in the cost of living. Between 2007 and 2008, Dibao lines in Shanghai and Wuxi increased by 14.3 and 16.7 per cent, respectively. Taking a simple average, therefore, the cost of living in the two cities increased by about 15.5 per cent, implying a 2.1 per cent increase in real per capita income in the two cities.

If, however, per capita expenditure is used as a measure, a much larger proportional increase can be observed. Per capita monthly expenditure increased from 744 yuan to 1,016 yuan in Shanghai and Wuxi, indicating growth of 36.6 per cent, or 21 percentage points, over and above the increase in the cost of living. Interpretation of this observation points to: 1) the inadequate

Table 6.5 Earnings and work-related welfare

	Pilot survey (4 cities)	RUMIC First wave (7 cities)	First wave (7 cities) weighted	Other urban-based migrant surveys (IE 1999)	(IE 2002)	(IPLE 2001)
Earnings per month (yuan)	1,486	1,524	1,553	716	785	962
Hourly earnings (yuan)	5.98	6.84	6.83	2.58	2.73	3.67
Wage-earner	5.92	6.50	6.90	2.49	2.76	3.10
Meal provided?				n.a.	n.a.	n.a.
3 meals	18.25	24.02	21.01			
2 meals	20.26	23.30	22.59			
1 meals	16.16	18.85	26.27			
No meal provided but subsidies given	8.65	6.56	5.92			
No meal, no subsidy	36.68	27.27	24.21			
Estimated food subsidy (yuan/month)	227	218	206			
Accommodation provided?				n.a.	n.a.	
Accommodation provided	46.81	57.94	49.98			37.31
No accommodation, but subsidies given	5.85	9.47	17.85			4.2
No accommodation, no subsidy	47.34	32.58	32.17			58.49
Estimated accomm. subsidy (yuan/month)	118	198	209			116
No unemployment insurance	88.92	82.82	77.74	96.36	94.68	n.a.
Self-employed	95.6	92.2	91.00	96.25	92.97	n.a.
Wage-earner	87.52	79.97	73.70	96.53	96.55	n.a.
No pension fund	82.96	77.4	73.61	94.95	91.63	n.a.
Self-employed	94.51	88.14	88.39	95.20	94.18	n.a.
Wage-earner	80.12	74.13	69.11	94.53	.86.79	n.a.

No work injury compensation	83.52	78.57	74.56	n.a.	n.a.	n.a.
Self-employed	98.17	93.45	92.38	n.a.	n.a.	n.a.
Wage-earner	80.03	74.04	69.13	n.a.	n.a.	n.a.
No housing fund	90.93	86.1	80.33	88.27	88.14	n.a.
Self-employed	95.6	93.6	92.31	92.96	92.97	n.a.
Wage-earner	89.76	83.82	76.68	80.44	79.54	n.a.

n.a. not applicable

Figure 6.5   **Earnings growth, 1991 – 2007** (first job, first month)

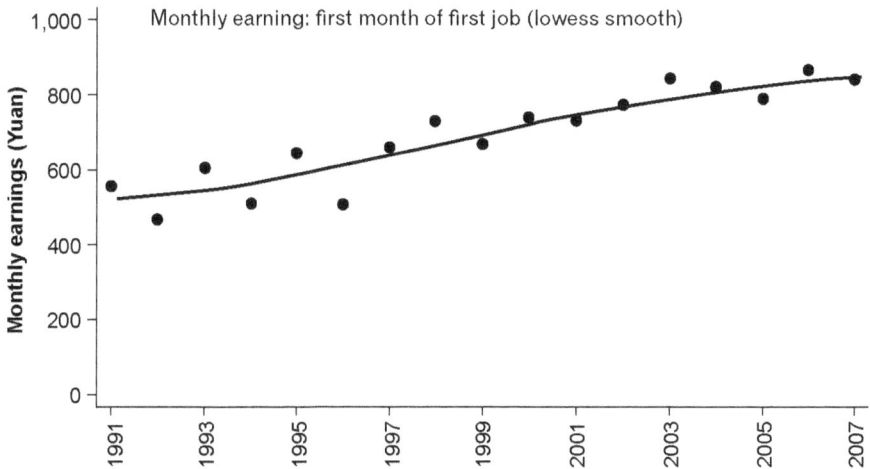

bandwidth = .8

reflection of increases in the cost of living from the change in Dibao lines over the two years; 2) the possible changes in migrants' attitude (the discount rate) as reflected in improved living standards. Both factors could contribute to the significant increase in per capita expenditure. A simple decomposition analysis of changes in expenditure suggests that items including food, clothing, housing and health care all register large increases, ranging between 37 per cent (for food) and 54 per cent (for clothing). Strictly speaking, despite the fact that China has been experiencing a period of high inflation since 2007, particularly rising food prices,[19] without controlling for the quality consumed, it is difficult to gauge the extent to which the increase in expenditure is due to price increases. Nevertheless, the RUMIC surveys provide some measures of housing quality, from which we observe the following: 1) the proportion of migrants living in suburban areas decreased by 5.5 percentage points; 2) the proportion of migrants living at their workplace decreased by 3.7 percentage points; 3) the proportion of migrants without private sanitation facilities remained essentially unchanged. In addition, living area per person has increased by 15 per cent. This observation suggests a small degree of improvement in housing conditions for migrant workers during 2007 and 2008. The degree of such development does not, however, seem to account for the dramatic increase in the cost of living. In contrast with the increase in other expenditure, education expenditure dropped by 20 per cent (the last column of Table 6.6), while the large increase in expenditure also drove down remittances significantly.

Even though relative to the pilot survey, per capita expenditure in the first wave of the survey increased much more than per capita income, migrants' savings rate remains as high as 39–48 per cent. This is measured by the difference between average per capita income and expenditure as a proportion of per capita income. Figure 6.6 indicates that almost all households (80–90 per cent) save, including those who might end up living below the Dibao line. The share of migrant households falling below local Dibao lines is 1–1.8 per cent and 9–13 per cent, by income and expenditure measures, respectively. A 1.2 percentage point fall in the poverty rate (expenditure measure) between 2007 and 2008 is identified for Shanghai and Wuxi. This result is consistent with the abovementioned fact that average household expenditure increased 16 percentage points over and above the increase in the cost of living.

In Table 6.6, poverty rates measured by two other poverty lines are also presented. According to the *China Development Report 2007* (CDRF 2007), China's official poverty lines are systematically biased downwards; however, the degree of the underestimation varies by region because of the different financial capacities of local governments. It is suggested in the *China Development Report 2007* that in 2004 the official poverty lines in the Eastern region were 71 per cent of the real minimum cost of living (MCL), while this ratio in the Central region was 64 per cent. The MCL lines are estimated using these ratios, together with the official Dibao lines in 2007 and 2008. The proportion of households in Shanghai and Wuxi living under this MCL line was 22.8 per cent in 2007 and 20.1 per cent in 2008 using expenditure measures, as opposed to 3.6 and 3 per cent for the two years using income measures. Additionally, information collected in the RUMIC surveys on the subjective cost of minimum living standards is used to formulate a 'subjective poverty line' for each city.[20] The proportion of migrants living below this subjective poverty line is 21.1 per cent (using expenditure) and 3.7 per cent (using income measures).

Migrants' health and mental health conditions are another important reflection of the overall well-being of migrants. The RUMIC surveys have collected relevant information and Table 6.7 presents some summary statistics. Based on migrants' self-assessment (row one), more than 85 per cent indicated excellent or good health. To supplement the subjective health measure, a number of objective health measures are also generated. The first important health indicator is whether the individual is under or overweight—the former is often an indication of malnutrition, and the latter suggests potential hypertension or

Table 6.6 Income, expenditure and housing conditions

	Pilot survey (4 cities)	First wave (7 cities)	First wave (7 cities) weighted	Pilot survey (Shanghai/Wuxi)	First wave (Shanghai/Wuxi)	Change (Shanghai/Wuxi)
	(1)	(2)		(3)	(4)	(5)=([4]/[3]-1)*100
Per capita income	1,740	1,634	1,682	1,418	1,668	17.63
Per capita expenditure	787	668	684	744	1,016	36.56
Per capita exp. as proportion of per capita income	45.23	40.88	40.64	52.47	60.91	..
Saving rate	54.77	59.12	59.36	47.53	39.09	..
Proportion of household saving	93.21	99.12	99.22	88.20	85.57	..
Annual food expenditure ([yuan/month]*12)	5,899	6,707	6921	5,151	7,053	36.92
Annual clothing expenditure ([yuan/month]*12)	2,094	2,110	2,125	1,898	2,929	54.32
Annual housing expenditure ([yuan/month]*12)	4,562	3,545	3,704	4,125	4,561	10.57
Monthly rent (yuan)	336	242	275	166	239	43.98
Annual total educational expenditure (yuan)	1,295	663	657.06	1,602	1,238	-22.72
Annual total health expenditure (yuan)	464	699	742.27	487	874	79.47
Annual remittances (yuan)	3,706	2,183	1,985	4,079	1,758	-56.90
Proportion living below Dibao line (income)	1.00	0.78	0.64	1.80	1.01	-43.89
Proportion living below Dibao line (expenditure)	13.37	11.68	9.88	11.60	10.40	-10.34
Proportion living below poverty line (income)	2.30	1.12	0.86	3.60	3.02	-16.11
Proportion living below poverty line (expenditure)	27.35	24.23	22.57	22.80	20.13	-11.41
Proportion living below subjective poverty line (income)	3.49	1.56	1.15	5.60	3.69	-34.11
Proportion living below subjective poverty line (expenditure)	36.73	29.78	27.71	31.80	21.14	-33.52
Living in suburban areas	46.72	49.40	52.60	54.28	48.82	..
Living at workplace	40.61	46.33	40.17	39.20	35.53	..
Size of housing per person (sq m/person)	12.08	11.16	11.93	10.84	12.53	15.59
Number with private bathroom	35.93	38.05	35.04	45.60	45.69	..

heart disease. The average body mass index (BMI) of RUMIC survey samples is similar to that found in the China Health and Nutrition Survey (CHNS) for 1989 and 1991, but the proportion of underweight migrants in the RUMIC surveys (15 per cent in the pilot and 10 per cent in the first-wave surveys) is much higher than that found in the CHNS (8.3 of rural and 9.2 per cent of urban samples). Similarly, the overweight rates are higher in the RUMIC surveys (Doak et al. 2002). While the proportion of underweight migrants is higher in the RUMIC surveys than in the CHNS 2004 data, the CHNS reports a higher proportion of overweight rural (25 per cent) and urban (30 per cent) samples. Part of the explanation for such differences could be the different measurement errors incurred when weight and height data are obtained in these surveys. Whereas the CHNS involved enumerators to measure respondents physically, the RUMIC surveys relied on self-reported information. Another objective measure of health is the proportion of migrants who were sick during the three months before the survey; 8.3 and 14 per cent of migrants reported being sick in the pilot and first-wave surveys, respectively. Among them, 20–24 per cent considered their illnesses serious. Seventy-five per cent of respondents in the first-wave survey did not see a doctor, of which 13 per cent had a serious illness. On average,

Figure 6.6   **Per capita monthly income and expenditure**

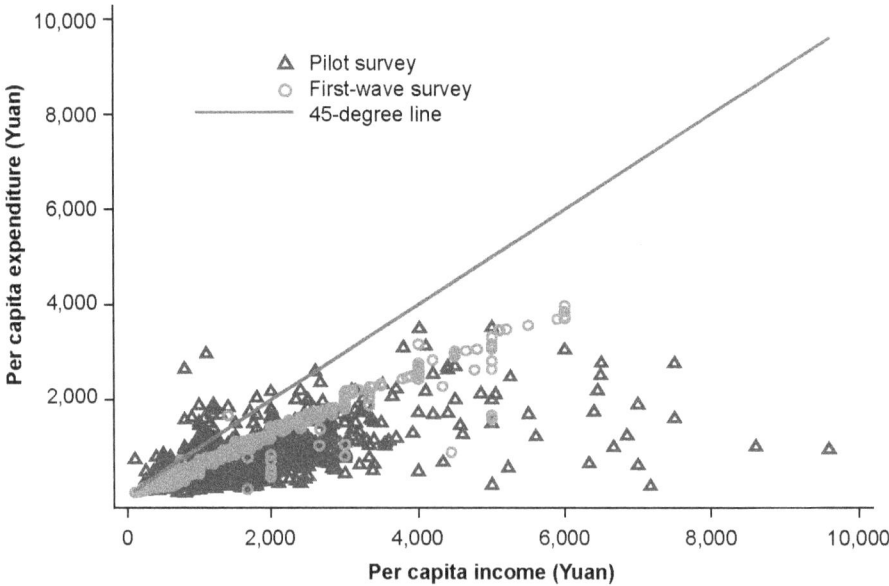

those who were sick spent 393 (pilot survey respondents) and 263 yuan (first-wave respondents) during the three months to treat their illness, and only 5.2–6 per cent of their total expenditure was reimbursed by insurance.

A unique contribution of the RUMIC surveys to the understanding of migrants' well-being is the collection of information on mental health conditions, measured by the 'General Health Question 12' (GH12).[21] The numbers reported in Table 6.7 indicate the proportion of migrants in the RUMIC samples suffering from mental health issues. The existing literature for comparison with these results is scarce; however, the last question, about happiness, has an equivalence in the 2002 IE survey, in which 11.2 per cent of migrants considered themselves to be not very happy or not happy at all—slightly lower than the figure suggested in the RUMIC surveys. Furthermore, using British survey data, Erens et al. (2001) reported GH12 for different minority groups in England, including ethnic Chinese. In that study, GH12 was summarised into three categorical variables: the proportion of individuals who answered 'no' to all of the 12 questions; the proportion who answered 'yes' to one to three of the above questions; and the proportion who answered 'yes' to four or more of the 12 questions. Erens et al. (2001) found that 69 per cent of the Chinese group reported 'no' to every one of the 12 questions, 28 per cent reported 'yes' to one to three of the questions, while 3 per cent reported 'yes' to four or more questions. Using the information for Chinese immigrants in the United Kingdom as an approximate benchmark, the RUMIC surveys' sample of respondents seem to suffer more from mental distress than other issues. In particular, 6.5–7.8 per cent of RUMIC survey sample migrants reported 'yes' to four or more questions—double the proportion reported in Erens et al. (2001).

Children of migrants

The unique nature of the guest-worker system for rural–urban migration in China means that many migrant children grow up with only one or neither parent living with them. Lack of parental care could have a significant impact on their current and future development. The RUMIC surveys collect a substantial amount of information from migrants about their children, including whether they migrated with them of were left behind in home villages. To our knowledge, this is the first attempt by researchers to examine the impact of migration on migrants' children with a relatively large sample survey. This subsection provides some first-cut summary statistics.

In the data for the first-wave RUMIC survey, there were 1,060 children recorded, including 893 children aged 16 years or younger.[22] The current living arrangements of the full sample of children and sub-sample of children aged 16 and below are reported in Table 6.8. Among children aged 16 and below (column two), almost

60 per cent were left behind in the rural hometown—among whom, 42 per cent lived with only one parent, 50 per cent lived with grandparents, 4.5 per cent lived with other relatives, and the remainder lived by themselves.[23]

When migrant parents were asked the main reason why their children did not live in cities with them, more than 50 per cent pointed directly to cost factors—high costs of living (37.5 per cent) and high school/childcare fees (16.4 per cent)—while 22 per cent indicated lack of a carer for the child, which, in turn, partly reflects the high cost of childcare.

Table 6.9 summarises characteristics of migrants' children in cities and those left behind. Among the children aged 16 and below, the average age of those left behind (8.4 years) is about 1.3 years older than those who migrated with their parents (7.4 years). This is understandable, as young children need more parental attention, and parents are therefore more likely to bring them to cities. Using the total sample of children, Figure 6.7 presents the age distribution of children by their current living arrangements. Those currently living with parents in cities are concentrated in the under-10 age group, while those left behind are more likely to be older then 10 years. Further, parents bring to cities slightly more (6 percentage points) girls than boys. Among children aged 16 and below in cities, 46 per cent were pre-school aged—10 percentage points higher than those children left behind. There are hardly any school drop-outs defined in the children above pre-school age and aged 16 or younger.[24]

Important research questions related to the impact of migration on children involve investigating the health, education and general development outcomes of migrants' children. Moreover, to study these questions requires information about these children's potential circumstances had their parents not migrated. While the collection of data to generate this information is currently under way, a simple comparison between children who migrated with their parents and those left behind is presented in Table 6.10, using parental-assessed indicators of outcomes of children's health, education and general development. Relative to those children left behind, migrant children living with their parents in cities seem to better off in terms of their health conditions. About 93 per cent of parents who had their children living with them in cities considered their children to be healthy or very healthy—5 percentage points higher than those children left behind. To examine how this health assessment differs by age group, Figure 6.8 plots the average proportion of children rated as healthy by age and current living arrangements. The figure shows that there is little difference in the proportion of children considered healthy for the very young age group and the over-16 age group. In the five-to-12 age group, however, the proportion of children rated healthy for those living in cities with their parents is almost 7–8 percentage points higher than their counterparts left behind in the countryside.

Table 6.7   Children's parent-assessed health and mental health

	Pilot survey (4 cities)	RUMIC First wave (7 cities)	First wave (7 cities) weighted	Other urban-based migrant surveys (IE 1999)	(IE 2002)	(IPLE 2001)
Healthy and very healthy	85.04%	86.41%	87.08%	98.02%	90.48%	97.48%
Have health insurance?	11.73%*	9.73%	14.36%	2.15%*	2.93%*	1.26%
Height (cm) (aged >=16)	165.65	165.78	165.86	n.a.	n.a.	165.96
Weight (kg) (aged >=16)	58.82	60.09	60.34	n.a.	n.a.	n.a.
BMI (aged >=16)	21.47	21.86	21.91	n.a.	n.a.	n.a.
Underweight (bmi <18.5)	14.65%	10.25%	9.54%	n.a.	n.a.	n.a.
Normal (bmi >=18.5 and bmi <25)	75.43%	76.83%	77.20%	n.a.	n.a.	n.a.
Overweight (bmi >=25)	9.92%	12.92%	13.25%	n.a.	n.a.	n.a.
Sick in previous three months	8.26%	15.65%	13.69%	n.a.	n.a.	8.36%**
Serious	20.44%	23.09%	24.07%	n.a.	n.a.	n.a.
Did not see a doctor		75.95%	75.02%	n.a.	n.a.	n.a.
Total health exp. (yuan/3 mth)	414	300	280	n.a.	n.a.	167
Out-of-pocket health exp. (yuan/3 mth)	393	282	263	n.a.	n.a.	167
Proportion paid by insurance	5.18	6.03	6.03	n.a.	n.a.	0.34
Mental health:						
Cannot concentrate	11.03%	12.32%	12.35%	n.a.	n.a.	n.a.
Have difficulty sleeping due to worry	8.14%	6.02%	6.56%	n.a.	n.a.	n.a.
Feel unable to play useful role in things	9.22%	9.96%	10.16%	n.a.	n.a.	n.a.
Feel unable to make decisions	7.23%	9.91%	10.02%	n.a.	n.a.	n.a.
Always feel under pressure	12.48%	9.75%	9.26%	n.a.	n.a.	n.a.
Feel unable to overcome difficulties	5.42%	4.57%	4.96%	n.a.	n.a.	n.a.
Feel life is meaningless	11.57%	15.53%	14.87%	n.a.	n.a.	n.a.
Cannot face up to problems	12.57%	12.68%	12.44%	n.a.	n.a	n.a.
Depressed	7.14%	5.62%	6.05%	n.a.	n.a	n.a.

Lack of self-confidence	4.25%	4.05%	4.54%	n.a.	n.a	n.a.
Feel worthless	2.98%	3.29%	3.78%	n.a.	n.a	n.a.
Not happy	14.20%	11.84%	11.49%	n.a.	11.20%	n.a.
GH12a score zero	70.29%	67.06%	67.63%	n.a.	n.a.	n.a.
GH12 score 1–3	21.95%	26.69%	25.90%	n.a.	n.a.	n.a.
GH12 score 4+	7.76%	6.24%	6.47%	n.a.	n.a.	n.a.

n.a. not applicable
* includes only employer-provided health insurance
** proportion of those sick who saw a doctor in the previous month
a General Health Question 12 (RUMIC survey)

Table 6.8   **Children's living arrangements**

	Total children	Children aged <=16
Total number	1,060	893
Percentage left behind	57.64	59.24
Among those left behind:		
Percentage living with one parent	37.45	42.26
Percentage living with grandparents	45.81	49.73
Percentage living with other relatives	3.98	4.50
Percentage living by him/herself	12.78	4.50
Why does the child not live with you? (%)		
City living costs too high	35.11	37.48
School/childcare fees are too high	16.30	16.40
No-one to look after the child in the city	19.08	22.16
The child is living with my spouse	9.54	11.17
It is better for the child to live in hometown	11.75	9.91
Other	8.22	2.88

The results for school performance, however, paint a quite different picture as to which group is doing better. A much higher proportion (14.4 per cent) of those children left behind enjoy excellent school performance compared with migrant children in cities (7.7 per cent). Figure 6.9 shows that the child's age is positively associated with parental assessment of the child's school performance, particularly for children older than 15. This is, perhaps, due to the fact that low performers are less likely to continue schooling after junior high school level. The large gap for school performance between those living in cities and those left behind is essentially constant across all age groups, except for the very young cohort (seven–eight years of age). Similarly, when parents are asked whether they are worried about their children's school performance, those who had children living with them recorded a higher proportion of positive answers ('yes') than those whose children had been left behind in rural villages (43 per cent versus 38 per cent). Figure 6.10 indicates that parents with children younger than nine worry most about their offspring, and especially those whose children have been left behind. For children aged 10 and above, parents appear to worry more about those who live with them in cities, compared with the ones left behind.

The RUMIC surveys, in addition to school performance, collect information from parents about other aspects of their children's behaviour that might cause them to worry—for example, skipping school, not finishing homework, watching too much television or playing too many Internet/computer games, being bullied, having 'bad' friends, or dating too early. While, on average, only 15–16 per cent of parents worry about these issues, the proportion is, again, higher for parents who are living with their children than for parents who left their children in rural villages.

There might be different reasons for why parents who live with their children do not consider their children to be doing very well, and to worry more about them. It is possible that pessimistic parents are more likely to take their children with them; or that parents are more likely to take children who are not doing well in the hope that parental supervision will help. The third possibility could be that the children who migrate with parents and those left behind are judged using different benchmarks. Parents with children living in cities tended to compare their children with city kids, whose education was often of a higher quality, whereas parents whose children were left behind were more likely to compare their children with other local rural children, whose general education quality was not as good as that offered in cities. Finding out exactly which explanation dominates is, however, beyond the scope of this preliminary study.

Finally, the RUMIC surveys also collect data for parental-assessed performance of pre-school-aged children. These data are presented in the bottom section of Table 6.10. In this group, few noticeable differences are observed between children who live with their parents in cities and those who are left behind.

Table 6.9 Basic characteristics of migrants' children

	Total children			Children aged 16 and below		
	Total	With parents	Left behind	Total	With parents	Left behind
Total number of children	1,060	379	681	893	338	555
Age	9.87	8.58	10.58	8.20	7.39	8.69
Males (%)	53.68	49.87	55.80	52.97	49.11	55.32
Education level (%)						
Pre-school	33.58	40.9	29.52	39.87	45.86	36.22
Primary	33.30	32.19	33.92	38.97	35.80	40.90
Junior high	15.00	15.04	14.98	16.80	16.27	17.12
Senior high	12.45	8.18	14.83	3.47	1.48	4.68
Three-year college	1.70	1.85	1.62	n.a.	n.a.	n.a.
University and above	2.83	0.26	4.26	n.a.	n.a.	n.a.
Drop-out	1.13	1.58	0.88	0.90	0.59	1.08

n.a. not applicable

Figure 6.7 Age distribution of migrant children by current living arrangements

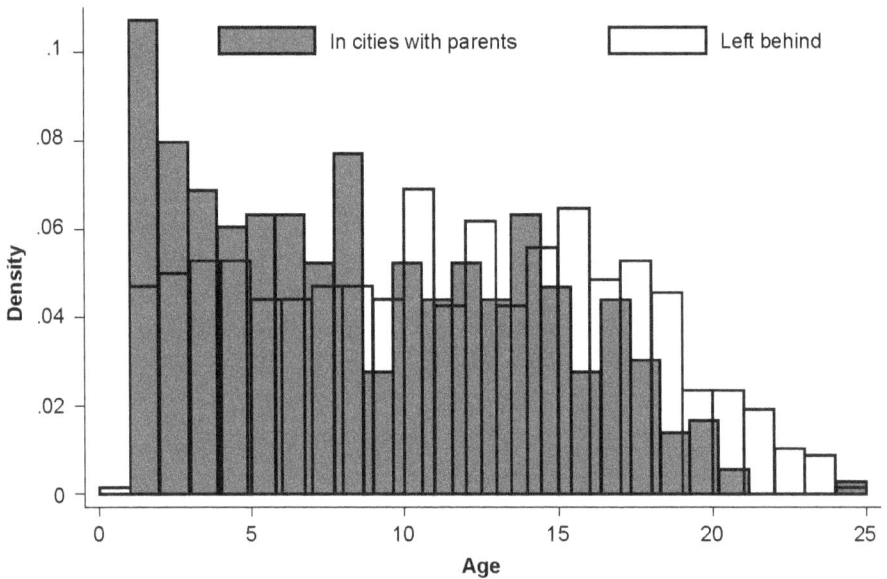

Table 6.10 **Parent-assessed children's health, education and other development outcomes by living arrangements**

	Total children			Children aged 16 and below		
Total children	Total	With parents	Left behind	Total	With parents	Left behind
Healthy and very healthy (%)	89.91	93.14	88.11	89.59	93.20	87.39
School-aged children	704	224	480	537	183	354
School quality (%)						
Best in the city/county	6.11	1.79	8.13	5.21	2.19	6.78
Pretty good in the city/county	33.52	37.05	31.87	29.80	36.61	26.27
Fair in the city/county	58.10	59.38	57.50	62.20	59.56	63.56
Bad in the city/county	2.27	1.79	2.50	2.79	1.64	3.39
School performance (%)						
Excellent	13.78	8.93	16.04	12.10	7.65	14.41
Good	40.34	44.64	38.33	40.22	45.36	37.57
Fair	43.47	44.64	42.92	44.69	44.81	44.63
Not good	1.56	1.34	1.67	2.05	1.64	2.26
Very bad	0.85	0.45	1.04	0.93	0.55	1.13
Do you worry about your child?						
Not at all	47.30	42.86	49.38	45.07	39.34	48.02
Worry about school performance	36.22	40.18	34.38	39.66	43.72	37.57
Worry about other things	16.48	16.96	16.25	15.27	16.94	14.41
School—total cost (yuan/year)						
Primary	1,582	2,295	1,207			
Junior high	2,713	2,937	2,593			
Senior high	5,066	4,787	5,152			
Pre-school-aged children				356	155	201
Not worried about language ability (%)				79.21	83.23	76.12
General development very good (%)				40.73	41.29	40.30
Going to childcare centre (%)				42.98	40.65	44.78
Monthly cost if going to childcare (yuan)				297	378	242

Figure 6.8 **Children's parent-assessed health status by age and current living arrangement**

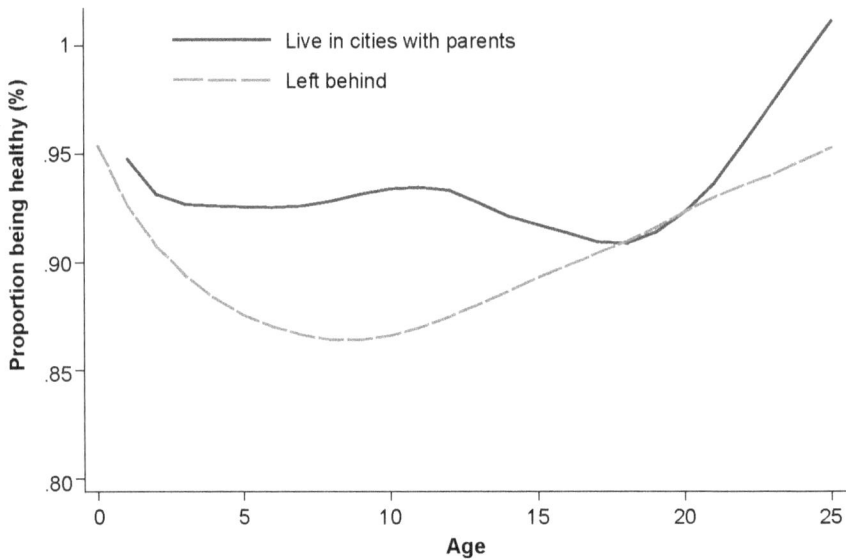

Conclusions

A large body of research has focused on the sources of China's rapid economic growth in the past three decades. A consensus emerging from the established literature indicates that the massive rural-to-urban migration has served as a crucial contributor to TFP growth by shifting labour from low-productivity rural sectors to modern industrial sectors in cities with significantly higher productivity. The main actors of this driving force—that is, the more than 120 million rural–urban migrants—constitute a research area that has long been understudied. The existing migration-studies literature is confined largely to the investigation of migration decisions and the short-term income impacts of migration on migrants and rural households. This chapter has outlined a host of fundamental questions concerning the patterns, the scope and the processes of migration as well as its extensive impact—all of which deserve much further in-depth investigation. Given the sheer size of population directly involved and indirectly affected by the migration process and the extensive and far-reaching impacts of this extraordinary movement, a better understanding of migration issues is needed.

Figure 6.9 **Children's parent-assessed performance at school by age and current living arrangement**

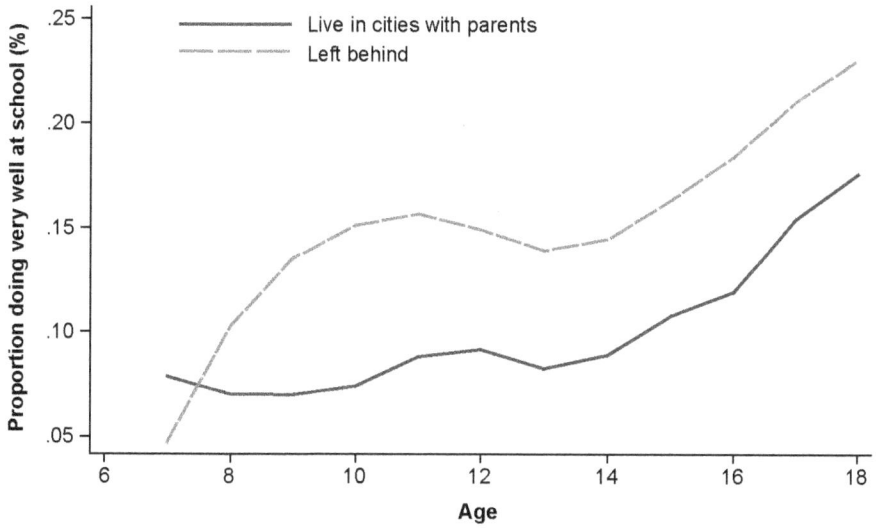

Figure 6.10 **Parents worried about child's school work by age and current living arrangement**

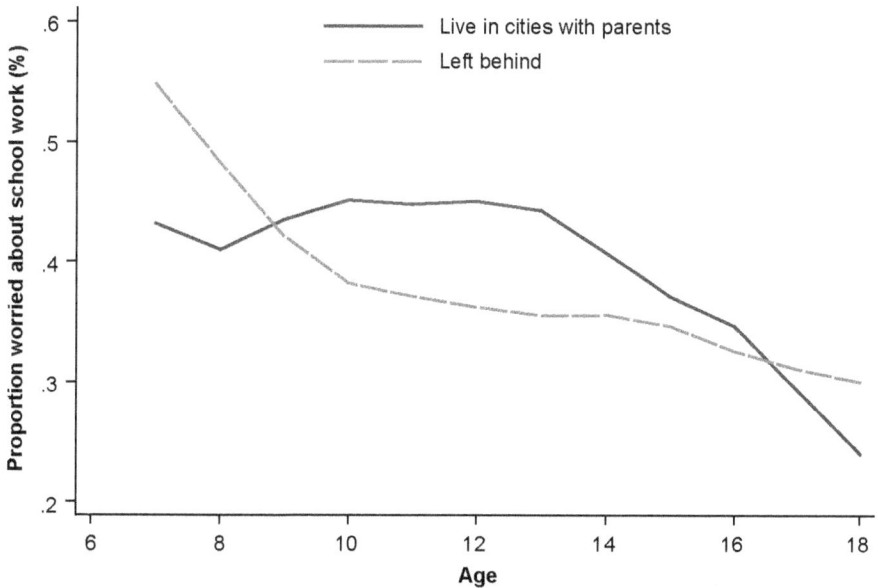

Admittedly, a major obstacle restricting the scope and depth of migration research is the extremely limited survey data for migrants and their comparative groups. The wide range of issues that require urgent further research are plagued by the unavailability of data with sound sampling strategies, broad scope and longer coverage. The newly launched RUMIC project embarked on its painstaking data-collection task with a sample size of 18,000 households across nine provinces and metropolitan areas. The RUMIC surveys identify and rectify the sampling biases of earlier urban-based migrants surveys and have developed a unique census-based sampling frame. With this improved sampling strategy, the RUMIC surveys are well situated to provide the information necessary for its ambitious research agenda. The RUMIC project intends to address questions concerning not just the impact of migration on migrants themselves, but the impact on their families and on the urban and rural communities that are integral parts of the grand migration process.

Using currently available data from the RUMIC surveys, this chapter offers a series of preliminary statistical analyses. The differences are noticeable when comparing the RUMIC census data for household structures and occupational industry distribution with a number of existing survey results. This outcome suggests that differences in sampling strategies and other aspects of survey arrangements can lead to major variations in survey results. In addition, using the RUMIC pilot-survey data and part of the first wave of survey data, we present a range of basic characteristics of migrants and their children. This exercise demonstrates the riches presented by the data and the possibility they bring for further in-depth migration studies. Information such as the health and education situation of migrants' children and the general well-being (including mental health) of migrants is the first of its kind to be made available. Furthermore, the potential of the RUMIC survey data will continue to unfold. Given time, the surveys will be able to develop their longitudial dimensions and enable important studies looking at income dynamics, long-term migration impacts on the second generation's health and development, as well as the processes of and channels for migrants' assimilation into urban communities.

Notes

1   The exact scale of internal migration in China remains a subject of considerable debate. Estimates differ depending on the data source and how migrants are defined in terms of geographic scope, the duration of their stay outside the source area and whether their movement was accompanied by a change in *hukou* (household registration system) status. Alternative estimates of the size of the floating population—those who live for more than six months outside the township where they have their *hukou*—indicate that it constituted nearly 150 million people in recent years, nearly 10 times the number in the late 1980s (NBS 2006).

2   The estimates are likely to underestimate the size of the rural migrant labour force because of the partial exclusion of migrant families.

3   Young (1995) argued that the 'East Asian miracle' was driven essentially by capital accumulation. As TFP growth accounted for only a small fraction of the growth of East Asian economies, the so-called miracle was merely a myth, as input accumulation-led growth was hardly sustainable. Krugman (1994) asserted a similar point.

4   As Young (2003) rightly pointed out, the data for most economies were filled with apparently inconsistent series. Choosing among them, one can produce almost any estimate of productivity growth imaginable. Therefore, it is not our intention to investigate the precise magnitude of productivity growth in China during the reform period. Rather, it is to note that productivity enhancement is an important contributor to China's economic miracle.

5   Similarly, Chow (1993) looked at a five-sector model of China's economy between 1952 and 1980. It was estimated that at the beginning of the reform period (1978), marginal products of labour (MPL) across agriculture, industry and commerce were 63 yuan, 1,027 yuan and 1,809 yuan, respectively (figures expressed in 1953 output values).

6   The efficiency index was calculated as the ratio of real GDP to efficient GDP, simulated by allocating labour and capital inputs such that the marginal returns to each factor were equal across all productive sectors.

7   An equally valid approach would be to investigate the comparison of incomes and welfare of non-migrants in home villages with what they would have earned had they migrated.

8   The urban samples are drawn from 15 cities, including Shanghai, Nanjing, Wuxi, Hangzhou, Ningbo, Guangzhou, Dongguan, Shenzhen, Chongqing, Chengdu, Wuhan, Hefei, Bangbu, Zhengzhou and Luoyang.

9   Based on the 2000 census data, Shanghai, Guangdong, Jiangsu and Zhejiang accommodated 66 per cent of migrants, while 47 per cent of migrants were from Sichuan, Chongqing, Anhui, Hubei and Henan Provinces.

10 The identification of a city boundary is based on the following principles. First, it should include as many workplaces where migrants are employed as possible. Second, it should not have large empty spaces. The cities defined by these rules are often smaller than the size of their boundary areas. For example, there were 18 administrative districts in Shanghai in 2005. Among them, nine districts are fully covered in the survey boundary and five are partly covered (Baoshang, Minhang, Jiading and New Pudong), while four districts (Songjiang, Jinshan, Qingpu and Fengxian), which are far from the city centre, are excluded. Similarly, Ningbo City comprises six districts and only two and half districts are included in our city boundary. For detailed information on the boundaries of the 15 cities, visit the RUMICI web site at http://rumici.anu.edu.au

11 Theoretically, this sampling strategy accounts for all working migrants who participate in any form of employment at the time of the census. While migrants who do not work at the time of the survey will be excluded from the sampling frame, we believe the number of non-working migrants is relatively small and does not constitute a bias significant enough to jeopardise the randomness of the sampling.

12 These include Beijing, Shenyang, Jinzhou, Nanjing, Xuzhou, Zhengzhou, Kaifeng, Pingdingshan, Chengdu, Zigong, Nanchong, Lanzhou and Pingliang.

13 The authors have access to all three surveys indicated here. The data presented are therefore from our own calculations.

14 The NBS survey asks rural households to report some basic information about their absent member(s) who have migrated to cities, such as the industry in which they work, their occupation and possible earnings.

15 We report unweighted and weighted means of the first wave of the RUMiC survey in all subsequent tables.

16 At first glance, there is some statistical evidence indicating an adverse effect on the relationship between husbands and wives due to living arrangements that separate them physically. Among those who live separately from their spouse, about one in every four consider there is some adverse effect, 2 per cent indicate a strong effect and the rest do not think there is any negative impact. Comprehensive evaluation of the impact of distant living arrangements, however, requires further analysis.

17 Once again, these differences could be related to the lack of randomness in the IE sample, which omits migrants who are not living in residential areas.

18 Part of this change might, however, reflect the sample bias inherent in the IE surveys, which include a much larger proportion of self-employed migrants who are less likely to have any insurance. Indeed, in the RUMiC survey, the proportion of self-employed workers without unemployment or work-injury compensation and pension insurance is almost 90 per cent for each category—much higher than for wage/salary earners.

19 China's CPI, the main gauge of inflation, reached an 11-year monthly high with a 7.1 per cent rise in January 2008. Huge increases in food prices pushed up the CPI by 4.8 per cent in 2007—the highest level since 1997.

20 Official poverty lines, Dibao lines, MCL lines and subjective poverty lines are described in Table A6.2.

21 GH12 covers a wide range of mental health issues, including lack of concentration, loss of sleep due to worry, feeling unable to play a useful role in things, feeling unable to make decisions, feeling under pressure all the time, feeling unable to overcome difficulties, feeling that life is meaningless, unable to face up to problems, depression, lack of self-confidence, feelings of worthlessness, and a general feeling of unhappiness.

22 The full sample of children includes all the children of respondent households who are aged 16 years or younger and those older than 16 but still at school. Therefore, the difference between the full sample of children and the younger sub-sample is the sub-sample of older children still at school.

23 Further investigation reveals that children living by themselves are aged seven to 16 and all are at boarding school.

24 The total number of school drop-outs is 8.

References

Borensztein, E. and Ostry, J.D., 1996. 'Accounting for China's growth performance', *American Economic Review*, 86(2):224–8.

China Development Research Foundation (CDRF), 2007. *China Development Report 2007* (in Chinese), China Development Research Foundation, Beijing.

Chow, G.C., 1993. 'Capital formation and economic growth in China', *Quarterly Journal of Economics*, 108(3):809–42.

——, 2002. *China's Economic Transformation*, Blackwell Publishing, London.

Consultative Group to Assist the Poor (CGAP), 2006. *China Labor: urban workers sent $30 billion to rural homes*. Available from http://www.cgap.org/portal/binary/com.epicentric.contentmanagement.servlet.ContentDeliveryServlet/Press/press_coverage31.php.

Doak, C., Adair, L., Bentley, M., Fengying, Z. and Popkin, B., 2002. 'The underweight/overweight household: an exploration of household sociodemographic and dietary factors in China', *Public Health Nutrition*, 5(1A):215–21.

Du, Y., Gregory, R.G. and Meng, X., 2006. 'Impact of the guest-worker system on the poverty and well-being of migrant workers in urban China', in R. Garnaut and L. Song (eds), *The Turning Point in China's Economic Development*, Asia Pacific Press and ANU E Press, The Australian National University, Canberra:172–202.

Du, Y., Park, A. and Wang, S., 2005. Migration and rural poverty in China, Mimeo., Department of Economics, University of Michigan.

Erens, B., Primatesta, P. and Prior, G., 2001. *Health Survey for England 1999: the health of minority ethnic groups*, The Stationery Office, London.

Fan, S., Zhang, X. and Robinson, S., 1999. *Past and future sources of growth for China*, EPTD Discussion Paper No.53, Environment and Production Technology Division, International Food Policy Research Institute, Washington, DC.

Fei, J.C.H. and Ranis, G., 1961. 'A theory of economic development', *American Economic Review*, September:533–65.

Giles, J., 2006. 'Is life more risky in the open? Household risk-coping and the opening of China's labor markets', *Journal of Development Economics*, 81:25–60.

Giles, J. and de Brauw, A., 2005. *Migrant opportunity and the educational attainment of youth in rural China*, IZA Discussion Paper No.2326, 12 September. Available from http://ssrn.com/abstract=757726.

Guo, F., Cai, F. and Zhang, Z., 2005. Rural migrants and shantytown communities in Chinese cities, Presentation to Migration and Poverty Conference, Beijing, October.

Hare, D., 1999. '"Push" versus "pull" factors in migration outflows and returns: determinants of migration status and spell duration among China's rural population', *Journal of Development Studies*, 35(3):45–72.

Hu, Z. and Khan, M.S., 1997. *Why is China's growth so fast?*, IMF Staff Papers, International Monetary Fund, Washington, DC.

Institute of Population and Labour Economics (IPLE), 2001. *China Urban Labor Survey*, Institute of Population and Labour Economics, Chinese Academy of Social Sciences, Beijing.

Islam, N. and Dai, E., 2007. *Alternative estimates of TFP growth in China: evidence from application of the dual approach*, Working Paper Series, International Centre for the Study of East Asian Development, Kitakyushu, Japan.

Krugman, P., 1994. 'The myth of Asia's miracle', *Foreign Affairs*, 73:62–78.

Lewis, W.A., 1954. 'Economic development with unlimited supplies of labour', *The Manchester School of Economic and Social Studies*, May:139–91.

Li, Q., 1995. 'Migrant workers' attitudes and the impact on social conflict' (in Chinese), *Sociology Study*, 4.

——, 2001. 'Research on remittances from rural–urban migrants' (in Chinese), *Sociology Study*, 4.

Lu, S., 2005. *Rural Left-Behind Children: psychological pressure* (in Chinese). Available from www.snzg.net/shownews.asp?newsid=7766

Ma, L. and Xiang, B., 1998. 'Native place, migration and the emergence of peasant enclaves in Beijing', *China Quarterly*, 155:546–81.

Meng, X., 2001. 'The informal sector and rural–urban migration—a Chinese case study', *Asian Economic Journal*, 15(1):71–89.

Meng, X. and Zhang, J., 2001. 'Two-tier labour markets in urban China: occupational segregation and wage differentials between urban residents and rural migrants in Shanghai', *Journal of Comparative Economics*, 29(3):485–504.

Murphy, R., 1999. 'Return migrant entrepreneurs and economic diversification in two counties in South Jianxi, China', *Journal of International Development*, 11:661–72.

National Bureau of Statistics (NBS), 1998–2003. *China Statistical Yearbook 1998–2003*, China Statistics Press, Beijing.

——, 2006. *China Yearbook of Rural Household Surveys*, China Statistics Press, Beijing.

——, 2007. *China Statistical Yearbook 2007*, China Statistics Press, Beijing.

Qi, C. and He, F., 2005. Rural–urban migrants' housing conditions in two urban villages of Ningbo City, Presentation to Migration and Poverty Conference, Beijing.

Rozelle, S., Taylor, J.E. and de Brauw, A., 1999. 'Migration, remittances, and productivity in China', *American Economic Review*, 89(2):287–91.

Shen, J., 2002. 'A study of the temporary population in Chinese cities', *Habitat International*, 26:363–77.

Sheng, L. and Peng, L., 2005. 'The population, structure and characteristics of rural migrant workers', *Research on Rural Migrant Labour in China*, China Statistics Press, Beijing.

State Council Research Group, 2006. *Report on Migrant Workers in China*, (in Chinese), China Yan Shi Publishing House, Beijing.

Taylor, J.E., Rozelle, S. and de Brauw, A., 2003. 'Migration and incomes in source communities: a new economics of migration perspective from China', *Economic Development and Cultural Change*, 52(1):75–101.

Wang, Y. and Yao, Y., 2001. *Sources of China's economic growth, 1952–99: incorporating human capital accumulation*, Working Paper, World Bank, Washington, DC.

Woo, W.T., 1997. 'Chinese economic growth: sources and prospects', in M.F. and F. Lemoine (eds), *The Chinese Economy*, Economica, London.

Xiang, B., 2005. *Migration and health in China: problems, obstacles and solutions*, Asian Metacentre Research Paper Series, National University of Singapore.

Young, A., 1995. 'Tyranny of numbers: confronting the statistical realities of the East Asian growth experience', *Quarterly Journal of Economics*, August:641–80.

——, 2003. 'Gold into base metals: productivity growth in the People's Republic of China', *The Journal of Political Economy*, 111(6):1220.

Zhang, L., 2005. 'Report on silicosis', (in Chinese), *Labour Protection*, 2005(4).

Zhang, X. and Tan, K.-Y., 2007. *Incremental Reform and Distortions in China's Product and Factor Markets*, International Food Policy Research Institute.

Zhao, Y., 1997. 'Labor migration and returns to rural education in China', *American Journal of Agricultural Economics*, 79:1278–87.

——, 1999a. 'Labor migration and earnings differences: the case of rural China', *Economic Development and Cultural Change*, 47(4):767–82.

——, 1999b. 'Leaving the countryside: rural-to-urban migration decisions in China', *The American Economic Review*, 89(2):281.

——, 2001. *The role of migrants' networks in labor migration: the case of China*, Working Paper E2001012, China Center for Economic Research, Peking University, Beijing.

——, 2002. 'Causes and consequences of return migration: recent evidence from China', *Journal of Comparative Economics*, 30:376–94.

Zheng, G., 2005. 'Physically damaged rural migrant workers', (in Chinese), *Sociology Study*, 3.

Zhu, L., 2001. 'Group discrimination—conflicts between migrant workers and urban natives', (in Chinese), *Jiang Hai Journal*, 6.

Zhu, N., 2002. 'The impact of income gaps on migration decisions in China', *China Economic Review*, 13(2–3):213–30.

Acknowledgments

The authors would like to thank the Australian Research Council, AusAID and the Ford Foundation for their financial support, and Dandan Zhang for her excellent research assistance.

Table A6.1 Different poverty lines, 2007 and 2008 (yuan per month)

	Dibao line		Poverty line		Subjective poverty line	
	2007	2008	2007	2008	2007	2008
Shanghai	350	..	..	..	..	..
400	493	563	536	565	..	..
Nanjing	300	350	423	493	464	493
Wuxi	300	350	423	493	..	..
521	..	..	..	..	..	..
Hangzhou	320	355	451	500	..	597
Ninbo	300	350	423	493	..	536
Wuhan	248	300	388	469	..	503
Chengdu	230	276	359	431	..	565
Guangzhou	390	..	549	..	637	..
Shenzhen	388	..	546	..	617	..

Note: Dibao data are from the Internet; poverty lines are adjusted based on Youjuan Wang's calculations of the ratio of the Dibao and poverty lines for the Eastern and Central regions.

Table A6.2 Different poverty lines

	Dibao		Poverty line		Subjective poverty line	
	2007	2008	2007	2008	2007	2008
Shanghai	350	400	493	563	536	565
Nanjing	300	350	423	493	464	493
Wuxi	300	350	423	493	..	521
Hangzhou	320	355	451	500	..	597
Ninbo	300	350	423	493	..	536
Wuhan	248	300	388	469	..	503
Chengdu	230	276	359	431	..	565
Guangzhou	390	..	549	..	637	..
Shenzhen	388	..	546	..	617	..

Note: Dibao data are from internet, poverty lines are adjusted based on Youjuan Wang's calculation of the ratio of dibao and poverty line for the Easten and Central regions.

Table A6.3   **Detailed and broad industry categories**

Code	72 industries	6 broad industry categories
1	Agriculture, forestry, animal husbandry and fishing	Construction
2	Mining	
3	Construction	
4	Food and drink manufacturing	Manufacturing
5	Tobacco manufacturing	
6	Textile manufacturing	
7	Clothing manufacturing	
8	Leather and shoes manufacturing	
9	Timber and furniture manufacturing	
10	Paper and paper product manufacturing	
11	Publishing, printing and video production	
12	Coking and oil refinary	
13	Medicine and other chemical manufacturing	
14	Rubber and plastic manufacturing	
15	Non-metalic mineral product manufacturing	
16	Basic metal product manufacturing	
17	Metal product manufacturing	
18	Machinary and equipment manufacturing	
19	Office equipment manufacturing	
20	Electrical and electronic equipment manufacturing	
21	Radio, TV and other communication equipment manufacturing	
22	Medical, optical, and other special equipment manufacturing	
23	Transportation manufacturing	
24	Machinery rental	
25	Government organisation	Education and government
26	Research	orgnisation
27	Education and training services	
28	Various type of agencies	Various types of agencies
29	Real estate agencies	
30	Consultancy services	Services
31	Security services	
32	Other public services	
33	Transportation services	
34	Post and communication services	
35	Bookshops	
36	Newspaper stands	
37	Financial service and insurance	
38	Hotels/motels	
39	Domestic services	
40	Entertainment/hairdressing/facial	
41	Childcare and care of old people	
42	Tourist centres	
43	Restaurants, cafés and fast food places	

Table A6.3 continued...

44	Small repair outlets	
45	Medical services	
46	Taxi company	
47	Bicycle or car parks	
48	Tailor shops	
49	Advertisment and other printing operations	
50	Dry cleaning stores	
51	Domestic equipment repairs	
52	Car repairs	
53	Wedding, funeral, and other religious operations	
54	Photo shops	
55	Recycling	
56	Sales of industrial products and tools	Wholesale and retail trade
57	Sales of metal products	
58	Street vendors	
59	Small shops	
60	Supermarket and department stores	
61	Chemists	
62	Shops to sell sports products	
63	Car sellers	
64	Bicycle and other domestic equipment stores	
65	Clothing, shoes, and bags stores	
66	Stationary stores	
67	Nursery and garden centres	
68	Construction material stores	
69	Home improvement stores	
70	Jewelry and other arts stores	
71	Musical equipment and other domestic equipment stores	
72	Second-hand stores	

7

Rethinking thirty years of reform in China
Implications for economic performance

Xiaolu Wang

Economic achievement after 30 years of reform

After three decades of economic reform, beginning in 1978, China has transformed itself from a centrally planned economy to a market economy. Prices for most commodities have been liberalised. The private sector, including private enterprises, shareholding companies and foreign-funded enterprises, has become the dominant part of the economy. Although the government is still playing an important role in the economy, the overall command system was abolished long ago.

During this period (1978–2007), the gross domestic product (GDP) growth rate in China was maintained at an average of 9.8 per cent annually—3.7 percentage points higher than that of the pre-reform period (1952–78) under the centrally planned regime. GDP in constant prices has increased by 14.8 times during the reform period, and achieved US$3,283 billion in 2007 by the yearly average exchange rate of CNY7.6 for US$1 (NBS various years; also below unless otherwise referenced). The size of the Chinese economy overtook Russia in 1992, Canada in 1993, Italy in 2000, France in 2005 and the United Kingdom in 2006, becoming the world's fourth largest in 2006. It is likely to surpass Germany in 2008. According to the purchasing power parity (PPP) measure of the World Bank, the Chinese economy is already the second largest after the United States (World Bank various years).

Due to its huge population size, China's GDP per capita is still low—only US$2,456 in 2007—although it is 10.8 times what is was in 1978. Using different PPP measures (for example, World Bank 2008), this per capita level could be expanded three to four times.

One of the most significant improvements during the reform period was a dramatic reduction in poverty levels. In 1978, 250 million rural people were living in poverty, according to the national poverty standard using constant prices. This figure reduced to 15 million in 2007. By the World Bank's higher standard of those living on 'one dollar a day', rural poverty in China reduced from 31.5 per cent in 1990 to 8.9 per cent of the population in 2005 (Gill et al. 2007).

China and Russia—two large countries and former centrally planned economies—shared many similarities before their reforms, but they adopted different reform methods and achieved very different results. In particular, there was a striking contrast in the economic performance of the two countries during their reform periods.

China introduced a family-based Household Responsibility System (HRS) in its agricultural sector and decentralised its central planning system at the beginning of its reform. Price control was gradually released. The non-state enterprise sector was encouraged to develop, and market competition gradually became the dominant mechanism in the economy during a long period of transition. In the first decade of reform, from 1978 to 1988, the size of the Chinese economy in real terms expanded by 2.6 times. The annual GDP growth rate during this period was 10 per cent. Urban and rural household income per capita in constant prices increased by 1.82 and 2.11 times, respectively. The rural population living in poverty was reduced by 60 per cent.

In Russia, some similar reform measures were adopted in the late 1980s, but were soon replaced by radical 'shock therapy' in the early 1990s. In 1992, price control was removed entirely and central planning was abolished. Most stated-owned enterprises were privatised in 1992 and subsequent years. If we regard the period 1990–2000 as the first decade of the reform period in Russia, GDP dropped by nearly 40 per cent—an annual decrease of 4.7 per cent. There was also hyperinflation and a dramatic reduction of people's incomes during this period.

Angus Maddison (2007) compared Chinese and Russian (initially, the Soviet Union) performance using a comparable PPP measure for a longer period. In 1978, the first year of Chinese reform, per capita GDP in China accounted for only 13 per cent of that in the Soviet Union. After 25 years of reform, it achieved 76 per cent of that in Russia in 2003. During this period, GDP per capita in China increased to 4.91 times its 1978 level, whereas GDP per capita in Russia shrank

Table 7.1   **Economic performance in China and Russia during the reform period**

	China	Russia (and USSR)	China/Russia (%)
GDP (US$ billion 1990 PPP)			
1978	935	1,018	92
2003	6,188	914	677
Growth (2003/1978)	662%	90%	
Per capita GDP (US$ 1990 PPP)			
1978	978	7,420	13
2003	4,803	6,323	76
Growth (2003/1978)	491%	85%	

Source: Maddison, A., 2007. *Chinese Economic Performance in the Long Run*, second edition, Organisation for Economic Cooperation and Development, Paris:Tables 4.4–4.5, p.102.

to 85 per cent of its 1978 level. In the same period and by the same measure, total Chinese GDP expanded from 92 per cent to 677 per cent of that in Russia (Table 7.1).

Russia experienced rapid economic growth in the post-Yeltsin period, although this was still a recovery towards its initial levels before the economic shock. In 2006, Russian GDP in constant prices was equal to only 97 per cent of its 1990 level, whereas the same ratio was 470 per cent for China (UN various years).

The next part of this chapter discusses the reasons for the different economic performances, and what lessons can be drawn from past reforms.

Chinese and Russian reforms: what made the difference?

Economists have offered various explanations for why the outcomes of the Chinese and Russian reforms were so different. It is commonly accepted that the speed and sequence of reforms are important. Unlike Russia's shock therapy, China made a step-by-step movement towards a market economy. This evolutionary approach not only smoothes the shock, it allows macroeconomic stabilisation and institutional building to be achieved. While the old central-control mechanism is gradually replaced by market mechanisms, rapid economic growth is achieved (for example, McKinnon 1993; Roland 2000; Maddison 2007).

Justin Lin (1995) provides a convincing, although incomplete, explanation for the Chinese–Russian contrast. He indicates that China and Russia had heavy resource misallocation under the centrally planned system, but reallocation of resources among sectors takes time. A sudden correction of price signals leads to production decreases in the sectors with over allocation of resources, but not to corresponding increases in the sectors in which resources are under allocated. According to Lin's explanation, the Chinese approach was to allow market-oriented growth in the under-allocated sectors first, so that misallocation was corrected in a relatively long period without drops in production (Lin 1995). Nevertheless, he did not explain why the real growth path in Russia in the reform period showed as an L-curve instead of a J-curve—that is, how can an expected temporary production decrease become a 10-year-long economic disaster?

Some other authors emphasise the different initial conditions in China and Russia. They argue that given these differences, the Chinese experience is not replicable, and the Russian recession is unavoidable. Among them, Popov (2000, forthcoming) shows, via a cross-country analysis, that the worse the economy distorted previously, or the higher the initial per capita GDP, the larger is the drop of output seen in liberalisation of a transitional economy.

Popov also indicates that the speed of liberalisation has a negative impact on output, although he does not treat the speed of liberalisation as a policy variable, but as an 'endogenously determined' variable, which is determined by the political situation. In this sense, all the policy changes already in place could be classified as being endogenously determined, although some are really imported. In addition, it is hard to believe that the Chinese economy was previously less distorted than other transitional economies.

One dimension of the Chinese and Russian reforms was never sufficiently discussed—that is, did the reforms lead to a process of 'Pareto improvement'? If not, who won and who lost? How were public interests affected in the reform? And how was economic performance related to the issue of interest redistribution during the reform and post-reform periods?

In the remaining parts of this chapter, I will show three points via a comparison of some reforms in China and Russia: first, most of the reforms in China have led to improved conditions for all groups of people; therefore, the Chinese reform has generally been a process of Pareto improvement. This was not the case in Russia because the reforms led to improvements for certain interest groups, but the majority of people were worse off in the long term. Second, most of the reforms in China were indigenously and endogenously determined, often after many policy debates and empirical experiments; therefore, they were usually in the interest of all the people or at least the majority of the people.

This explains why the reforms could be in a path of Pareto improvement. In Russia, some major reform measures were determined externally and by a small group of élites, although they were undertaken in the name of public interest. Third, the reform outcomes in China and Russia indicate a strong link between economic performance and interest redistribution—that is, a reform benefiting all groups of people is usually a process of Pareto improvement, leading to better economic performance. This implies that whether the reform policy takes the public interest as its priority is crucial for the economy.

The following sections of the chapter comprise reviews of three major reforms in China compared with those in Russia.

Agricultural reform

Agricultural reform in China, beginning in 1978, was the first step in the country's economic reforms. As a replacement for the old and inefficient commune system, the rural HRS was not a government-designed reform measure. It was an innovation by farmers originating from the mid 1950s, and brought consistently better outcomes for increasing output and reducing rural poverty in pilot practices from the 1950s to the 1970s. It also incurred many political attacks, and was several times suppressed by top leaders for ideological reasons (RGCRD 1981).

During 1978–80, farmers in different regions reintroduced this system and achieved remarkable success in increasing agricultural output and incomes; the system therefore spread automatically to broader areas. In Anhui and Sichuan Provinces, it was supported by provincial leaders Wan Li and Zhao Ziyang,[1] but was attacked by conservative leaders at the central and local government level as a serious crime of 'anti-socialism'. The following is an example of a typical debate from 1980 between Wan Li and an anonymous senior official, 'A', both of whom were responsible for the central government's agricultural policy at the time (Zhao 2007):

A: HRS does not fit the socialist character, therefore, it should not be widely applied.

Wan: Why not? This is what people want; they only want to get enough food.

A: It deviates from the socialist direction; it is not a road towards 'common prosperity'.

Wan: Socialism and the people, which one would you choose?

A: I choose socialism!

Wan: I choose the people!

The real outcome of the HRS was, however, more convincing than any ideology. After many conflicts, the HRS was accepted and promoted formally by the central government in 1982. Until 1984, more than 97 per cent of Chinese villages adopted the HRS. The commune system was abolished.

The HRS is a household-based farming system that does not alter collective land ownership, but distributes farm land to households under long-term contracts. Unlike the old commune system, it provides adequate incentives and autonomy to farmers. In 2006, all land contract levies, together with agricultural taxes, were abolished; land use is now free for farmers.

With the introduction of the HRS, the pricing mechanism of agricultural products was also changed. The government first increased the State's purchasing prices for grain by 20–50 per cent in 1979, and then gradually liberalised the grain market. Similar things happened to other agricultural products. The HRS reform, plus price increases, led to remarkable increases in agricultural output and farmers' incomes in the early 1980s. The longstanding problem of food shortages was resolved. In real terms, farmers' per capita incomes in 1984 increased by 2.5 times its 1978 level. The proportion of the rural population living in absolute poverty was reduced by half—from 250 million to 128 million (Table 7.2). Urban–rural income disparity was also reduced.

Another important condition for the success of the agricultural reform was that, in the initial stage, reformers entered and dominated the core group of the Chinese leadership. They had rich working experiences and good knowledge of agricultural issues, they understood the needs of rural people and made rural development and the improvement of people's lives a target for reform.

Table 7.2   **Farm output and income before and after agricultural reform in China**

	1978 (before HRS)	1984 (after HRS)	1984/1978 (%)
Grain (Mt*)	304.8	407.3	134
Cotton (Mt)	2.2	6.3	289
Oil-bearing crops (Mt)	5.2	11.9	229
Fruit (Mt)	23.8	47.8	201
Rural income per capita (yuan)	134	355	250[a]
Rural poverty (million people)	250	128	51

* Mt = mega-tonnes
[a] income levels in current prices, but the relative change is calculated in constant prices
Source: National Bureau of Statistics (NBS), 2005. *China Compendium of Statistics 1949–2004*, China Statistics Press, Beijing.

Agricultural reform in Russia was very different. In the early 1990s, the leadership accepted the advice of the International Monetary Fund (IMF) and set a target for agricultural reform to follow the US agricultural model, and scheduled to develop one million private farms. To achieve this, they privatised agricultural land, disbanded state farms and collective granges and withdrew all state subsidies for agricultural production. This type of reform did not have the support of the majority of rural people. According to a survey (Qiao 2002), only 32.2 per cent of Russian farmers supported land privatisation without reservation, while 39.7 per cent opposed it. Only 18.1 per cent of farmers supported the free trade of land, while 60 per cent opposed it.

Prices for agricultural products were liberalised completely, together with all other products in 1992, and this immediately caused hyperinflation. Input prices for farming increased much faster than output prices, resulting in rapid decreases in farmers' incomes. From 1990 to 1998, grain production dropped by 46 per cent, and the gross value of agricultural output dropped by 47 per cent. In 1999, 53 per cent of Russian rural residents lived below the official poverty line. Agriculture in Russia has improved in recent years, although it is still recovering from recession.

In general, the agricultural reforms in China were a 'bottom-up' process. It adopted a model preferred by most farmers, which led to quality-of-life improvements for nearly all the 790 million rural residents—who accounted for 82 per cent of the total population in 1978. The urban population also benefited because the food-supply situation was much improved. The reforms also made significant contributions to overall economic performance. In contrast, agricultural reforms in Russia followed a 'top-down' approach: the leadership imposed an imported agricultural model on farmers without considering indigenous needs and local situations. The outcome of the reform was undesirable, and most people suffered because of it.

Price reform

Price reform in China was a crucial part of its transition to a market economy. A 'dual-price system' was formed gradually in the early and mid 1980s. As a transitional measure towards a market economy, this system allows market prices to work while planning prices remain; this smoothes economic shocks and maintains economic growth in the early stages of reform.

During 1978–80, a large number of state-owned enterprises (SOEs) were brought into the experimental schedule of 'expanding enterprise autonomy'. They were allowed to sell any excess above their state quota of products outside the state plan, at flexible prices, and were also allowed to purchase inputs outside the plan at flexible prices, when the state supply was not available. This experiment was started in Sichuan Province in 1978 and achieved good outcomes, so it soon expanded to cover 6,600 SOEs across China in 1980, and then applied to all SOEs. This formed a market system that coexisted with the planned system. The market prices indicated the direction of demand, gave enterprises incentives and played a role in supply–demand adjustment at the margin.

After the initial success, debates about price reform were still going on in the mid 1980s, focusing on a few questions: 1) should planning prices or market prices play the dominant role in the economy; 2) how could the imbalanced economy be brought into equilibrium—that is, first via administrative adjustment of controlled prices or by liberalising price controls; 3) should these adjustments or liberalisation be done as fast as possible or over a relatively long period (for example, Chinese Institute for Economic System Reform 1987)?

It became clear that controlled prices were not functioning well, and further market-oriented price reforms were needed. Price reforms that were too radical, however, would exceed the economy's capability to bear the shock. As a result, the government persisted with a dual-price system in the 1980s. The scope of state planning and price control was gradually reduced or liberalised in some industries, where there was equilibrium or excess supply, but was kept essentially at the same level in other fields where there was a serious shortage of supply, such as in the steel industry.[2]

This was because, in these fields, the gaps between the controlled and market-determined prices were large (for example, by then, the market price for steel was three times higher than the controlled price); a sudden abolition of controlled prices would have led to violent changes in input prices and serious shocks to firms in downstream industries, and to consumers. Serious inflation, unemployment, enterprise bankruptcy and economic declines would be expected.

Nevertheless, for basic inputs, even though price controls remained, prices for products were increasingly market oriented. The higher market prices provided incentives to firms to meet additional demand. In the 1980s, the output of steel increased from 37 to 66 mega-tonnes (Mt) (NBS 2005), most of which was produced by SOEs, but was promoted by market prices. Along with economic growth, the relative importance of market prices increased and that of controlled prices shrank.

In 1988, top leaders attempted to launch a 'price-reform storm' to liberalise the remaining portion of controlled prices. This scheme resulted in massive panic purchasing of consumer goods and bank squeezes, so the leadership soon decided to abandon it.

Another important contributing factor to price reform was the rapid growth of market-oriented non-state enterprises. This substantially increased the scope for market prices to work. The non-state enterprises gradually became dominant in the economy, and finally led market prices to play the dominant role. Table 7.3 shows how the price mechanism transformed.

While prices were gradually marketised, price levels during the transitional period were basically stable, and hyperinflation was avoided. The consumer price index (CPI) as an annual average for the three decades from 1978–2007 was 5.5 per cent. During this period, the CPI exceeded 10 per cent in only five years: 1988–89 and 1993–95. The CPI reached its highest level—24.1 per cent—in 1994. The annual CPI remained below 3 per cent in 16 of the 30 years.

A decade after the beginning of price reforms, China had essentially grown out of the 'shortage economy' of the early 1990s. Serious supply bottlenecks were eliminated. The economic structure became more balanced. Most importantly, most commodity prices were determined via market competition, which substantially promoted efficiency increases.

In Russia, all prices were suddenly liberalised in January 1992 by the Yeltsin administration without sufficient discussion or preparation. The main 'theoretical' base was an analogy: 'You cannot leap over a ditch in two jumps.' Radical price reforms led immediately to hyperinflation. Using 1991 as the base year (100 per cent), the CPI reached 1,629 per cent in 1992, 15,869 per cent in 1993 and 64,688 per cent in 1994. The majority of people's savings disappeared. The serious inflation forced the enterprise sector to increase nominal wages for workers to survive, the enterprise sector then forced the central bank to issue money to finance the unavoidable increases in nominal wages, and the rapid increases in monetary supply further fuelled hyperinflation. In 2000, the CPI

Table 7.3 **Proportion of products subject to market prices, 1978–2005** (per cent)

	1978	1997	2005
Controlled price	>90	11.9	7.2
Guided price		3.6	
Market price	<10	84.4	92.8

Note: Calculated as weighted average of retail commodities, production inputs and agricultural products in gross value.
Source: Fan, G., Wang, X. and Zhu, H., various years. *NERI Index of Marketization of China's Provinces*, Economic Sciences Press, Beijing.

reached 9,344 times the 1991 level (UN various years). The Russian experience of price reform presented a real challenge to the 'Washington consensus', because its radical price liberalisation did not allow macroeconomic stability to be achieved—although both were necessary, according to the Washington consensus.

Even though nominal wages increased dramatically, price increases were twice as high as wage increases. Most people suffered from income drops and the unemployment rate increased substantially during this period. Production collapsed, and GDP decreased year by year. Until 1998, GDP dropped to only 57 per cent of its 1990 level in real terms. Sixteen years later, in 2006, GDP in Russia had not fully recovered to its 1990 level.

Price reform is not itself an objective in former centrally planned economies; rather, the objectives are better incentives, improved efficiencies and optimised resource allocation. As indicated by their economic performance, these were achieved in China, but not in Russia. Table 7.4 compares inflation, unemployment and economic growth data for China and Russia during the first decade of their price reforms (1980–90 for China and 1990–2000 for Russia).

To summarise the price reforms in China, the dominant position of market prices in the economy was finally established via a gradual process of evolution. This reform strategy was based mainly on the consideration of macroeconomic stability as well as protecting consumers, workers, enterprises and the State from the shocks of inflation, unemployment, production shrinkage and reductions in budgetary revenue. These goals were achieved. In Russia, all the negative impacts were considered to be a necessary cost to pay for the ultimate objective of liberalisation. Radical price liberalisation resulted in serious disadvantages for the majority of people, as well as heavy economic losses.

Ownership structure transformation

State-owned enterprises dominated the Chinese non-agricultural industries before the reforms. They shared 78 per cent of the gross output value of industry in 1978; the remaining 22 per cent was shared by collective enterprises. Transformation of the ownership structure of the economy began with development of Township and Village Enterprises (TVE) in rural areas, and foreign-funded enterprises (FEs) and enterprises with investment from Hong Kong, Macao and Taiwan (HMTEs) in four special economic zones in the 1980s. After these enterprises achieved good results, the development of FEs, HMTEs, private enterprises and joint-stock companies was officially promoted.

Table 7.4 **Price reform in the first decade: inflation, unemployment and GDP growth** (per cent)

	China, 1980–90	Russia, 1990–2000
CPI (annual average)	7.0	276.2[a]
GDP growth (annual average)	9.3	−4.2
Unemployment rate	4.9, 2.5[b]	5.4, 13.4[c]

[a] 1991–2000
[b] urban unemployment rate for 1980 and 1990, respectively
[c] total unemployment rate for 1990 and 1998, respectively
Sources: United Nations (UN), various years. *Monthly Bulletin of Statistics*, online database, United Nations; National Bureau of Statistics (NBS), 2005. *China Compendium of Statistics 1949–2004*, China Statistics Press, Beijing; National Bureau of Statistics (NBS), 2007. *China Statistical Yearbook 2007*, China Statistics Press, Beijing. World Bank, 2007. *2007 World Development Indicators*, World Bank, Washington, DC.

There was a lot of debate about non-state enterprises in the 1980s. An influential opinion in the government was that these enterprises competed with SOEs for inputs and markets, 'undermining the socialist economy', and they should therefore be banned or restricted; others argued that these enterprises provided new employment opportunities, lifted people's incomes, produced various consumer goods that were needed in the market and contributed to the State's revenue, and should therefore be developed further. The latter view dominated central government policy after the mid 1980s, and became more and more convincing when non-state sector development performed better than the state sector.

During 1980–90, TVE employment increased from 30 million to 93 million, and the share of TVEs in the value of gross industrial output (GIOV) increased from about 5 per cent to 20 per cent. Private enterprises, HMTEs, FEs and joint-stock companies also grew rapidly; they shared only 0.75 per cent in GIOV in 1980, but this increased to 9.8 per cent in 1990. They became the main engine of economic growth in the 1990s and shared 68 per cent of GIOV in 2007. In 2007, the entire non-state enterprise sector shared 70.5 per cent of industrial output. Urban non-state enterprises employed 229 million people in 2006—representing 77.3 per cent of urban employment. The formerly state-dominated economy has basically transformed into a private sector-dominated economy. Table 7.5 shows the changes in the industrial sector in the past three decades.

Table 7.5   **Ownership structure of the industrial sector (share in gross output value)** (per cent)

Year	SOE	Non-state enterprises	Collective	Private[a]
1978	77.6	22.4	22.4	.
1990	54.6	45.4	35.6	9.8
2000	47.3[b]	52.7[c]	13.9	38.8
2007	29.5[b]	70.5[c]	2.7	67.8

. insignificant
[a] private sector, including private enterprises, foreign-funded enterprises and joint-stock companies
[b] SOE data for 2000 and 2007, including joint-stock companies with a controlling state share
[c] data exclude small non-state enterprises with annual sales below RMB5 million
Sources: National Bureau of Statistics (NBS), 2005. *China Compendium of Statistics 1949–2004*, China Statistics Press, Beijing; National Bureau of Statistics (NBS), 2007. *China Statistical Yearbook 2007*, China Statistics Press, Beijing. National Bureau of Statistics (NBS), 2008. *2007–2008 Collection of Statistics*, China Economic Monitoring and Analysis Center of the National Bureau of Statistics, CEMAC Print, Beijing.

The transposition of the state and non-state enterprises in the economy was due basically to two elements: continued robust growth of the private and foreign enterprises during the past three decades, which was much faster than the growth of the SOEs, and privatisation of SOEs, mainly from the late 1990s onwards.

Whether the SOEs should be privatised was a controversial issue. Although SOEs became market oriented to a certain extent, their performance was generally undesirable. From 1984 to 1996, the net value of fixed assets in SOEs in the industrial sector increased from CNY340 billion to CNY2,386 billion, whereas their total profits decreased from CNY71 billion to CNY41 billion (all in current prices). Their profit/assets ratio was only 0.8 per cent, and the profit margin (profit to total sales) was only 1.5 per cent in 1996—both significantly lower than the average (1.7 per cent and 2.6 per cent, respectively). During the same period, the SOEs' total losses increased from CNY2.7 billion to CNY79 billion (in current prices), which led to a big deduction in their profits (NBS 1997, 1998). At the same time, non-performing bank loans built up to the trillion-yuan level, mainly due to SOEs. The 1997 East Asian financial crisis further sharpened the pain.

In 1997, the central government formally adopted a new policy called 'seize the big and free the small' (State Council 1997), allowing small SOEs to be sold. The policy for large and medium-sized SOEs emphasised improving their management and transforming them into 'modern enterprise systems'—that is, joint-stock corporations and limited-liability companies.

The number of SOEs in the industrial sector reduced from 113,800 in 1996 to 24,961 in 2006 (the latter figure included pure SOEs and joint-stock corporations with a controlling state share; hereafter, called state-controlled enterprises, or SCEs). Their employment in industry reduced from 43 to 18 million people. Most small SOEs were fully privatised.

The remaining SCEs in the industrial sector produced less then 30 per cent of the total industrial output in 2007. Their performance, however, was much improved. Their total profit increased from CNY41 billion to CNY849 billion during the period 1996–2006, and the profit/asset ratio increased from 0.8 per cent to 6.3 per cent, which was close to the industrial average of 6.7 per cent.[3] Non-performing loans reduced substantially; they accounted for 23.6 per cent of total loans in 2002, and reduced to 6.7 per cent in 2007. This implies that previous policies for improving SOE management and non-state share participation have been effective.

While a large number of SOEs have been privatised, the general performance of the private sector has improved significantly, implying better achievement of the privatised former SOEs.

Privatisation of the SOEs was not, however, a painless operation. Twenty million SOE workers were laid off in 1998, with another 26 million in later years until 2006. The lack of a social-security system in the first few years meant that laid-off workers received only limited financial support and some of them immediately experienced hardship. This was a real shock to a large part of the population, although it was less serious than what happened in Russia. The development of the non-state sector meant it was able to provide many job opportunities for the re-employment of laid-off workers, especially in the coastal areas where the non-state sector was better developed.

Another negative outcome was low transparency and unfair distribution of the former state assets in some regions, where the process of privatisation was not well regulated or monitored and many under-the-table deals benefited mainly a small group of people.

To rethink the process of SOE reform, the negative effects might have been reduced if the social-security systems had been built earlier and the process of reform had been better regulated, more transparent and introduced step by step.

Although there were some similarities between China and Russia in SOE privatisation, it was a more painful process in Russia than in China. Russia's SOE privatisation began in 1992, when a large proportion of state assets was distributed to all Russian residents in the form of warrant stocks worth 10,000 rubles per person. The privatisation program was designed by a small group of élites, who regarded privatisation itself as being the supreme objective, even above the public interest, although such a design seemed fair because everyone had equal shares. Because of hyperinflation caused by the radical price reform, the value of the warrant stocks soon shrank to the price of a pair of shoes. This provided an opportunity for those who were astute enough to collect the warrants from the public at extremely low prices. The enterprise shares soon became highly concentrated and a small number of people became rich owners of the former SOEs (for example, Freeland 2000).

The remaining state assets were sold later, usually at token prices to insiders and those who had relationships with government officials. The second phase of privatisation was even less fair, due to manoeuvring behind the scenes and so on. In a number of industries, privatisation did not bring about market competition; instead, the state monopoly was replaced by private monopolies or oligopolies. In particular, those state companies holding the most valuable natural resources, such as oil and natural gas, were privatised in a non-transparent and very unfair way. This allowed a small group of oligarchs to pay nothing to become the owners of half of Russia's national wealth. This was why the Putin administration had to carry out a re-nationalisation program of oil companies.

In the Forbes world-100 rich list for 2007, Russia had 13 of the world's richest people, most of whom were oil oligarchs, and China had none.

As well as being unfair, radical privatisation in Russia, together with radical price reform, caused heavy economic losses instead of real economic growth. As indicated earlier in this chapter, GDP in Russia in 1998 dropped to only 57 per cent of its 1990 level. The only comparable scale of economic disaster in modern Russian history was during World War II. Up to 2006, GDP in Russia had recovered to 97 per cent of its 1990 level, but this recovery should be attributed partly to increasing world oil prices, as Russia is an important oil exporter.

The experiences and lessons of privatisation in China and Russia are meaningful. They show that, when the public interest is taken as the first priority of reform objectives, the outcome is more likely a Pareto improvement and this fertilises economic growth and development. This is represented by the results of non-state sector development in China. On the contrary, when a certain ideological objective—whether a communist or a capitalist one—or the interests of a certain group of people are placed above the public interest, an inefficient redistribution of resources or wealth is likely to occur. This was the case in Russian privatisation, and to some extent in China's SOE privatisation program in the late 1990s.

Implications for further reform

The theory of public choice shows that even a publicly elected government in a democratic country can make decisions contrary to the public interest, and an election can itself lead to a result not in accordance with the majority's interests (Buchanan and Tullock 1962; Arrow 1963). This explains what happened in Russia in the reform period. The theory also leads to the issue of 'government failure' against 'market failure', and therefore to a solution of 'small government'. The reform experiences in China and Russia indicate, however, that the size of government should not be the only concern—or even the major concern. A more important issue is how to impel the government to behave in the public interest with a long-term perspective, and to play a role as an impartial arbitrator in social conflicts between interest groups.

Although reform in the past three decades has achieved great success in China, the above issue remains unresolved in an institutional base. After the establishment of a basic market framework, a lot of new problems appeared— relating mainly to the role of the government.

Income inequality is widening. According to the World Bank (2006), the Gini coefficient increased from 0.32 to 0.45 during the period 1980–2001. It is likely to be even greater because unreported income (including illegal income) is huge, and is concentrated to a small proportion of high-income earners (Wang 2007). Rent-seeking behaviour and corruption in the public sector are increasing; there is evidence of unjustified distribution of returns from land, natural resources and financial resources.

It is natural that income inequality increases to a certain extent when the economy is in a market-oriented transition. The continued income divergence indicates that other factors are also at work—due mainly to institutional defects or incompletion, which leads to distortions in income distribution, challenges for social justice, and creates uncertainty for social stability and growth sustainability in the long run. This is because the old central-planning framework is withering away, but a new institutional framework has not been completed. This indicates a need for further reform: mainly institutional innovations towards a more transparent, better monitored government system, including better regulated public resource management and better managed public services, and also an effective legal framework and better law enforcement. These are the necessary conditions for sustainable long-run growth and development towards the middle of the twenty-first century.

Notes

1   At the time, Wan and Zhao were the provincial party secretaries of Anhui and Sichuan, respectively. Wan became a vice-prime minister and the director of the State Agriculture Committee in 1980, and the chairman of National People's Congress Standing Committee between 1988 and 1993. Zhao became the prime minister (1980–87) and then the general secretary of the Communist Party (1987–89).
2   Much of the information on the process of reform in China was based on the author's personal experience while working for the State Committee for Economic Restructuring of China. This is the case below unless otherwise referenced.
3   This improvement should be discounted to some extent because 43 per cent of the SCE profit came from oil companies, which benefited mainly from increasing oil prices. Nevertheless, after deduction of this, the improvement is still significant.

References

Arrow, K.J., 1963. *Social Choice and Individual Values*, second edition, Yale University Press, New Haven.

Buchanan, J. and Tullock, G., 1962. *The Calculus of Consent: logical foundations for constitutional democracy*, University of Michigan Press, Ann Arbor.

Chinese Institute for Economic System Reform, 1987. *China: development and reform (1984–1985)*, Spring and Autumn Press, Beijing.

Fan, G., Wang, X. and Zhu, H., various years. *NERI Index of Marketization of China's Provinces*, Economic Sciences Press, Beijing.

Freeland, C., 2000. *Sale of the Century* (Chinese translation, 2004), CITIC Publishing House, Beijing.

Gill, I. and Kharas, H., 2007. *An East Asian Renaissance: ideas for economic growth*, World Bank, Washington, DC.

Lin, J.Y., 1995. 'Chinese economic reform and development of economics', in J.Y. Lin, G. Yi, W. Hai, W. Zhang, F. Zhang, and M. Yu (eds), *Economics and Chinese Economic Reform* (in Chinese), Shanghai People's Publisher, Shanghai.

Maddison, A., 2007. *Chinese Economic Performance in the Long Run*, second edition, Organisation for Economic Cooperation and Development, Paris.

McKinnon, R., 1993. *The Order of Economic Liberalization: financial control in the transition to a market economy* (second edition), Johns Hopkins University Press, Baltimore.

National Bureau of Statistics (NBS), 1997. *China Statistical Yearbook 1997*, China Statistics Press, Beijing.

——, 1998. *China Statistical Yearbook 1998*, China Statistics Press, Beijing.

——, 2005. *China Compendium of Statistics 1949–2004*, China Statistics Press, Beijing.

——, 2007. *China Statistical Yearbook 2007*, China Statistics Press, Beijing.

——, 2008. *2007–2008 Collection of Statistics*, China Economic Monitoring and Analysis Center of the National Bureau of Statistics, CEMAC Print, Beijing.

——, various years. *China Statistical Yearbook*, China Statistics Press, Beijing.

Popov, V., 2000. 'Shock therapy versus gradualism: the end of the debate (explaining the magnitude of transformational recession)', *Comparative Economic Studies*, 42(1):1–57.

——, forthcoming. 'Shock therapy versus gradualism reconsidered: lessons from transitional economies after 15 years of reforms', *Comparative Economic Studies*. Available from http://ssrn.com/abstract=918226.

Qiao, M., 2002. 'A comparison of Chinese–Russian agricultural reforms', *East Europe and Middle Asia Research*, 6.

Research Group for China's Rural Development (RGCRD), 1981. *Selected Materials on Household Responsibility System. Volume I*, Research Group for China's Rural Development, Beijing.

Roland, G., 2000. *Transition and Economics: politics, market and firms*, MIT Press.

State Council, 1997. *The State Council notice on endorsing the State Economic and Trade Commission report for SOE reform and development in 1997*, 19, 23 May.

United Nations (UN), various years. *Monthly Bulletin of Statistics*, online database, United Nations.

Wang, X., 2007. 'Grey income and income inequality in China', *Comparative Studies*, 31, China CITIC Press, Beijing.

World Bank, 2007. *2007 World Development Indicators*, World Bank, Washington, DC.

——, 2008. *World Development Report: agriculture for development*, World Bank, Washington, DC.

——, various years. *World Development Report*, World Bank, Washington, DC.

Zhao, L., 2007. 'Du Runsheng: the Chinese reform should be on guard against dignitary capitalism', *Southern Weekend*, 17 May.

8

China's rapid emissions growth and global climate change policy

Ross Garnaut, Frank Jotzo and Stephen Howes

The world has entered a period of exceptionally fast economic growth, with rapid economic development especially in China, followed by India and many other low-income countries. Early twenty-first century rates of economic growth have been even higher than the average in the 'Golden Age' of the 1950s and 1960s, so the current period could be called the 'Platinum Age' (Garnaut and Huang 2007). This rapid economic growth goes hand in hand with increasing resource use and pressure on the environment, including the build-up of greenhouse gases and resulting climate change.

In most of its first two centuries, modern economic growth was located in a small number of countries, in Western Europe, North America and Oceania, and in Japan (Maddison 2001). In the third quarter of the twentieth century, it extended into a number of relatively small economies in East Asia. A new era began in the fourth quarter of the past century, with the rapid extension of the beneficent processes of modern economic development into the heartland of the populous countries of Asia, including China, India and Indonesia. Incomes are now growing rapidly in a large proportion of the developing world.

In the absence of a major dislocation of established trends, fast growth is likely to continue for a considerable period. The contemporary slow-down in the United States and some other industrialised countries will reduce average total global growth for a while, but is unlikely to break the momentum of strong Chinese, developing-country and global growth. The first two decades of the twenty-first century will see a greater absolute increase in annual human output and consumption than was generated in the whole previous history of our species, and then almost that much again in the next decade to 2030 (Garnaut 2008b).

This growth is heavily dependent on energy use. Rising global energy prices can be expected to reduce substantially the growth in petroleum consumption, but not necessarily the rate of expansion of total fossil-fuel emissions, given the widespread availability of coal—the most emissions-intensive fuel. This is true for most countries that do not have stringent greenhouse gas control policies in place, and certainly for China.

China is poised to be the main engine of world growth in the next two decades. China is also one of the world's countries most reliant on coal for energy use, and therefore one of the world's most carbon-intensive countries—that is, it has one of the highest ratios of carbon dioxide emissions to energy use. China has already overtaken the United States as the largest global emitter (MNP 2008). The combination of China's large, rapidly growing economy and its carbon intensity means that in the coming years it will have an influence on greenhouse gas emissions unmatched by any other country.

Action will be needed by all major economies to limit and reduce greenhouse gas emissions to levels that limit the risk of climate change to acceptable levels. Given China's unique position, its policies will be crucial for global climate change prospects.

In this chapter, we analyse recent trends in China's growth including for energy and carbon dioxide emissions and present business-as-usual projections to 2030 for China and the world. Then we summarise recent policy developments in China and ask what kind of commitments China could and would need to undertake in a world of comprehensive climate change mitigation.

Recent trends

Carbon dioxide emitted from the combustion of fossil fuels is the largest and fastest growing greenhouse gas; it is the focus of this chapter. Through the 'Kaya identity' (Kaya and Yokobori 1997),[1] carbon dioxide emissions growth can be decomposed to changes in economic growth, energy intensity (of gross domestic product, GDP) and carbon intensity (of energy)

$$\Delta CO_2 = \Delta GDP * \Delta(energy/GDP) * \Delta(CO_2/energy) \qquad (1)$$

Summary data for these variables are presented in Table 8.1, for the world, for China and for the world excluding China. There has been a world-wide acceleration this decade in the growth of all three of GDP, energy and emissions (see also Raupach et al. 2007).[2] Table 8.1 shows how of much of the recent acceleration in growth in global emissions is due to China. Global emissions fell in the 1990s due largely to the collapse of the transitional economies. Apart from China, emissions growth this decade (2000–05) is lower than it was in the 1970s and 1980s: 1.5 per cent compared with 1.8 per cent. With China, however, total global emissions growth this decade is 2.9 per cent—well up from the 1971–90 global average emissions growth rate of 2.1 per cent. China's carbon dioxide emissions grew by 10.6 per cent on average between 2000 and 2005—more than three times the growth rate of the 1990s.[3] Fifty-five per cent of the growth

Table 8.1   **A comparison of GDP, energy and carbon dioxide emissions growth rates and elasticities for the world and China, 1971–2005**

World	1971-90	1990-2000	2000-05
Emissions growth (per cent)	2.1	1.1	2.9
GDP growth (per cent)	3.4	3.2	3.8
Energy growth (per cent)	2.4	1.4	2.7
Energy/GDP growth (per cent)	-1.0	-1.8	-1.1
CO_2/energy growth (per cent)	-0.3	-0.2	0.3
Energy/GDP elasticity	0.71	0.43	0.69
Emissions/energy elasticity	0.87	0.82	1.10
China			
Emissions growth (per cent)	5.5	3.2	10.6
GDP growth (per cent)	7.8	10.2	9.4
Energy growth (per cent)	4.3	2.5	9.1
Energy/GDP growth (per cent)	-3.3	-6.9	-0.2
CO_2/energy growth (per cent)	1.2	0.7	1.4
Energy/GDP elasticity	0.55	0.25	0.97
Emissions/energy elasticity	1.29	1.27	1.16
World excluding China			
Emissions growth (per cent)	1.8	0.8	1.5
GDP growth (per cent)	3.2	2.6	3.0
Energy growth (per cent)	2.3	1.2	1.7
Energy/GDP growth (per cent)	-0.9	-1.3	-1.3
CO_2/energy growth (per cent)	-0.4	-0.4	-0.2
Energy/GDP elasticity	0.70	0.48	0.56
Emissions/energy elasticity	0.81	0.68	0.90

Notes: Emissions growth is carbon dioxide from the combustion of fossil fuels (excluding industrial processes). Energy growth is total primary energy supply measured in million tonnes of oil equivalent (Mtoe). GDP growth is measured using 2000 US$ purchasing power parity (PPP).
Source: International Energy Agency (IEA), 2007b. *CO$_2$ Emissions from Fuel Combustion*, International Energy Agency, Paris.

in global emissions between 2000 and 2005 occurred in China.

Figure 8.1 uses the same data as Table 8.1 to explore the differences in carbon dioxide, energy and GDP relationships in China and globally. World-wide (excluding China), the ratios of carbon dioxide emitted to energy used and energy to GDP have fallen gradually and fairly smoothly. In China, in contrast, the ratio of carbon dioxide to energy has risen over time. The ratio of energy to GDP in China was stable until the late 1970s, fell sharply until about 2000, and has been roughly stable since. World-wide, carbon dioxide to GDP has fallen more sharply than either the carbon dioxide/energy or energy/GDP ratio (since both these have been in decline)—but not in China since carbon dioxide/energy has been increasing. Since 2000, GDP and energy have been growing at the same rate in China, and emissions slightly faster than both.

What has been driving the unusual trends observed in China? China's energy intensity in the 1970s was very high. During the 1980s and 1990s, major improvements were achieved in the energy efficiency of the Chinese economy. This can be traced to strong improvements in industrial energy efficiency driven by government regulation, including the shut-down of small inefficient power plants (motivated in part by concerns about local air pollution), changes

Figure 8.1 **Carbon dioxide emissions/GDP, energy/GDP and carbon dioxide emissions/energy for the world and China, 1971–2005** (1971=100)

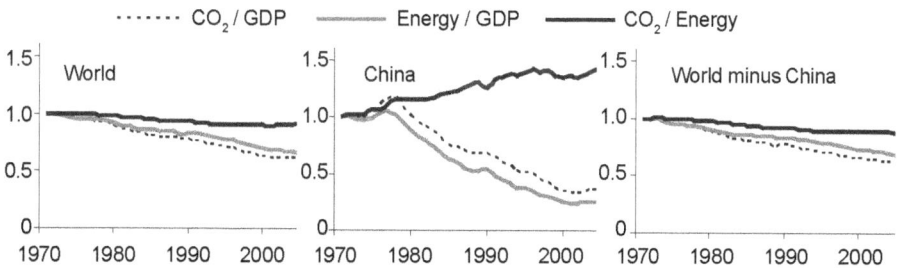

Notes: Emissions growth is carbon dioxide from the combustion of fossil fuels (excluding industrial processes). Energy growth is total primary energy supply measured in million tonnes of oil equivalent (Mtoe). GDP growth is measured using 2000 US$ purchasing power parity (PPP).
Source: International Energy Agency (IEA), 2007b. CO_2 *Emissions from Fuel Combustion*, International Energy Agency, Paris.

in ownership of state-owned enterprises, rising energy prices and structural change (Fisher-Vanden et al. 2004; Wu et al. 2005; Sinton and Fridley 2000). Many of the efficiency gains in the 1990s were, however, one-offs, and the move towards greater private-sector control of the economy weakened the emphasis on energy-saving measures. Quadrelli and Peterson (2007) report that investment in energy conservation as a share of total energy investment in China declined from 13 per cent in 1983 to 7 per cent in 1995 and to 4 per cent in 2003. China's energy-intensive industries have boomed in recent years. Between 2000 and 2006, crude steel production in China grew by an annual average of 22 per cent, pig iron grew by 21 per cent and cement grew by 13 per cent (NBS 2007a). All this has resulted in rapid energy growth in China in recent years (Figure 8.2). Between 2000 and 2005, total energy consumption grew by more than half in just five years. There has been a slight slow-down in energy growth in 2006 and 2007, but rates are still close to 10 per cent. After remaining unchanged between 2000 and 2005, energy intensity fell in 2006 by 2 per cent.[4] First-half figures for 2007 indicated a fall in energy intensity of 2.8 per cent.[5]

The average carbon intensity of China's energy use has kept increasing. This is due largely to a changing energy mix, with growth concentrated in fossil fuels.

Figure 8.2   **Energy consumption in China, levels and growth, 1978–2006**

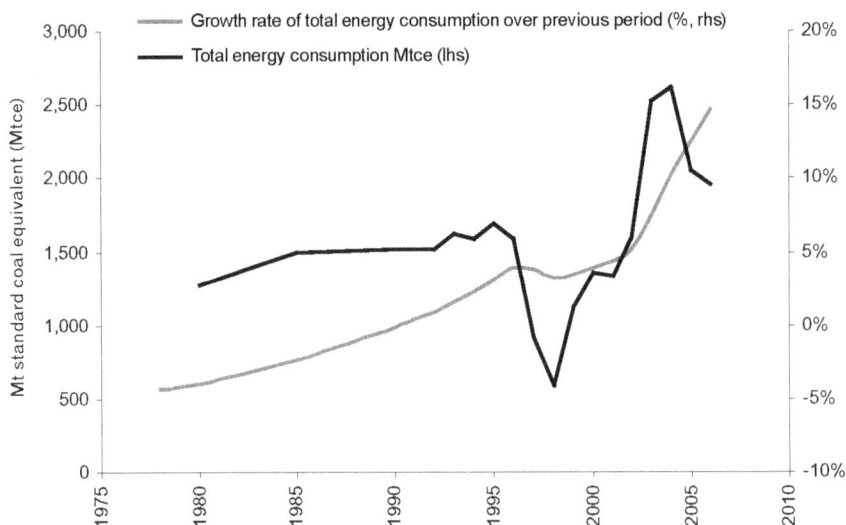

Source: National Bureau of Statistics, 2007a. *2007 China Statistical Yearbook*, China Statistics Press, Beijing.

From 1990 to 2005, coal energy demand doubled and the share of coal in total energy demand increased from 61 to 63 per cent (IEA 2007a). Increased coal use is predominantly for use in power generation, which is almost 80 per cent coal based and booming; in 2006, nearly 90 per cent of new electricity-generation capacity was coal fired. Lower or zero-emissions sources (renewable sources including hydroelectricity, nuclear power and gas) have been growing fast but from low bases or, in the case of biomass, have stagnated in absolute terms. Oil use, especially for transport, almost trebled from 1990 to 2005, increasing its share in total energy use from 13 to 19 per cent.

A business-as-usual projection

This section projects fossil fuel-related carbon dioxide emissions to 2030 under a business-as-usual, constant-policy approach.[6] Policies already in place to reduce emissions are assumed to continue, but it is assumed no new ones are put in place, even if a government has committed to do so. We start with projections from the most recent International Energy Agency (IEA) *2007 World Energy Outlook* (IEA 2007a), which make use of extensive information on energy systems in a partial-equilibrium framework. Using an emissions-growth decomposition framework, we then make adjustments, based on the analysis presented in this chapter, to selected macroeconomic assumptions—namely, GDP growth in non-Organisation for Economic Cooperation and Development (OECD) countries and the intensity of energy use with regard to GDP in China. The strength of this approach is that it builds on the specialist knowledge of the IEA, and makes clear what assumptions might need rethinking. Its limitation is that it does not capture the general-equilibrium effects that would derive from the changes in assumptions.

GDP

We review and adjust IEA (2007a) growth rates for the three most-populous developing countries—China, India and Indonesia—and for other developing and transitional regions.

Our growth forecasts for China draw on the growth-accounting framework of Perkins and Rawski (2008). As Table 8.2 shows, we accept the Perkins–Rawski projections for education-enhanced labour, and assume a figure of 3.1 per cent total factor productivity (TFP) growth for the entire period, which is the rate of TFP growth in the past decade. Perkins and Rawski assume a slow-down in the rate of capital formation, but investment rates are rising, and Garnaut and Huang (2005) argue that these are in fact likely to rise even higher than current levels. We assume investment stays at 45 per cent of GDP until 2015

Table 8.2   **Growth-accounting projections for China, 2005–25**

Annual average	2005–2015		2015–2025	
growth (%)	Perkins–Rawski	Platinum age	Perkins–Rawski	Platinum age
Labour growth (%)	2.0	2.0	1.0	1.0
Capital growth (%)	9.8	11.0	5.6	7.3
Capital share	0.43	0.43	0.43	0.43
TFP growth (%)	3.6	3.1	3.0	3.1
GDP growth (%)	9.0	9.0	6.0	6.8

Notes: All 'Platinum Age' (current chapter) assumptions are, unless otherwise stated, from Perkins, D. and Rawski, T., 2008. 'Forecasting China's economic growth over the next two decades', in L. Brandt and T.G. Rawski (eds), *China's Great Economic Transformation*, Cambridge University Press. Note that Perkins and Rawski's 3.6 per cent total factor productivity growth figure for 2005–15 is not presented as a realistic estimate, but derived by the authors to show what it would take, given their projected capital and labour growth, to achieve 9 per cent GDP growth.
Source: Garnaut, R., Howes, S., Jotzo, F. and Sheehan, P., 2008 (forthcoming). *Emissions in the Platinum Age: the implications of rapid development for climate change mitigation*, Background Working Paper for the Garnaut Climate Change Review, *Oxford Review of Economic Policy*.

and then falls to 40 per cent by 2025. Embedding these assumptions into the Perkins–Rawski framework results in projected growth of 9 per cent from 2005 to 2015 and 6.8 per cent for 2015–25 (Table 8.2).[7] Considered against China's recent performance, and its good prospects for continued double-digit growth (Garnaut and Huang 2005, 2007), we consider this projection to be relatively conservative. China has averaged about 10 per cent GDP growth per annum since 1990. The latest figures for 2006 and 2007 are for 11.6 per cent and 11.9 per cent growth, respectively (NBS 2008a). Our growth projections are, however, well above widely used international projections, including that of the IEA, which in its reference scenario has China's GDP growth at 7.7 per cent for 2005–15 and 4.9 per cent for 2015–30.

The recent acceleration of growth in the developing world has extended well beyond China. The growth acceleration is most evident from the period 2004–07, during which all developing-country regions as well as the group of transitional economies grew at 5 per cent per annum or more. We see this acceleration of growth in developing countries as owing much to better policy settings and to the spill-over effect of rapid Chinese growth—and therefore as sustainable. Based on growth accounting (Garnaut et al. 2008), we use, for India, 7.5 per cent for 2005–30 as a GDP growth projection, and for Indonesia 6.5 per cent. For developing countries other than China, India and Indonesia, we use a weighted average of IEA projections (two-thirds) and performance in the past four years (one-third).

Under the assumptions deployed, all developing and transitional countries are projected to be growing faster than they were in the latter decades of the past century, but slower than at rates observed in the past four years—with the exception of Indonesia.

Energy intensity

The IEA (2007a) projects significant falls in the energy intensities of developing countries from current levels.[8] Although this is inconsistent with historical experience (Figure 8.3), we accept it on the grounds that high energy prices might induce greater efforts to improve energy efficiency.

Future trends in energy intensity in China are a matter for debate. This is not surprising given the variability of past trends (Figures 8.1, 8.2 and 8.3). Will the downward trend in energy intensity observed in the 1980s and 1990s resume, or is the recent cessation of that trend a permanent break?

Figure 8.3 **Energy intensity in China and other developing countries, 1971–2005**

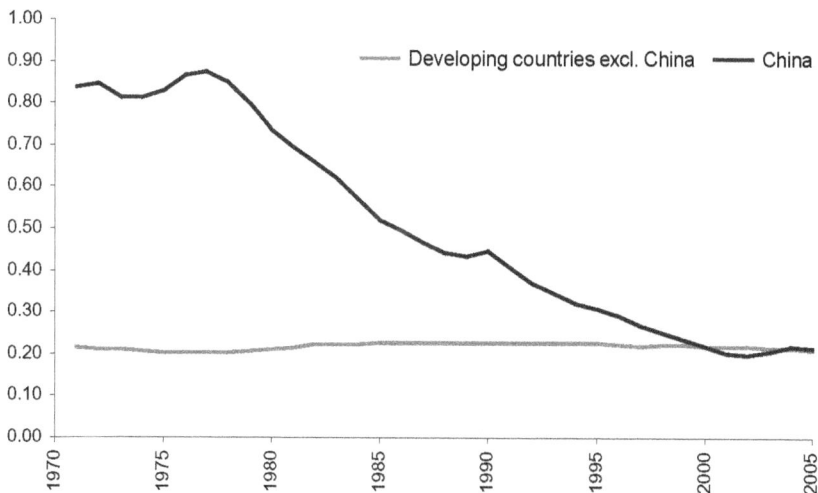

Note: Energy intensity is defined as the ratio of primary energy supply (in million tonnes of oil equivalent) to GDP (in billions of 2000 US$ purchasing power parity).
Source: International Energy Agency (IEA), 2007b. *CO$_2$ Emissions from Fuel Combustion*, International Energy Agency, Paris.

The IEA assumes that the downward trend will resume, and that China will reduce energy intensities by 2.5 per cent per annum until 2015, and then by 2.8 per cent to 2030. This projection is consistent with the position of some observers that China uses energy inefficiently and has plenty of room to improve (Cooper 2004).

China's energy intensity is high if market exchange rates are used to compare economies, but less so if purchasing power parities (PPPs) are used. Moreover, the international experience does not suggest that energy intensity falls in the course of development. As Figure 8.3 shows, China has been the exception rather than the rule. In most developing countries, and as an average, the energy intensity of the economy has been remarkably constant in the past three and a half decades.

A closer examination of economies whose development path China might be thought to be following—Japan, Korea and Taiwan—similarly belies the notion that energy intensity falls as countries develop (Figure 8.4). Analysis of the historical experience of Japan, Taiwan and Korea shows only small decreases in energy intensity as per capita income increases, with energy

Figure 8.4   **Energy intensity in China, Japan, Korea and Taiwan, 1971–2005**

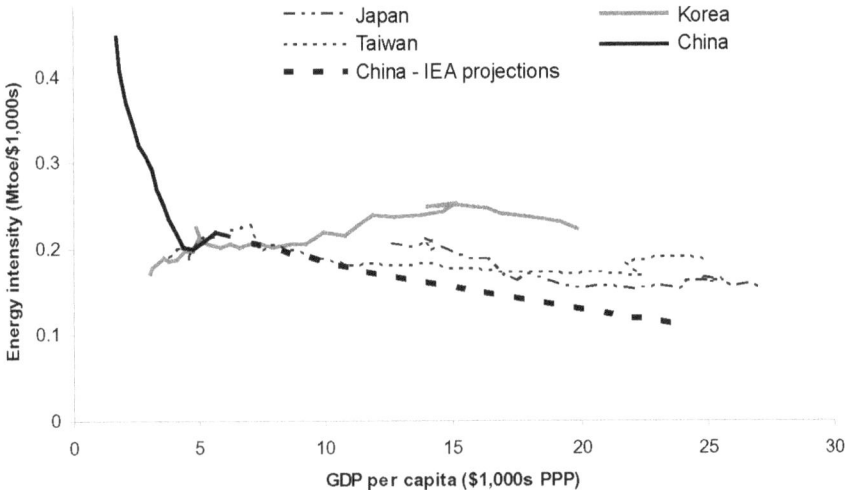

Source: Data from IEA (2007b). *CO₂ Emissions from Fuel Combustion*, International Energy Agency, Paris.

intensity roughly flattening out at levels well above that assumed by the IEA for China in coming decades.

The IEA (2007a) argues that energy intensity will fall as the structure of the Chinese economy shifts from heavy to light industry. It is true that China has put in place a number of policies to slow the growth of energy-intensive sectors: for example, through taxes on energy-intensive exports (see the next section). The shift to heavy industry is, however, no more than would be expected as an economy develops with a high level of investment (extraordinarily high in China's case) underpinning a shift in comparative advantage towards more capital-intensive manufacturing. Energy-intensive heavy industry has been growing so fast in China that its growth will have to fall very significantly for its share in the economy to fall.

We assume that energy intensities decline at two-thirds the rate assumed by the IEA. This gives China a rate of energy-intensity reduction around the developing-country average. The adjustment gives energy elasticities for China of 0.8 for 2005–15 and 0.7 for 2015–30, which is consistent with or below the work of Sheehan and Sun (2007), but well above the IEA's projections for the two periods of 0.66 and 0.4, respectively.

Carbon intensity

We assume that the carbon intensities of energy use (carbon dioxide/total primary energy supply) stay broadly constant, in line with IEA projections. This is a conservative approach on two counts. First, in recent years, emission intensities have in fact been increasing in the developing world, due to the shift to coal, with a pronounced increase in carbon intensity in China. The tighter supply constraints on oil will continue to force substitution to other fuels. The prices for traded coal have increased in recent times also, but this reflects purely short-term supply constraints in mining and shipping rather than resource scarcity, and an expanded coal supply is expected to balance current excess demand.

Second, if energy use does turn out to be higher than projected by the IEA (as we argue it will), a disproportionate amount of the extra demand will be met by (emissions-intensive) coal. In particular, our projections have emissions growing at about the same rate as energy in China. If, however, the growth of coal use continues to stay high, emissions will grow faster than energy. Between 2000 and 2005, coal use in China increased on average by 11.7 per cent. In 2006, China's coal consumption grew by 11.9 per cent and, in 2007, according to preliminary estimates, it grew by 7.8 per cent (NBS 2007a, 2008b).

Emissions

Putting these assumptions together results in annual average growth in China's carbon dioxide emissions of 7.5 per cent from 2005–15, and 4.7 per cent from 2015–30. This is significantly above mainstream projections. IEA (2007a) projects emissions growth for China in these two periods of 5.4 per cent and 1.9 per cent. The IEA (2007a) 'rapid growth' scenario projects somewhat higher emissions growth for China of 6.4 per cent for 2005–15 and 2.7 per cent for 2015–30. On the other hand, the Platinum Age projections are not above the range of China-specific studies. Auffhammer and Carson (2008), in dynamic statistical models for China's emissions based on regional data up to 2004, forecast annual average carbon dioxide emissions growth of 11–12 per cent for 2001–10 (see also Sheehan and Sun, this volume).

Under business-as-usual, largely because of China's rapid emissions growth, global emissions growth will stay high. The Platinum Age projections predict 3.1 per cent annual average growth in carbon dioxide emissions from 2005–30 (Garnaut et al. 2008:Table 9). These are again above the range of official global projections. For example, the IEA (2007a) projects 2.1 per cent annual average growth in carbon dioxide emissions for this period.

Under the Platinum Age projections, China would almost double its share in global carbon dioxide emissions, from 19 per cent in 2005 to 37 per cent in 2030 (Figure 8.5). China's emissions in this scenario would be about three times as large as the United States' at the end of the period. The share of current non-OECD countries in global emissions would be more than 70 per cent.

After 2030, China's emissions growth will slow considerably, not least because its population growth will probably turn negative thus further slowing GDP growth. However, by 2030, it will be too late for the world to start acting on climate change. If the world does reach the levels of emissions projected in Figure 8.5 it will have locked in dangerous levels of temperature increase in subsequent decades. In any case, if the world is still not on a climate change mitigation path, other developing countries are set to take over from China after 2030 as the prime drivers of rapid emissions: first India (on its way to overtaking China as the world's most populous country and, eventually, the largest economy), and then other regions including sub-Saharan Africa.

Clearly, China's growth path in the next two decades will be decisive for global climate change. Its policies and stance in international negotiations will be increasingly important influences on the global response to climate change.

Figure 8.5   **Historical and projected carbon dioxide emissions levels, 1990–2030**

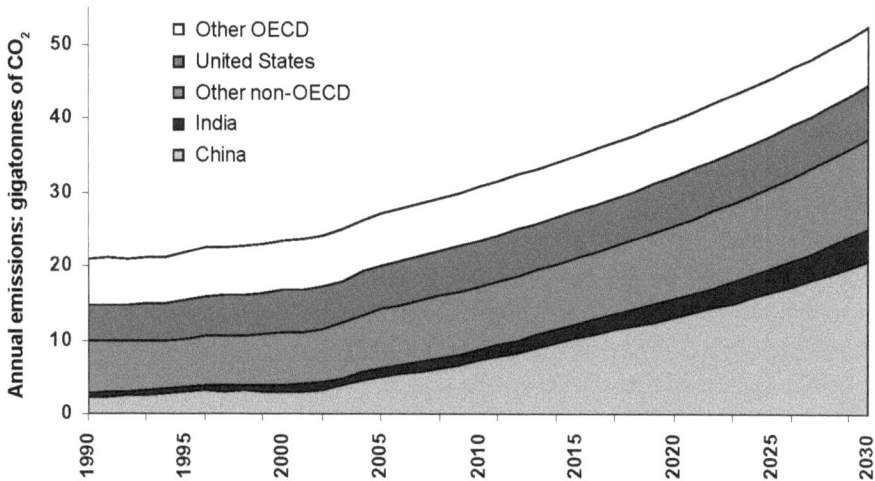

Sources: Data for 1990–2005 from International Energy Agency (IEA), 2007b. *CO$_2$ Emissions from Fuel Combustion*, International Energy Agency, Paris; Platinum Age projections from Garnaut, R., Howes, S., Jotzo, F. and Sheehan, P., 2008. *Emissions in the Platinum Age: the implications of rapid development for climate change mitigation*, Background Working Paper for the Garnaut Climate Change Review, forthcoming in *Oxford Review of Economic Policy*.

China's climate change policies and goals

Recall that the projections in the previous section are constant-policy ones, assuming no further policy action to reduce greenhouse gas emissions. In reality, however, climate change is moving up the policy agenda in China.

China's National Development and Reform Commission, in its National Climate Change Program (NDRC 2007), noted a range of observed and projected climatic changes in China, including temperature increases, changes in precipitation patterns, increases of arid areas and risks of desertification, sea-level rise, the possibility of more frequent extreme weather or climate events and the accelerated retreat of glaciers. The NDRC put forth a set of objectives, principles and policies to deal with climate change.

The principles include an emphasis on the UN Framework Convention on Climate Change's principle of 'common but differentiated responsibilities', active participation in international cooperation, equal emphasis on the mitigation of and adaptation to climate change and integration with other policies such as those for energy conservation and agricultural productivity.

Various policies and programs to reduce energy consumption have been launched or announced (NDRC 2007; Pew Center 2007). They include closing inefficient power plants to the tune of about 8 per cent of current generating capacity, closing small or outdated industrial plants, agreements for incentives for the largest 1,000 enterprises, improving end-use efficiency through standards and labelling and mandatory fuel-economy standards for cars that are more stringent than those in the United States. Specific advanced processes are to be used in energy-intensive industries.

Policies also aim to change the composition of the economy. The service sector's share is to be increased, and the scale of energy-intensive industries reduced in favour of high technology and information industries. Export taxes are already levied on energy-intensive industries—at 15 per cent for a range of metals and 10 per cent for primary steel products—while import tariffs are to be reduced to 0–3 per cent (Pew Center 2007).

Reducing the energy intensity of the economy is motivated not predominantly by climate change objectives, but importantly by China's concerns about energy security (Downs 2004), and by concerns about local environmental impacts such as air pollution. It is unclear to what extent comprehensive policies to limit greenhouse gas emissions will in fact be implemented—and, even if fully implemented, they will fall a long way short of stopping the growth in China's emissions. The NDRC (2007:19) cautions that 'with [the] current level of technology development, to reach the development level of the industrialized countries, it is inevitable that per capita energy consumption and CO_2 emissions will reach a fairly high level'.

On the energy-supply side, a range of policies is aimed towards lowering carbon intensity (NDRC 2007). The share of renewable energy sources is to be increased under the Renewable Energy Law, predominantly through a doubling of hydropower capacity. Nuclear power capacity is to be quadrupled. Within thermal power generation, the development of high-efficiency coal power plants is to be accelerated, and methane arising from coal-mining is to be utilised for power generation to a greater extent. China is also involved in initiatives with the United States, Europe and Australia on carbon capture and storage.

Programs to reduce emissions are also planned or are under way beyond the energy sector, such as accelerated reforestation and the development of rice varieties that are low in methane emissions (NDRC 2007).

The target contained in China's eleventh Five-Year Plan is to reduce the energy intensity of GDP by 20 per cent from 2005 to 2010. This is against the backdrop of a broader strategy of quadrupling GDP from 2000 to 2020, while doubling energy consumption (Development Research Center 2005, see also Sinton et al

2005), thus limiting the growth rate in energy use to approximately half the GDP growth rate. The 2005–10 target implies an average annual reduction in energy intensity of 4.4 per cent, starting from 2005 levels. Such reductions would be a drastic turn around from almost unchanged intensity in the previous five years, and modest reductions in 2006 and 2007 (see above). Analysis of energy-saving options (Lin et al. 2007) concludes that with vigorous policy action it could be possible to meet the 20 per cent energy intensity reduction target, but time clearly is running out for 2010.

Nevertheless, the target of constraining growth in energy use to half the rate of GDP growth could be a useful yardstick in years to come, and could underpin a possible international commitment on national emissions by China.

Possible commitments by China in an effective global mitigation regime

Strong domestic policy action in China is likely to eventuate only if there is strong action in other major countries, especially the United States. Political change is under way in the United States, with growing support in Congress for domestic policies to control greenhouse gas emissions. Both presidential candidates for the upcoming election are running on platforms that include goals of reducing US emissions by 60 per cent (John McCain) and 80 per cent (Barack Obama) by 2050, compared with 1990 levels.

How far the negotiations under the UN 'Bali road-map'—or indeed any alternative negotiating framework—can take the world towards effective climate change mitigation depends in large measure on the commitments that China and the United States are prepared to make. China's dominance in global emissions and emerging importance as a global economic force mean that it is no longer credible to claim that mitigation has to be undertaken exclusively in industrialised countries. Effective mitigation of climate change will require global emissions to peak in the near future, and then decline to well below current levels (IPCC 2007). Given fast emissions growth in China and other developing countries, and their rapidly rising share in global emissions, emissions need to be limited and then reduced in all countries.

As argued by the Garnaut Climate Change Review (Garnaut 2008a), broad international acceptance of national emissions limits will require a heavy emphasis on per capita measures, leading ultimately to equal per capita emissions rights across all countries. This would have to be implemented gradually, and fast-growing countries close to the world average per capita emissions would need to be given some headroom to allow their emissions to go above the global average for a limited time. China is in this situation, with emissions per person

Figure 8.6 **China's future emissions or emission entitlements under different scenarios, 2000–2030**

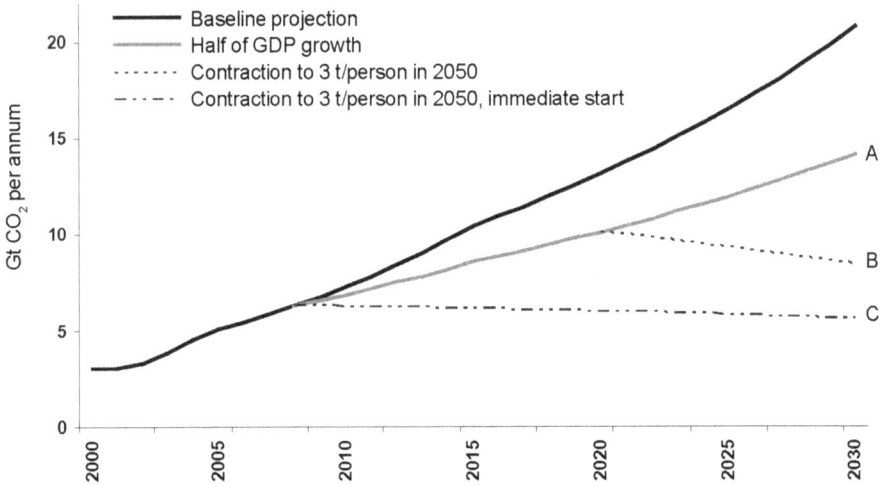

soon to reach the global average on a fast-rising curve. Headroom could take the form of the emission-intensity target applying for a number of years, before shifting to a trajectory of reductions in per capita emissions.

Consistent with this, a first and significant step for China would be to hold the growth in its carbon dioxide emissions to half the rate of its GDP growth, as shown by Trajectory A in Figure 8.6. This would be consistent with China's announced policy for halving its energy intensity by 2020, and increased use of renewable and other low-emissions energy to counterbalance increases in high-emissions coal. Over time, adherence to this trajectory would drive emission levels significantly below business-as-usual levels. In 2030, China's fossil-fuel emissions would be almost one-third lower than in the business-as-usual case—and the same as they would be in 2021 without policy action (Figure 8.6).

At some point, China's scope for headroom would come to an end, and China's per capita emissions would need to start to fall. The second trajectory (B) in Figure 8.6 illustrates such a contraction scenario, which takes over from the 'half GDP growth' trajectory in 2020. The scenario shows China's emissions for a linear reduction in per capita emissions—or emissions entitlements, if there were international trade in entitlements—towards 3 tonnes per person in 2050 (3 tonnes per person is mentioned as a 2050 global goal by Stern [2008] and is used here to illustrate the sort of long-term emissions reductions that will be required). Trajectory B implies an approximately 2 per cent reduction rate in annual absolute emissions entitlements in the 2020s. It would bring emissions (entitlements) back to 2012 levels by 2030—a 60 per cent reduction relative to business as usual.

Even this level of mitigation by China might not suffice. Garnaut et al. (2008) analyse what would be required to adhere to a limited stabilisation path (aiming not to exceed an atmospheric concentration of 550 parts per million of carbon dioxide). Assume that industrialised and transition countries (the so-called 'Annex 1' countries of the United Nations Framework Convention on Climate Change) reduce emissions by 30 per cent by 2020 (the upper bound of the range of emissions reductions by 2020 proposed by the European Union). This gives room for the emissions of developing countries to increase by an annual average of only 1.4 per cent to 2020. The figure of 1.4 per cent is not much more than the rate of population growth for the developing world, and well below the 4 per cent annual growth in emissions assumed for China to 2020, under the intensity target.

Trajectory C in Figure 8.6 shows the emissions path for China in a case in which per capita emissions contract from their current level towards the 3 tonnes of carbon dioxide target for 2050, starting in 2009. While it is not realistic to assume that China does not need any headroom and can start to cut emissions immed-iately, this trajectory provides a useful illustrative bound on its possible mitigation effort.

Whether and where China lies between its business-as-usual baseline and the most stringent mitigation trajectory (C) will depend on whether and to what extent it chooses to participate in the global mitigation regime. This in turn will be a function of industrial-country participation and pressure, China's assessment of the political and economic benefits of participation and non-participation and the give and take of international negotiations. Most importantly, it will depend on China's own assessment of the importance of avoiding high risks of dangerous climate change.

Realising any of mitigation Trajectories A, B or C would require strong policy action, including for fuel switching away from coal, expansion of renewable and/or nuclear power, geosequestration for carbon dioxide (carbon capture and storage) when and where available, and accelerated structural change and improvements in energy efficiency throughout the economy. It could, however, be less difficult to achieve than it appears at first sight. Rapid projected emissions growth presents great opportunities, because much of China's infrastructure is not yet in existence. The additional costs of building new power supply systems, manufacturing industries or transport infrastructure to a low-carbon standard are much lower than refurbishing or replacing existing stocks, which would be needed to achieve similar reductions in slower-growing economies.

Comprehensive emissions pricing, through a tax or an emissions trading scheme, would be the most economically efficient way of achieving this outcome. International harmonisation of marginal abatement costs would allow emissions reductions at least at cost. Linking to international emissions markets, which are emerging or expanding in the European Union, the United States, Japan, Australia, New Zealand and probably in other countries in future, could achieve this for China. Depending on the allocation of emissions entitlements, this could benefit China economically even without taking into account the benefits from lesser climate change impacts and risks. The International Monetary Fund's *World Economic Outlook* (IMF 2008), for example, shows simulations in which China's consumption increases significantly, if allocations are made on the basis of current emissions or population. China could sell excess emissions entitlements if remaining below its agreed trajectory of emissions entitlements, or dampen compliance costs by buying entitlements elsewhere—a realistic scenario once developing countries start coming into an international system.

Emissions trading will be critical to providing incentives for China and other developing countries to participate in the international mitigation regime, but it is only one part of the story. Increased industrial-country financing of the research, development and commercialisation of low-emissions technology, as well as the financing of the transfer of this technology to developing countries, will be critical.

Conclusion

It is in China more than anywhere else that global climate change mitigation will be decided. If China does not participate in the global mitigation effort, its emissions will continue to grow rapidly and will account for a rising share of global emissions. Many 'mainstream' analyses underestimate likely economic growth in years to come, and could be too optimistic about reductions in the energy intensity and carbon intensity of China's energy system, unless there are strong and comprehensive policies to reduce emissions. Indeed, with high oil prices, there might be a strong shift to coal, which will further increase the rate of growth in emissions.[9]

Moreover, if China does not participate, or only participates marginally, in the global mitigation effort, there will be an indirect effect that will see other countries reduce their levels of ambition and effort since they will know that, without China, their efforts cannot avert climate change risks. On the other hand, if China does participate fully in global mitigation efforts, these two effects can be reversed. China's own emissions will be much lower, and other countries—industrialised and developing—are likely to do more.

China is starting to put a range of ambitious climate change policies in place, and has announced goals of reducing the energy intensity of the economy and increasing the share of low-emissions energy. If implemented, these policies will limit emissions growth to well below the business-as-usual trajectory. China's goal of limiting the growth in energy use to half the growth rate of the economy could become the basis for a near-term emissions target that will allow it to play a leading role in international discussions. Constraining the growth of China's emissions in this way could be compatible with the start-up phase of an international system of per capita emissions converging to a common and lower level, with some headroom provided in the interim for fast-growing developing countries.

Notes

1   The Kaya identity further decomposes economic growth into population growth and growth in income per capita.
2   Since energy intensity and carbon intensity are declining, acceleration for them means that they are declining less rapidly.
3   Growth in other greenhouse gas emissions is thought to be much slower. Emissions of methane and nitrous oxide, which account for about one-sixth of China's emissions in carbon dioxide equivalent terms, are reported to have grown by only 0.6 per cent per annum from 1994 to 2004 (NDRC 2007).
4   In 2006, energy consumption grew by 9.6 per cent, and GDP by 11.6 per cent (NBS 2007a, 2008a).
5   See NBS (2007b).
6   This section draws heavily on Garnaut et al. (2008).
7   We extend the latter projection to 2030.
8   The IEA projects an annual average decline of 1.8 per cent for energy intensity in the developing world excluding China in the next 25 years.
9   China is investing in coal-to-liquid plants and is expected to start operating the largest such facility outside South Africa later in 2008. See Nakanishi and Niu (2008).

References

Auffhammer, M. and Carson, R.T., 2008. 'Forecasting the path of China's CO_2 emissions using province-level information', *Journal of Environmental Economics and Management*, doi:10.1016/j.jeem.2007.10.002.

Cooper, R., 2004. *A carbon tax for China*. Available from http://www.economics. harvard.edu/faculty/cooper/papers_cooper (accessed 20 June 2008).

Downs, E.S., 2004. 'The Chinese energy security debate', *The China Quarterly*, 177:21–41.

Fisher-Vanden, K., Jefferson, G.H., Liu, H. and Tao, Q., 2004. 'What is driving China's decline in energy intensity?', *Resource and Energy Economics*, 26(1):77–97.

Garnaut, R. and Huang, Y., 2005. 'Is growth built on high investment sustainable?', in R. Garnaut and L. Song (eds), *The China Boom and its Discontents*, Asia Pacific Press and ANU E Press, The Australian National University, Canberra:1–18.

——, 2007. 'Mature Chinese growth leads the global Platinum Age', in R. Garnaut and Y. Huang (eds), *China: linking markets for growth*, Asia Pacific Press and ANU E Press, The Australian National University, Canberra:9–29.

Garnaut, R., 2008a. *Interim Report to the Commonwealth, State and Territory Governments of Australia*, February. Available from www.garnautreview. org.au.

——, 2008b. Measuring the immeasurable: the costs and benefits of climate change mitigation, The Sixth H.W. Arndt Memorial Lecture, The Australian National University, Canberra, 5 June.

Garnaut, R., Howes, S., Jotzo, F. and Sheehan, P., 2008 (forthcoming). *Emissions in the Platinum Age: the implications of rapid development for climate change mitigation*, Background Working Paper for the Garnaut Climate Change Review, *Oxford Review of Economic Policy*.

International Energy Agency (IEA), 2007a. *World Energy Outlook 2007: China and India insights*, International Energy Agency, Paris.

——, 2007b. *CO_2 Emissions from Fuel Combustion*, International Energy Agency, Paris.

International Monetary Fund (IMF), 2008. *World Economic Outlook 2008*, International Monetary Fund, Washington, DC.

Intergovernmental Panel on Climate Change (IPCC), 2007. *Fourth Assessment Report, Working Group III*, Intergovernmental Panel on Climate Change, Cambridge University Press, Cambridge, Mass.

Kaya, Y. and Yokobori, K. (eds), 1997. *Environment, Energy, and Economy: strategies for sustainability*, United Nations University Press, Tokyo.

Lin, J., Zhou, N., Levine, M.D. and Fridley, D., 2007. *Achieving China's Target for Energy Intensity Reduction in 2010: an exploration of recent trends and possible future scenarios*, China Energy Group, Lawrence Berkeley National Laboratory, Berkeley.

Maddison, A., 2001. *The World Economy: a millennial perspective*, Organisation for Economic Cooperation and Development, Paris.

MNP (Netherlands Environmental Assessment Agency), 2008. Global CO_2 emissions: increase continued in 2007, Netherlands Environmental Assessment Agency, Bilthoven. Available from http://www.mnp.nl/en/publications/2008/globalco2emissionsthrough2007.html.

Nakanishi, N. and Niu, S., 2008. 'China builds plant to turn coal into barrels of oil', *Reuters*, 4 June.

National Bureau of Statistics, 2007a. *2007 China Statistical Yearbook*, China Statistics Press, Beijing.

National Bureau of Statistics, 2007b. *Communiqué on National Energy Consumption for Unit GDP in the First Half of 2007*, 2007-07-31. http://www.stats.gov.cn/english/newsandcomingevents/t20070731_402422194.htm

National Bureau of Statistics (NBS), 2008a. Announcement on verified GDP data in 2006 and 2007, National Bureau of Statistics, Beijing. Available from http://www.stats.gov.cn/english/newsandcomingevents/t20080410_402473201.htm (accessed 10 April 2008).

——, 2008b. Statistical communique of the People's Republic of China 2007, 28 February. Available from http://www.stats.gov.cn/english/newsandcomingevents/t20080228_402465066.htm.

Development Research Center, 2005. *China National Energy Strategy and Policy 2020*, Development Research Center of the State Council, Beijing. Available from http://www.efchina.org/FReports.do?act=detail&id=155.

National Development Research Commission (NDRC), 2007. *China's National Climate Change Program*, National Development and Reform Commission, Beijing.

Perkins, D. and Rawski, T., 2008. 'Forecasting China's economic growth over the next two decades', in L. Brandt and T.G. Rawski (eds), *China's Great Economic Transformation*, Cambridge University Press, Cambridge, Mass.

Pew Center, 2007. *Climate Change Mitigation Measures in the People's Republic of China*, Pew Center on Global Climate Change, Arlington.

Quadrelli, R. and Peterson, S., 2007. 'The energy–climate challenge: recent trends in CO_2 emissions from fuel combustion', *Energy Policy*, 35(11):5938.

Raupach, M.R., Marland, G., Ciais, P., Le Quere, C., Canadell, J.G., Klepper, G. and Field, C.B., 2007. 'Global and regional drivers of accelerating CO_2 emissions', *Proceedings of the National Academy of Sciences*, 0700609104.

Sheehan, P. and Sun, F., 2007. *Energy use and CO_2 emissions in China: interpreting changing trends and future directions*, CSES Climate Change Working Paper No.13, Centre for Strategic Economic Studies, Victoria University, Melbourne.

Sinton, J.E. and Fridley, D.G., 2000. 'What goes up: recent trends in China's energy consumption', *Energy Policy*, 28(10):671–87.

Stern, N., 2008. *Key Elements of a Global Deal on Climate Change*, London School of Economics, London.

Wu, L., Kaneko, S. and Matsuoka, S., 2005. 'Driving forces behind the stagnancy of China's energy-related CO_2 emissions from 1996 to 1999: the relative importance of structural change, intensity change and scale change', *Energy Policy*, 33(3):319–35.

9

China can grow and still help prevent the tragedy of the CO_2 commons

Warwick J. McKibbin, Peter J. Wilcoxen and Wing Thye Woo

On the road to prosperity

China and India have finally embarked on the path of modern economic growth. China's economy has grown at an average annual rate of almost 10 per cent for the past 30 years, and India's has grown more than 8 per cent every year since 2004. Just like the experiences of post-1868 Japan and post-1960 South Korea and Taiwan, China and India are now on the trajectory of catch-up growth that will bring them in the long run to the same living standard as Western Europe, Japan and the United States. At that point, the share of global income produced by China and India will equal their share of global population (which is anticipated to be about 35 per cent).

This projected parity in living standards in the long run will represent a return to the global economic situation that persisted in the first 1,600 years of the Gregorian calendar (Table 9.1). In year zero, China and India had 58 per cent of the global population and 59 per cent of global gross domestic product (GDP); and the respective numbers in 1600 were 53 per cent and 52 per cent (despite the growing divergence in GDP per capita in Western Europe from 1500 onwards). The relatively slow growth of China and India in the past 400 years changed the situation dramatically. By 1973, China and India's share of global GDP had fallen to only 7.7 per cent although the two countries accounted for 37 per cent of the world's population. China's economic deregulation and integration into the world trade and financial systems since 1978 and India's since 1991 raised their share of world GDP to 20.6 per cent in 2003.[1] Given the

Table 9.1 Global economic and demographic changes, 0–2003

Year	0	1000	1500	1600	1700	1820	1870	1913	1950	1973	1998	2003
Part A: GDP per capita (1900 international $)												
Western Europe	450	400	774	894	1,024	1,232	1,974	3,473	4,594	11,534	17,921	19,912
United States			400	400	527	1,257	2,445	5,301	9,561	16,689	27,331	29,037
Japan	400	425	500	520	570	669	737	1,387	1,926	11,439	20,413	21,218
China	450	450	600	600	600	600	530	552	439	839	3,117	4,803
India	450	450	550	550	550	533	533	673	619	853	1,746	2,160
World	444	435	565	593	615	667	867	1,510	2,114	4,104	5,709	6,516
Part B: Share of world GDP (per cent of world total)												
Western Europe	10.8	8.7	17.9	19.9	22.5	23.6	33.6	33.5	26.3	25.7	20.6	19.2
United States			0.3	0.2	0.1	1.8	8.9	19.1	27.3	22.0	21.9	20.6
Japan	1.2	2.7	3.1	2.9	4.1	3.0	2.3	2.6	3.0	7.7	7.7	6.6
China	26.2	22.7	25.0	29.2	22.3	32.9	17.2	8.9	4.5	4.6	11.5	15.1
India	32.9	28.9	24.5	22.6	24.4	16.0	12.2	7.6	4.2	3.1	5.0	5.5
Part C: Share of world population (per cent of world total)												
Western Europe	10.7	9.5	13.1	13.3	13.5	12.8	14.8	14.6	12.1	9.2	6.6	6.3
United States	0.3	0.5	0.5	0.3	0.2	1.0	3.2	5.4	6.0	5.4	4.6	4.6
Japan	1.3	2.8	3.5	3.3	4.5	3.0	2.7	2.9	3.3	2.8	2.1	2.0
China	25.8	22.0	23.5	28.8	22.9	36.6	28.2	24.4	21.7	22.5	21.0	20.5
India	32.5	28.0	25.1	24.3	27.3	20.1	19.9	17.0	14.2	14.8	16.5	16.7
Memo items												
World GDP (in billion)	103	117	247	329	371	694	1,101	2,705	5,336	16,059	33,726	40,913
World population (in million)	231	268	438	556	603	1,041	1,270	1,791	2,525	3,913	5,908	6,645

Sources: data for 0–1998 from Maddison, A., 2001. *The World Economy: a millennial perspective*, Organisation for Economic Cooperation and Development, Paris; data for 2003 from Maddison, A., 2007. 'horizontal-file_03-2007.xls', File downloaded from Internet. Available from www.ggdc. net/maddison/Historical_Statistics/horizontal-file_03-2007.xls

still large gap between the average income in China and Western Europe in 2003—$4,803 and $19,912, respectively (measured in 1990 international Geary-Khamis dollars)[2]—continued high growth in China could continue for the next two decades.

The very likely return of China to the centre stage of the global economy has given rise to immense optimism on some fronts, and intense pessimism on a number of other fronts. Optimistic analysts have predicted that China's re-emergence as an independent growth pole will create a new web of synergistic relationships that will unleash greater global prosperity. On the other hand, pessimistic analysts have pointed out that the major new rising powers in the twentieth century came into conflict with the existing powers: Germany in World War I, the Japanese–German axis in World War II and the Soviet Union in the Cold War.

The important lesson from the history of the twentieth century is not, however, that conflict is inevitable but that rising powers and existing powers should work together to avoid past mistakes—to falsify Karl Marx's quip that 'history repeats itself, first as tragedy, second as farce'. It is really not naive to think that conflict is preventable because the most important power to rise and prevail in the twentieth century was the United States, and it has, in general, been a stabilising force in the international order. Averting the pessimistic outcome requires adherence to the multilateralist principle of the existing powers accommodating rising powers, and the latter becoming responsible stakeholders in the international system.

The dialogue between the existing and rising powers must necessarily be comprehensive because the range of global public goods that must be supplied is very broad (ranging from the maintenance of the universal postal system to the peaceful use of outer space), and the nature of some of these global public goods is highly complicated (for example, a scheme to control the emission of greenhouse gases). In this chapter, we will confine discussion to an economic issue in which the need to engage China in constructive dialogue is important for sustainable global growth. The issue is the protection of the global environmental commons by addressing China's emissions of carbon dioxide.

The chapter is organised as follows: section two makes the case that climate change could be a key obstacle for China. It shows that even under conservative assumptions, the business-as-usual growth path might cause an environmental collapse before China achieves parity in living standards with the countries in the OECD. Section three reviews the history of energy production and consumption in China, and then uses a dynamic multi-country general-equilibrium model (the G-cubed model) to project a realistic business-as-usual trajectory for carbon dioxide emissions. Section four proposes a novel hybrid policy as an alternative to the commonly discussed cap-and-trade mechanism to control carbon dioxide emissions. Section five employs the G-cubed model to examine the economic consequences of the different instruments to reduce carbon dioxide generation. Section six concludes with recommendations for the form of future international climate agreements and how China can be encouraged to participate.

The fallacy of composition in modern economic growth?

We began this chapter with the optimistic projection that China and India will achieve parity in living standards with Western Europe, which leads immediately to the question of when this convergence will occur. Between 1913 and 2003, when Japan was on its catch-up growth trajectory, the annual growth rate of average income was 3.1 per cent in Japan and 1.9 per cent in Western Europe and the United States. It is possible to use this information to undertake a very crude back-of-the-envelope calculation to see what stresses might begin to emerge over time. Suppose we assume that

- Western Europe grows 1.5 per cent annually from 2003 onwards
- China and India grow 3.1 per cent annually from 2003 until reaching parity with Western Europe, and then 1.5 per cent annually.

Under these assumptions, China will achieve income parity with Western Europe by 2100, and India will achieve parity by 2150.[3] The common GDP per capita in 2150 will be about $180,000.

This extrapolation might fail to be realised, however, not because of political reasons, as commonly feared, but because of environmental reasons. It will not be wars that will derail the catch-up growth; rather, the growth process could prove to be unsustainable because of the fallacy of composition. Specifically, it is possible that a continual improvement in living standards might be achievable for a small subset of large countries, but not for all large countries together. A global equilibrium with a common living standard, which existed in the first millennium, might not be replicable in 2150 because the earlier situation was an agriculture-dominated equilibrium in which the average income was stagnant at $440. In contrast, the envisaged global equilibrium will have an average income of $180,000, which will be growing at 1.5 per cent annually.

The difference is between the vicious circle of Malthusian growth and the process of what Simon Kuznets (1966) has labelled 'modern economic growth (MEG)'. In MEG, society is urbanised, the economy is industrialised and increasingly service oriented, and human capital rivals physical capital in its contribution to economic growth. A key ingredient, so far, in this historically unprecedented sustained growth in prosperity has been energy from fossil fuels. The result is that the concentration of carbon dioxide in the earth's atmosphere has risen from 280 parts per million (ppm) in the pre-industrial age to 379 ppm in 2005 (IPCC 2007:37).

Under existing energy technologies, the scale of growth in China and India will be associated with a very large increase in global carbon dioxide emissions and with rapidly rising carbon dioxide concentrations. There is now a substantial literature suggesting that the increase in carbon dioxide concentrations has contributed substantially to global warming and climate change.[4] According to the Intergovernmental Panel on Climate Change (IPCC 2007), climate change has

- *very likely* contributed to sea-level rise during the latter half of the twentieth century
- *likely* contributed to changes in wind patterns, affecting extra-tropical storm tracks and temperature patterns
- *likely* increased temperatures of extremely hot nights, cold nights and cold days
- *more likely than not* increased the risk of heat waves, areas affected by drought since the 1970s and the frequency of heavy precipitation events
- led to the ocean becoming more acidic, with an average decrease in pH of 0.1 units.[5]

There is serious concern expressed in IPCC reports that there could be severe and irreversible problems resulting from climate change.[6] What is the level of the threshold carbon dioxide concentration that would unleash calamity on the world economy and human life? The truth is that we do not know. David King, the chief scientific advisor to the British government, suggested that 'we should prevent atmospheric CO_2 [concentrations] going beyond 500 ppm' (Kirby 2004),[7] and Michael Raupach, an Australian atmospheric scientist, advocated a limit of 550 ppm (Beer 2007).[8] It has become quite common to adopt the position that the threshold carbon dioxide concentration for dangerous consequences is 560 ppm—a doubling of the pre-industrial value of 280 ppm. Of course, the possibility that the threshold is 500 ppm or even 840 ppm cannot be ruled out definitively on *a priori* grounds.

At the present incremental rate of 2 ppm of atmospheric carbon dioxide annually, the 560 ppm mark would be breached by 2100, just when China is about to reach parity in living standards with Western Europe.[9] If there were indeed a catastrophic threshold of carbon dioxide concentration of 560 ppm,

China and India could achieve income parity with Western Europe, Japan and the United States in 2150 only because the environmental collapse triggered by the growth of China and India brought down the incomes of Western Europe, Japan and the United States! This new equilibrium of income parity produced by the 'fallacy of composition' could well be characterised by global acrimony and strife.

The crucial point is that one does not have to accept the existence of a catastrophic threshold level of carbon dioxide concentration in order to conclude that unless there are future revolutionary breakthroughs in green technology or fundamental shifts in the nature of economic growth, China and India could achieve income parity with the rich countries only by creating serious global environmental problems. Clearly, China is one of the key countries that needs to be brought into the global framework with a clear commitment to take action on greenhouse gas emissions.

Figure 9.1　**Comparison of projections of energy consumption in China, 1990 – 2030** (quadrillion [10^{15}] BTU)

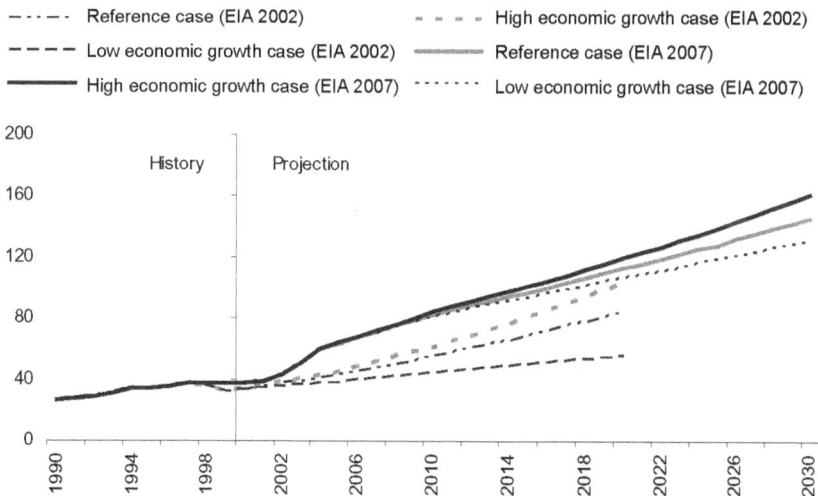

Note: The base years for projections reported in EIA 2002 and 2007 are 1999 and 2004, respectively.
Sources: Energy Information Administration / International Energy Outlook 2002 and 2007

Moreover, it is important to bring China quickly into an international agreement because its dramatic recent rises in energy use and greenhouse gas emissions have been unanticipated by most analysts, and the potential for further upside surprises on emissions remains as China's strong growth could be more durable than anticipated. For example, the Energy Information Administration (EIA) of the US Department of Energy provides projections of carbon dioxide emissions by major countries in its annual *International Energy Outlook*. The EIA makes projections for Chinese energy consumption for three scenarios: high economic growth, the reference case and low economic growth.

Figure 9.1 reports projections from the *2002 International Energy Outlook* (EIA 2002) and the *2007 International Energy Outlook* (EIA 2007). The shocking fact is that for the future years that overlap in both reports, in every case China's projected energy consumption in the low-growth scenario in the 2007 report is above the projected energy consumption in the high-growth scenario in the 2002 report. The 2002 high-growth forecast for 2020 was 102.8 quadrillion British thermal units (BTU) and the 2007 low-growth forecast for 2020 was 106.6 quadrillion BTU. The 2002 'reference-case' forecast was 84.4 quadrillion BTU in 2020, and the 2007 reference-case forecast was 112.8 quadrillion BTU in 2020—an upward revision of 33.6 per cent. Even more important, carbon dioxide emissions in 2005 were 50 per cent higher than the forecast made in 2002.

Past and future patterns of energy use and carbon dioxide emissions in China

China is now the second largest user of energy in the world after the United States, and the EIA (2007) projects China will become the largest by 2025 (Table 9.2), when China will consume 19.6 per cent of the world's supply of energy and the United States will consume 19 per cent. China will, however, become the world's biggest emitter of carbon dioxide earlier than 2025. In 2015, China will account for 20.7 per cent of global carbon dioxide emissions while using 17.4 per cent of global energy, and the same figures for the United States will be 19.4 per cent and 20.1 per cent, respectively. This is partly because it is anticipated that China will expand its use of fossil fuels.

The fuel composition of energy consumption in China is shown in Figure 9.2. Much of the recent rise in energy consumption took the form of an increased use of coal. Coal has been the major energy source in China throughout the period of growth since the reforms in the early 1990s. The surge in energy use since 2002 is obvious from the figure, and it results from a number of factors including rising GDP growth since 1998 (Figure 9.3) and a recent rise in the energy intensity of GDP (Figure 9.4). The shift in the energy intensity of the

Table 9.2 China's share of global energy consumption and carbon dioxide emissions, 1990–2030

	1990	2003	2004	2010	2015	2020	2025	2030
Energy consumption								
United States	24.4	23.1	22.5	20.8	20.1	19.5	19.0	18.7
OECD Europe	20.1	18.7	18.2	16.5	15.3	14.2	13.4	12.7
Japan	5.3	5.2	5.1	4.6	4.3	4.1	3.8	3.6
Australia/New Zealand	1.3	1.4	1.4	1.3	1.3	1.3	1.2	1.2
Other OECD	5.8	6.7	6.5	6.6	6.4	6.4	6.3	6.2
China	7.8	11.7	13.3	16.2	17.4	18.6	19.6	20.7
India	2.3	3.4	3.4	3.6	3.9	4.1	4.4	4.5
Other Non-OECD	33.1	29.8	29.5	30.5	31.4	32.0	32.3	32.2
World total	100.0	100.0	100.0	100.0	100.0	100.0	100.0	100.0
CO_2 emissions								
United States	23.5	22.7	22.0	20.1	19.4	18.8	18.7	18.5
OECD Europe	19.3	16.9	16.3	14.6	13.4	12.4	11.6	10.9
Japan	4.8	4.9	4.7	4.1	3.8	3.5	3.3	3.0
Australia/New Zealand	1.4	1.6	1.6	1.5	1.4	1.4	1.4	1.3
Other OECD	4.8	5.7	5.4	5.4	5.2	5.2	5.1	5.0
China	10.5	12.8	13.2	19.0	20.7	22.1	23.5	25.0
India	2.7	4.0	3.8	4.4	4.7	4.9	5.0	5.1
Other Non-OECD	33.1	28.8	28.4	29.1	29.8	30.1	30.1	29.9
World total	100.0	100.0	100.0	100.0	100.0	100.0	100.0	100.0
Memo items								
Energy used (Quadrillion BTU)	26.2	32.1	33.2	40.4	43.4	46.5	50.1	53.5
CO_2 emitted (Million metric tonnes)	21,246	25,508	26,922	30,860	33,889	36,854	39,789	42,880

Source: Energy Information Agency (EIA), 2007. *International Energy Outlook*, Department of Energy, Washington, DC.

Chinese economy was due to a number of factors driving structural change, including: increased electrification, greater energy demand from manufacturing, greater energy demand by households and greater use of cement and steel as infrastructure spending rose.

Perhaps more interesting than the historical experience of Chinese energy use are future trends in energy and greenhouse gas emissions, particularly since an increasingly worrying picture of the global climate has emerged in the past half-decade. Projecting future energy use and greenhouse gas emissions in China, especially over horizons of more than a decade, is very difficult. It is tempting to construct future projections by simple extrapolation of recent trends. A somewhat more sophisticated approach is to apply the 'Kaya identity' (Kaya 1990), which decomposes emissions growth into four components: changes in emissions per unit of energy, changes in energy per unit of per capita GDP, growth of per capita GDP, and population growth. The four components are then projected separately. This is the approach taken, for example, in many of the studies cited by the IPCC (2007) and by Garnaut et al. (2008).

The Kaya identity is a useful historical decomposition but it is not an ideal forecasting framework. Each of its components is really an endogenous outcome resulting from a wide variety of individual decisions, and they cannot be

Figure 9.2   **Energy consumption in China by source, 1980–2005**

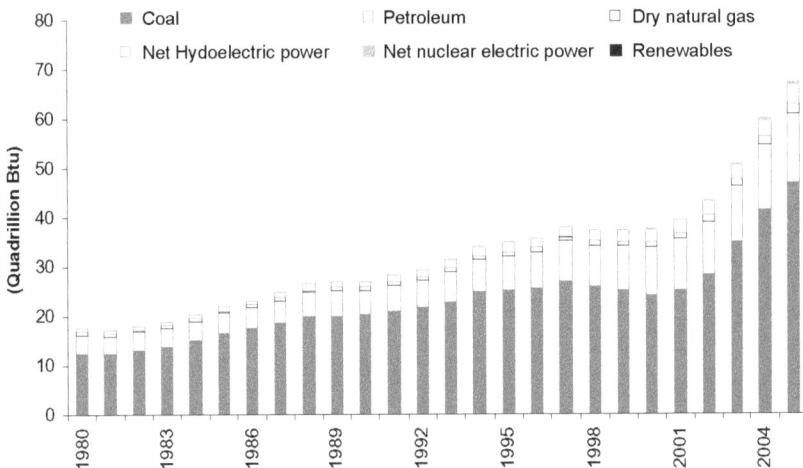

Source: Energy Information Agency (EIA), 2007. *International Energy Outlook*, Department of Energy, Washington, DC.

Figure 9.3   **Chinese GDP growth in purchasing power parity, 1980 – 2005**

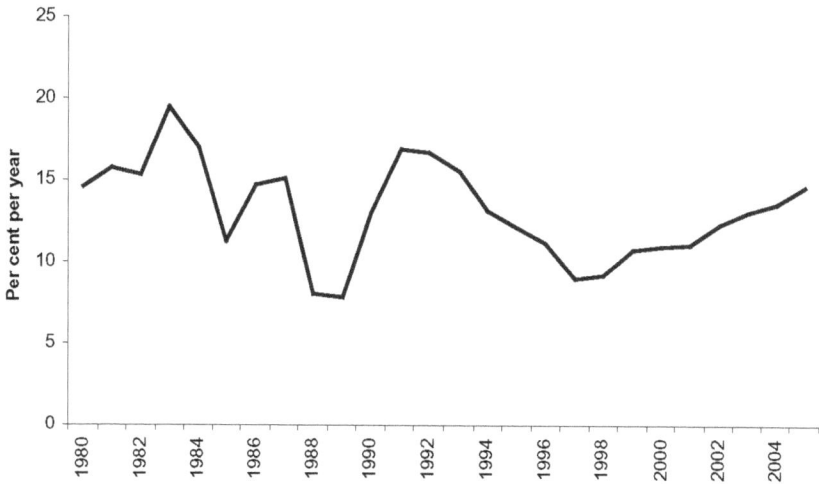

Source: International Monetary Fund (IMF), 2008. *World Economic Outlook*, April, International Monetary Fund, Washington, DC.

Figure 9.4   **Energy** (1,000 BTU) **per unit of GDP (purchasing power parity)** (1980=100), **1980 – 2005**

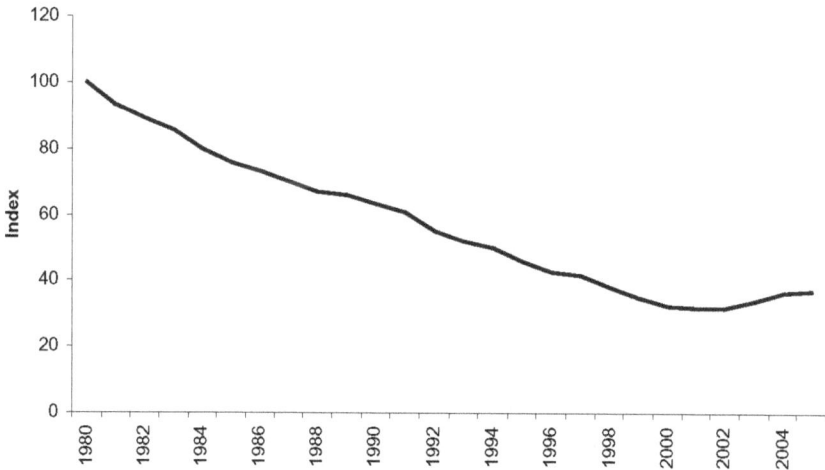

Source: Energy Information Agency (EIA), various years. *International Energy Outlook*, Department of Energy, Washington, DC.

assumed to remain constant in the future. As shown by Bagnoli et al. (1996) and McKibbin et al. (2007), overall economic growth is not the only important determinant of energy use. Identifying and understanding the underlying sources of economic growth is critical, and it is particularly important to understand how the structure of an economy evolves in response to changes in energy prices.

Figure 9.5 shows EIA projections for carbon dioxide emissions by energy source in China for the reference-case scenario. It is clear that coal is the overwhelming source of carbon dioxide emissions in China—historically and in these projections. It is expected to be the major source of energy, and therefore emissions, in the foreseeable future. This is not surprising given the large quantity of low-cost coal available in China and the assumptions of unchanging relative energy prices in these projections. Over time, the share of emissions from petroleum is projected to rise with greater use of motor vehicles and other transportation. These types of projections are dependent on assumptions about the relative price of energy to other goods and the relative price of alternative energy sources.

Figure 9.5 **Projections of carbon dioxide emissions by fuel type in China, 1990–2030** (mega-tonnes of carbon dioxide)

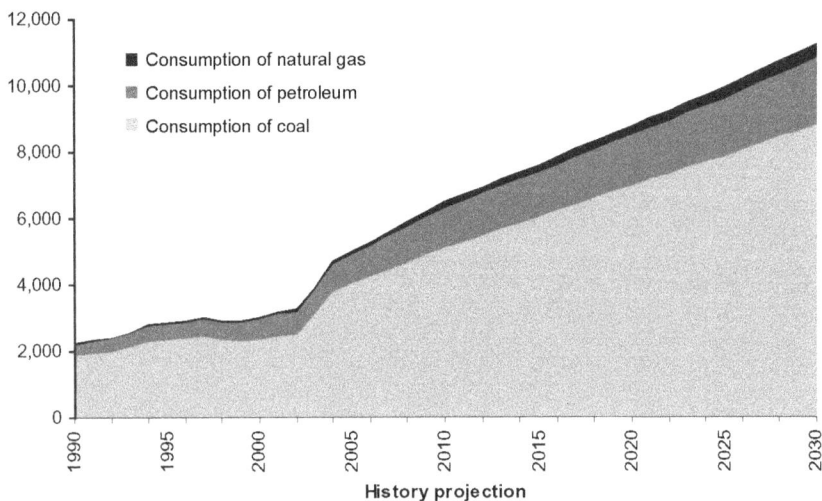

Source: Energy Information Agency (EIA), 2007. *International Energy Outlook*, Department of Energy, Washington, DC.

Figure 9.6 shows the global sources of carbon dioxide from burning fossil fuels, by region, in 1990 and those projected in the *2007 International Energy Outlook* (EIA 2007) for 2030. Not only is China currently an important source of carbon dioxide emissions, it is expected to grow quickly. Its absolute size, shown in Figure 9.6, and its share in global emissions (shown in Table 9.1) emphasise that China is a critical country in the debate about policies to deal with climate change.

We now present our own projections of carbon dioxide emissions from the G-cubed multi-country model (McKibbin and Wilcoxen 1998; gcubed.com). A summary of the approach is provided here but further details on the technique used in the G-cubed model can be found in McKibbin and Wilcoxen (2007). In the following discussion, the sources of economic growth are labour-augmenting technical change at the industry level and population growth. The population growth assumptions are based on the 2006 UN population projections (mid-scenario). In order to simplify the discussion, labour-augmenting technical change is referred to as 'productivity growth' throughout the remainder of this chapter.

In the G-cubed model, a productivity catch-up model is assumed to be the driver for productivity growth by sector and by country. The United States is assumed to be the technological leader in each sector. Other countries are allocated an initial productivity gap by sector and a rate at which this gap is

Figure 9.6   **Global carbon dioxide emissions from fossil fuels, 1990 and 2030**

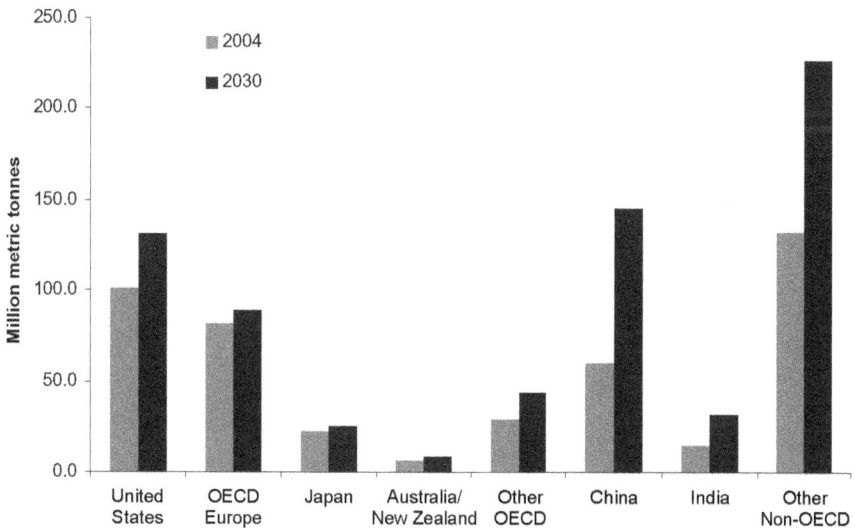

closed. For industrial countries and China, this is assumed to be a time-varying rate, which on average is 2 per cent per annum from 2006. For other developing countries, it is assumed to range between 2 per cent per annum and 1 per cent per annum, depending on the region. In this chapter, initial Chinese productivity is assumed to vary across sectors, and averages about 20 per cent of the productivity in the equivalent sector in the United States in 2002.

The results from the G-cubed model for Chinese carbon dioxide emissions are shown in Figure 9.7. This has a business-as-usual baseline as well as two other lines that will be discussed in section five below, which involve different assumptions about policy interventions. The business-as-usual projections from G-cubed are higher than the projections in EIA (2007). Carbon dioxide emissions in

- the EIA low-growth scenario rose from 6,400 mega-tonnes (Mt) in 2010 to 10,143 Mt in 2030
- the EIA reference-case scenario rose from 6,497 Mt in 2010 to 11,239 Mt in 2030
- the EIA high-growth scenario rose from 6,615 Mt in 2010 to 12,500 Mt in 2030
- the G-cubed model rose from 7,855 Mt in 2010 to 14,114 Mt in 2030.

The difference between the EIA's projections and those of G-cubed is reminiscent of the difference between the EIA's 2002 and 2007 projections. The higher projections by G-cubed come from it forecasting a higher economic growth rate in China than the EIA (2007) and a smaller change in the energy intensity of GDP in China (the latter being an endogenous result of the assumptions imposed for sectoral productivity growth in China). It must be stressed, however, that our G-cubed projections (like projections by others) are highly uncertain and change quite significantly if assumptions about the rate of catch-up are varied.

The principles to guide reduction in carbon dioxide emissions

There are many vexing, fundamental issues in deciding how to prevent catastrophic climate change. These issues include the following.

- There is still much about the science of climate change that we do not fully understand. Is climate change a linear or an abrupt discontinuous function of carbon dioxide concentration?[10] Is there are saturation point in the absorptive capacity of the earth's sinks for atmospheric carbon dioxide?
- There are immense difficulties in computing the costs and benefits of climate change. How should we value irreversible events such as species

extinction? How should we value the benefits to the present generation and the costs to the unborn future generations?

- There are serious challenges to designing effective implementation and oversight mechanisms for the carbon dioxide reduction process. How can national carbon dioxide caps be enforced? How can we build in incentives for mutual policing among the polluters dispersed around the world?

- The reduction of carbon dioxide emissions will only delay, not stop, the increase in carbon dioxide concentrations towards the danger level. The only long-term solution is likely to be shifting to non-fossil fuel energy. It is, however, impossible to know when these alternative fuels will be available at commercially viable costs, and at the vast scales that will ultimately be required. If the carbon dioxide reduction mechanism is designed to buy time for this development, how long will we need?

- There is unlikely to be an amicable way to distribute the burden of reducing carbon dioxide. Should the existing polluters be 'grandfathered' into the international treaty? What should be the relative burden for the rich, middle-income and poor nations? Alternatively, should the cap be based on carbon dioxide allowances per person?

The world, obviously, cannot afford to continue on the business-as-usual path until there is broad consensus about most of the above issues. Rates of carbon dioxide emissions are increasing, the tangible consequences of climate change are already evident and there is the real possibility that 'projections from climate models have been too conservative' (Gulledge 2008:56).[11] The sense of urgency is real, and this is why many countries signed the Kyoto Protocol on 11 December 1997 as a pragmatic way to effect at least a temporary improvement on the business-as-usual situation. The signatories from industrial countries agreed to reduce their carbon dioxide emissions in the period 2008–12 to 95 per cent of their 1990 levels, on average (that is, 5 per cent below their 1990 emissions), and to allow the permits for carbon dioxide emissions to be tradable internationally. China was not required to undertake any reduction because it was a developing country. The United States signed the treaty but never ratified it because it exempted large developing countries, particularly China and India. Since the United States and China are the world's two largest carbon dioxide emitters,[12] the Kyoto Protocol was rendered grossly inadequate as a carbon dioxide reduction mechanism. Nordhaus (2008:92) has estimated that global emissions in 2010 under the Kyoto Protocol will be only 1.5 per cent lower than under the business-as-usual outcome.

To be effective, any carbon dioxide reduction scheme must include as many of the large emitters as possible and it should move them towards substantial long-term reductions in emissions. There are three classes of market-based mechanisms that could put the world on this agreed global carbon dioxide emissions path

- mechanisms that do not specify the carbon dioxide emissions path for each country: for example, a global carbon tax
- mechanisms that specify an immediately binding carbon dioxide emissions path for each country: for example, a domestic cap-and-trade scheme or an international cap-and-trade scheme
- mechanisms that specify a carbon dioxide emissions path that is not immediately binding: for example, a domestic carbon tax or the McKibbin–Wilcoxen hybrid (MWH) approach.

In practice, actual emissions are unlikely to hit target emissions at every point in time. We label the quantity target 'immediately binding' if the emissions above the target are explicitly penalised. The quantity target is labelled 'not immediately binding' when the above-target emissions pay the same carbon tax as the below-target emissions, and the carbon tax is later adjusted to bring anticipated emissions to the target path. Naturally, the global and national target paths, and the level of international and domestic carbon taxes are modified over time to take into account how close the actual emissions have been to target emissions, revelations in abatement costs and developments (and anticipated developments) in areas such as technology.

The global carbon tax

Given a desired time path of global carbon dioxide emissions, it could be possible to identify a time-varying common carbon tax that would motivate the private sector in each country to hold collective carbon dioxide emissions to the target amount in the absence of unexpected developments. A global carbon tax would have to be revised at fixed periods in light of its performance, improvements in technology, advances in scientific knowledge and new information and ideas. A global carbon tax has the virtue of not distorting the comparative advantage of the different countries.

Since much of the increase in atmospheric carbon dioxide concentrations since the Industrial Revolution has been due to industrialised countries, perhaps developing countries could be exempted from the global carbon tax for a period or until they have reached a certain level of income.

The domestic carbon tax

A carbon tax could also be applied at the domestic level. Given a time profile of desired carbon dioxide emissions for a country, it would be possible to identify the carbon tax required to achieve this. This approach is, however, likely to be inefficient in the global sense because it will not guarantee that the marginal cost of emissions reductions will be the same across countries. The probable outcome would be a distortion of comparative advantage. Again, developing countries might be exempted temporarily from having to impose this domestic carbon tax.

Domestic cap and trade

A country could issue emissions permits to match a national target emissions path. The permits could be given free to existing carbon dioxide emitters or auctioned to the general public, and would be tradable within the country but not across borders. This approach, like the domestic carbon tax, is unlikely to produce a globally efficient pattern of abatement. Developing countries might be given ceilings on carbon dioxide emissions that are binding only when they attain a particular income level.

International cap and trade

An international treaty that establishes a global carbon dioxide emissions path and allocates carbon dioxide emissions among countries could also allocate internationally tradable emissions permits to those countries; the Kyoto Protocol falls under this category. Developing countries could be given more permits than they would need for their current emissions, and they could then sell the excess and use the revenue to accelerate development and buy green technology. This approach would equate the costs of abatement at the margin and would not distort comparative advantage.

The McKibbin–Wilcoxen hybrid (MWH) approach

McKibbin and Wilcoxen (2002a, 2002b) have proposed a hybrid approach that combines
* an internationally determined path for emissions reductions for each country, which is translated into a limited supply of long-term national permits
* sales of annual national permits (in order to accommodate deviations from a national path) sold at a price that is determined by international negotiations: say, every five years.

Both types of permits would be valid only in the country of issue: there would be no trade across borders.[13] Every year, firms would be required to hold a portfolio of permits equal to the amount of carbon they emit.[14] The portfolio could include any mix of long-term and annual permits. The long-term permits could be owned outright by firms, or they could be leased from other permit owners. Except for the case of developing countries, which we will discuss in detail later, the amount of long-term permits for each country would intentionally be set lower than the anticipated amount of emissions (for example, set below the target emissions path). If the target turns out to be sufficiently tight, there will be demand for the annual permits, which will impose an internationally fixed upper bound on the short-term price of carbon emissions.

Each country would manage its own domestic hybrid policy using its own existing legal system and financial and regulatory institutions. There would be no need for complex international trading rules, for the creation of a powerful new international institution or for participating governments to cede a significant degree of sovereignty to an outside authority. The international dimension of the MWH consists of two actions: 1) setting a notional (or 'aspirational') greenhouse gas emissions trajectory for each country; and 2) harmonising the price of annual permits across participating countries.[15]

The number of long-term permits would be guided by the international negotiations over the target emissions path for the country. For example, the international treaty establishing the MWH mechanism could suggest that signatories distribute no more long-term permits than their allotments under the Kyoto Protocol. The number of long-term permits would be set when a country joined the scheme, but the country's government would have considerable flexibility in how the permits were used. A government that wished to tackle climate change more aggressively could choose to distribute few long-term permits;[16] and a government that preferred a carbon tax could distribute no long-term permits at all.[17] The treaty would not need to specify rigid allocations of long-term permits because emissions would generally be controlled at the margin by the price of annual permits. The number of long-term permits affects only the distribution of permit revenue between the private sector and the government; it does not affect the country's total emissions. Distributing a small number of long-term permits means the government will earn a lot of revenue from annual permit sales, but it might also lead to significant political opposition. Distributing a larger number means less government revenue but the permits would be very valuable to the private sector and permit owners could be expected to form a powerful lobby in support of the policy. In either case, one country's decision has little effect on other signatories.

Long-term permits

A 100-year permit would be akin to a book of 100 coupons, with each coupon corresponding to a particular year and stating the amount of greenhouse gas emissions the holder is entitled to emit. In line with a declining level of target emissions, the coupon for each year would allow a smaller amount of greenhouse gas emissions than the previous year. Once distributed, the long-term permits could be traded among firms, or bought and retired by environmental groups. The permits would be very valuable because: 1) there would be fewer available than needed for current emissions; and 2) each permit would allow annual emissions over a long period. As a consequence, the owners of long-term permits would form a private-sector interest group that would greatly enhance the long-term credibility of the policy: permit owners would have a clear financial interest in keeping the policy in place.

When initially distributed, the long-term permits could be given away, auctioned or distributed in any other way the government of the country saw fit. One option would be to distribute them for free to industry in proportion to each firm's historical fuel use—for example, a firm might receive permits equal to 90 per cent of its 1990 carbon emissions. Such an approach would be relatively transparent and would limit the incentives for lobbying by firms. Although the allocation would be based on historical emissions, the tradability of the permits would mean that they were not tied in any way to the original recipient or any particular plant, and hence would not create differences in marginal costs across firms or plants. Moreover, the existence of annual permits limits the ability of incumbent firms to create entry barriers by keeping their long-term permits off the market: entrants could simply buy annual permits. Incumbent firms would benefit financially from the initial distribution of permits, but unless they were previously liquidity constrained, they would not be able to use their gains to reduce competition.[18]

Another alternative would be to auction the permits. Auctioned permits would be exactly like a carbon tax except that the industry would have to pay the entire present value of all future carbon taxes up front. As the number of long-term permits was intentionally kept below the target path of emissions, at least a few annual permits would be sold in every year. The price of a permit during the auction would be bid up to the present value of a sequence of annual permit purchases.

Annual permits

The government would sell annual permits for an internationally agreed price—say, for $20 per tonne of carbon. There would be no restriction on the number of annual permits sold, but each permit would be good only in the year it was issued. Annual permits give the policy the advantages of an emissions tax: they provide clear financial incentives for emissions reductions but do not require governments to agree to achieve any particular emissions target regardless of cost. The existence of the annual permits introduces a degree of flexibility in the target. Over time, the global carbon price would be readjusted if either the global target were not being met as well as desired or if the global target were changed because of new information about climate science or marginal abatement costs.

Treatment of developing countries

To be effective in the long run, the agreement will eventually need to include all countries with significant greenhouse gas emissions. It is unlikely, however, that all countries will choose to participate at the beginning. Developing countries, for example, have repeatedly pointed out that industrialised countries are overwhelmingly responsible for current greenhouse gas emissions, and that those countries should therefore take the lead in reducing emissions. As a result, an international climate policy will need to cope with gradual accessions taking place over many years. Its design, in other words, must be suitable for use by a small group of initial participants, a large group of participants many years in the future and all levels in between. One important role for the treaty's long-term permit guidelines would be to distinguish between industrialised and developing countries. For example, a country such as China would be allowed to distribute more long-term permits than needed for its current carbon emissions. In that case, it would be committing itself to slowing carbon emissions in the future, but would not need to reduce its emissions right away. As the country grows, its emissions will approach the number of long-term permits. The market price of long-term permits would gradually rise, and fuel users would face increasing incentives to reduce the growth of emissions. Once the long-term target becomes a constraint, annual permits would begin to be sold and would smooth out the evolution of annual carbon costs.

A generous allotment of long-term permits would reduce the disincentives to join faced by developing countries, but that alone might not be enough to induce widespread participation. If stronger incentives were needed, it would be possible to augment the treaty with a system of foreign-aid payments or with programs for technology transfer to participating developing countries.

The firewall of separate markets under MWH

Because the permit markets under this policy are separate between countries, shocks to one permit market do not propagate to others: for example, accession by a new participant has no effect on the permit markets operating in other countries.[19] Likewise, the collapse of one or more national permit systems would be unfortunate in terms of emissions control, but it would not cause permit markets in other countries to collapse as well. In contrast, under the Kyoto Proto-col, shocks in one country—ineffective enforcement or withdrawal from the agreement, for example—would cause changes in permit prices around the world. For permit owners and permit users, investments in emissions reductions would be more risky under the Kyoto Protocol than other systems.

Compartmentalisation is especially important for a climate change agreement because of the uncertainties surrounding climate change: the agreement must survive through intervals in which warming seems to be proceeding more slowly than expected, which could create political pressure to abandon the agreement on the grounds that it is not necessary. Such intervals could arise because of random fluctuations in global temperatures from year to year, or because the policy is succeeding in reducing the problem. The latter point is worth emphasising: if a climate regime is successful at reducing warming and preventing significant damage, it will be easy for complacency to arise: many people might interpret the absence of disasters to mean that the risks of climate change were overstated.

Another advantage of multiple national permit markets, rather than a single international one, is that the incentives for enforcement are stronger. Individual governments would have little incentive to monitor and enforce an international market within their borders—and it is easy to see why: monitoring polluters is expensive, and punishing violators imposes costs on domestic residents in exchange for benefits that accrue largely to foreigners. There would be a strong temptation for governments to look the other way when firms exceeded their emissions permits. For a treaty based on a single international market to be effective, therefore, it will need to include a strong international mechanism for monitoring compliance and penalising violations. National permit markets reduce the problem substantially because monitoring and enforcement become a matter of enforcing the property rights of a group of domestic residents—the owners of long-term permits—in domestic markets.

Incentives for investments in carbon dioxide reduction under MWH

Some commentators argue that the MWH mechanism is more complex than an emissions tax or conventional permit system, but it is more likely to encourage private-sector investment in capital and research that will be needed to address climate change. To see why, consider the incentives faced by a firm after the policy has been established. Suppose the firm has the opportunity to invest in a new production process that will reduce its carbon emissions by 1 tonne every year. If the firm is currently covering that tonne by buying annual permits, the new process will save it $20 per annum. If the firm can borrow at a 5 per cent real rate of interest, it will be profitable to adopt the process if the cost of the innovation is $400 or less. For example, if the cost of adoption were $300, the firm would be able to avoid buying a $20 annual permit every year for an interest cost of only $15; adopting the process, in other words, would eliminate 1 tonne of emissions and raise profits by $5 per annum.

Firms owning long-term permits would face similar incentives to reduce emissions because doing so would allow them to sell their permits. Suppose a firm having exactly the number of long-term permits needed to cover its emissions faced the investment decision in the example above. Although the firm does not need to buy annual permits, the fact that it could sell or lease unneeded long-term permits provides it with a strong incentive to adopt the new process. To keep the calculation simple, suppose that the permits are perpetual and allow 1 tonne of emissions per annum. At a cost of adoption of $300, the firm could earn an extra $5 per annum by borrowing money to adopt the process, paying an interest cost of $15 per annum, and leasing the permit it would no longer need for $20 per annum.

The investment incentive created by MWH rises in proportion to the annual permit fee as long as the fee is low enough to be binding—that is, low enough that at least a few annual permits are sold. For example, raising the fee from $20 to $30 raises the investment incentive from $400 to $600.

The upper limit on incentives created by the annual fee is the market-clearing rental price of a long-term permit in a pure tradable permit system. Above that price, there would be enough long-term permits in circulation to satisfy demand and no annual permits would be sold. For example, if long-term permits would rent for $90 a year under a pure permit system, the maximum price of an annual permit under the hybrid model would be $90.

The critical importance of credibility becomes apparent when considering what would happen to these incentives if firms were not sure the policy was going to remain in force. If the policy were to lapse at some point in the future, emissions permits would no longer be needed. At that point, any investments made by a firm to reduce its emissions would no longer earn a return. The effect of uncertainty about the policy's prospects is therefore to make the investments it seeks to encourage substantially more risky.

Since the incentives created by the policy increase with the price of an annual permit, a government might try to compensate for low credibility by imposing higher annual fees. For example, suppose a government would like a climate policy to generate a $400 incentive for investment but firms believe that there is a 10 per cent chance the policy will be abandoned each year. For the policy to generate the desired incentive, the annual permit price would have to be $60 rather than $20. That is, the stringency of the policy (as measured by the annual permit fee) must triple in order to offset the two-thirds decline in the incentives arising from the policy's lack of credibility. In practice, the situation is probably even worse. Increasing the policy's stringency is likely to reduce its credibility further, requiring even larger increases in the annual fee. For example, suppose that investors believe that the probability the government will abandon the policy rises by 1 per cent for each $20 increase in the annual fee. In that case, maintaining a $400 investment incentive would require an annual fee of $70 rather than $60, which would be accompanied by an increase in the perceived likelihood of the policy being abandoned from 10 per cent to 12.5 per cent.

The general lesson is that a low-cost but highly certain policy generates the same incentives for action as a policy that is much more expensive but less certain. A hybrid policy with a modest annual permit price would generate larger investment incentives than a more draconian, but less credible, emissions target imposed by a more conventional system of targets and timetables. The MWH proposal is more credible than a carbon tax because it builds a political constituency with a large financial stake in preventing backsliding by future governments. It is therefore likely to provide more incentive to the private sector to make investments to reduce greenhouse gas emissions.

Coping with new information

Over time, more information will become available about climate change, its effects and about the costs of reducing emissions. If it becomes clear that emissions should be reduced more aggressively, the price of annual permits can be raised. The political prospects for an increase would be helped by the fact that raising the price of annual permits would produce a windfall gain for owners of long-term permits, since the market value of long-term permit prices would rise as well.[20]

If new information indicates that emissions should drop below the number allowed by long-term permits, raising the price of annual permits would need to be augmented by a reduction in the stock of long-term permits. One option would be for each government to buy and retire some of the long-term permits it issued. Other approaches would be possible as well: for example, accelerating the expiration date of the permits.

Comparing methods for reducing China's carbon dioxide emissions

Three market-based mechanisms

The Clean Development Mechanism (CDM) of the Kyoto Protocol allows industrialised countries to use credits for emissions-reducing actions taken in China to help meet their obligations under the protocol. This approach cannot be scaled up sufficiently to have the required effect of significantly reducing China's carbon emissions because it is project based and has proven very complex and costly to administer.

In this section, we show some results for alternative policy regimes and discuss what they imply for emissions and economic growth in China. Figure 9.7 contains various paths of greenhouse gas emissions from energy use in China under three different policy regimes

- a domestic carbon tax
- an international cap-and-trade scheme
- the MWH approach.

The business-as-usual line in Figure 9.7 is the projection of Chinese emissions from energy use from the G-cubed model under the assumptions already discussed above.

In order to compare the key aspects of the three policy regimes, we assume that all countries take on the emissions reduction path that is contained in the recent *World Economic Outlook* of the International Monetary Fund (IMF 2008). Emissions in each country, and for the world as a whole, rise along the business-as-usual path for a number of years, gradually peaking in 2028, falling back to 90 per cent of the 2002 emissions level about 2050 and then dropping to 40 per cent of the 2002 level by 2100. Along this business-as-usual trajectory, China and other developing countries would take on the same commitment as industrial countries but initially with a more gradual reduction target.[21]

In the first policy option, labelled 'Country target' in Figure 9.7, China reaches its target by implementing a domestic carbon tax. All other countries are assumed to follow a similar strategy and achieve their targets through domestic actions only. This country-by-country targeting achieves a common global outcome but with a wide variety of costs across countries.

Figure 9.7   **China's carbon dioxide emissions from energy, 2008 – 2050**

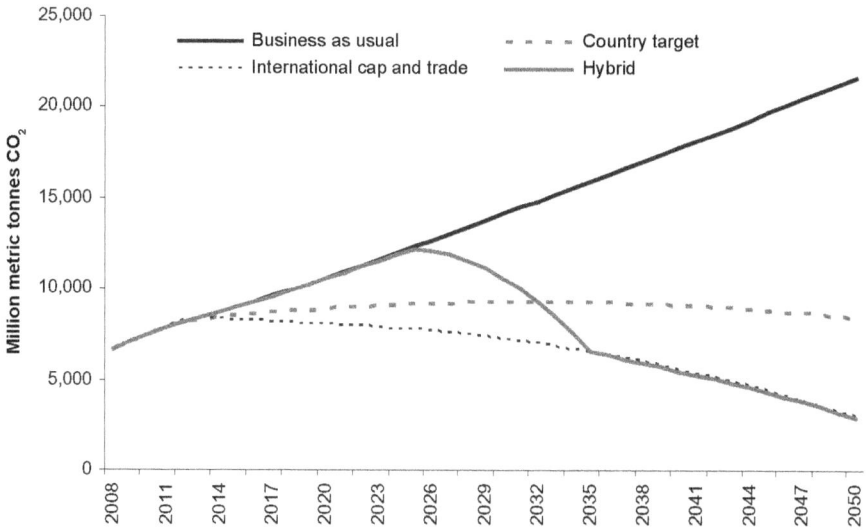

Source: G-cubed model in McKibbin, W. and Wilcoxen, P., 2008. Building on Kyoto: towards a realistic climate change agreement, Paper presented to the Brookings Institution High-Level Workshop on Climate Change, Tokyo, 30 May.

Figure 9.8   **China's GDP change from emissions reduction, 2008 – 2050**

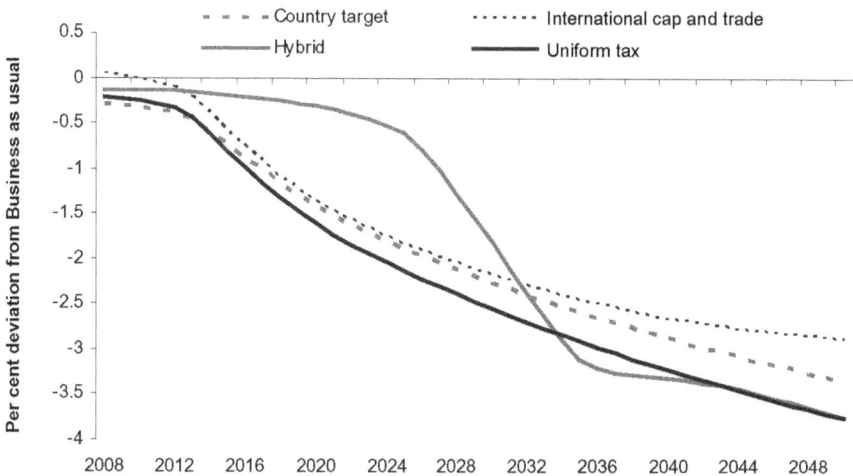

Source: G-cubed model in McKibbin, W. and Wilcoxen, P., 2008. Building on Kyoto: towards a realistic climate change agreement, Paper presented to the Brookings Institution High-Level Workshop on Climate Change, Tokyo, 30 May.

The results indicated by the 'International cap and trade' (triangles) line in Figure 9.7 are the emissions outcomes when China is given a permit allocation based on its target emissions and is then allowed to buy or sell emissions permits on international markets. China can therefore change its emissions outcome by selling permits at the world price (which is common to all countries). In the G-cubed model, under this allocation of permits, China has among the lowest marginal abatement costs in the world—that is, it is much cheaper to reduce a unit of carbon in China than in most other countries, reflecting the energy infrastructure and sources of emissions in China. The outcome is that emissions fall more quickly in China as China cuts its emissions domestically to sell permits abroad. Eventually, the marginal cost of abatement in China rises enough to reach equilibrium with the rest of the world.

The third policy option shown in Figure 9.7 is the MWH approach, in which China is allocated an amount of long-term permits equal to twice its 2008 emissions (which is more than the real amount of emissions in the first few years of its accession to the international climate treaty) but declining over time at the same rate as other countries.[22] These permits cannot be used outside China and therefore do not directly affect emissions in other countries. In this case, China's short-term carbon price is zero for a number of years because there are more permits available than needed, and emissions in China continue to rise along the business-as-usual path. When China grows enough to reach its emissions constraint, it starts to sell annual permits at the price stipulated by international agreement. Eventually, the carbon emissions path begins to fall until it reaches the emissions outcome under the international cap-and-trade system. This is not surprising since the uniform price under the MWH is designed to be almost the same as the price that would be delivered under the cap-and-trade policy.[23] The results are the same because the model is run under conditions of complete certainty about future events. With uncertainty, it would be necessary to refine the carbon price iteratively over time to try to reach the desired global target in a 'learning-by-doing' fashion. Under the cap-and-trade system, however, the target would be reached but at the cost of potentially very high volatility in carbon prices, and therefore economic costs.

Figure 9.8 shows the GDP outcome for China under the three different policies. The results are expressed as a percentage deviation from the business-as-usual path. Under the 'country target' and 'cap-and-trade' regimes, GDP begins to fall from the beginning of the regime in 2013. By 2025, the GDP loss to China from the carbon policy is about 1.8 per cent per annum. The international cap-and-trade policy leads to slightly lower GDP loss than the no-trading case because China is able to sell permits to raise income, which slightly offsets the

GDP loss for deeper cuts. The MWH model delays the significant GDP losses until China reaches the binding permit constraint, which begins about 2028.

Advanced technology diffusion

Another policy approach that is often advocated as a means of enhancing emissions reductions world-wide is the deployment of advanced energy technology in China. In this section, we present some results from McKibbin and Wilcoxen (2008) in which this policy is explored. The business-as-usual path discussed above is based on the assumption that energy technologies in each economy gradually improve at rates similar to those seen in recent historical data. Many policies now under discussion are, however, explicitly intended to accelerate the development and deployment of advanced technologies that would reduce greenhouse gas emissions. Some of these technologies, such as the integrated gasification combined cycle (IGCC) process to generate electricity from coal, reduce carbon dioxide emissions by substantially improving the efficiency of fossil-fuel combustion. Other technologies, such as carbon capture and sequestration, would reduce emissions by removing carbon dioxide from the exhaust stream after combustion. Yet other technologies, such as hybrid engines or carbon-fibre components for automobiles, would reduce emissions by lowering the fuel required per unit of service demanded (vehicle kilometres travelled, for example). Finally, advanced technology for non-fossil fuel sources of electricity, including nuclear power and renewable energy, would reduce carbon dioxide emissions by shifting the overall fuel mix. In this section, we examine the potential for accelerated deployment of advanced technology to reduce carbon dioxide emissions associated with electric power generation.

Since improved technology will allow more electricity to be produced from any given input of fossil fuel, we represent advanced technologies in the model via fuel-augmenting technical change. In essence, this approach captures the fact that new technology allows the same outcomes (output produced, distance travelled, and so on) to be produced with less physical energy. Factor-augmenting technical change introduces a distinction between physical inputs of energy (kilowatt hours, for example) and the effective value of those inputs to energy users. For example, increasing the efficiency of a coal-fired power plant from 41 per cent to 49 per cent using ultra-supercritical boiler technology would allow 19.5 per cent more electricity to be produced from a given amount of coal (an 8 per cent gain on a base of 41 per cent). In effect, the technology allows a new plant using 1 tonne of coal to produce the same amount of electricity that would have required 1.195 tonnes of coal in an older plant. The technology, in effect, serves to augment the physical fuel used.

Because the G-cubed model aggregates all electric power technologies into a single electric sector in each country, shifts of the fuel mix away from fossil fuels towards nuclear and renewable energy can also be modelled as fossil fuel-augmenting technical change. For example, a country increasing the share of non-fossil fuel generation in its fuel mix from 40 per cent to 55 per cent, and hence reducing its fossil fuel share from 60 per cent to 45 per cent, is effectively generating 33 per cent more electricity for any given input of fossil fuel.

Using industry projections of the rate of diffusion of a range of innovations in electricity generation between 2008 and 2030, we produced the augmentation factors shown in Table 9.3. The values shown include both effects mentioned above: improvements in the efficiency of fossil-fuel combustion, and shifts in the fuel mix away from fossil fuels. By 2030, for example, the 1.66 shown for Japan indicates that advanced technology and fuel switching will mean that the ratio of total electricity produced to fossil-fuel input will be 1.66 times that ratio today. We assume that technology and fuel switching continue beyond 2030, although at a diminishing rate. By 2045, for example, the augmentation factor for Japan increases to 2.09. The augmentation factors vary considerably by country. Improvements are very limited in developing countries other than China and India: the 2030 augmentation factor is only 1.13. India's augmentation factors are quite high, reflecting the fact that India currently relies heavily on coal burned in boilers with very low efficiency. Better technology therefore improves India's performance considerably. In contrast, Europe's augmentation factors are relatively low: it currently relies least on fossil fuels of all of the regions, and its current technology is relatively efficient. It therefore has less room for improvement.

Figure 9.9 shows the effect of the advanced-technology scenario on carbon emissions in China. For comparison, the business-as-usual results are shown as well. The business-as-usual trajectories are indicated with diamonds and the advanced-technology trajectories are indicated with triangles and labelled 'high innovation'. By 2050, emissions are lowered by 500 Mt per annum. This is a significant reduction from focusing only on electricity generation, but interestingly it is not as large as might be expected given the substitution we have assumed. This result is seen because in a rapidly growing economy such as China, the introduction of enhanced technology results in greater wealth and this higher wealth is spent partly on greater energy consumption. When we reduce the amount of carbon per unit of electricity, therefore, we also raise the amount of electricity used. This rebound effect of technological deployment on income growth is sufficient in China to partly offset the reduction in emissions from the new technology. This suggests that a combination of policies to deploy

Table 9.3   **Fossil-fuel augmentation factors i.e. productivity in electricity generation relative to business-as-usual**

Region	2030	2045
United States	1.67	2.1
Japan	1.66	2.09
Australia	1.73	2.19
Europe	1.49	1.8
Rest of OECD	1.67	2.09
China	1.67	2.1
India	1.8	2.31
Other developing countries	1.13	1.22
Former Soviet Union	1.71	2.16
OPEC	1.22	1.35

Note: Each number represents the ratio of electricity per unit of fossil fuel consumed in the advanced technology simulation to electricity per unit of fossil fuel consumed in the business-as-usual simulation.

Figure 9.9   **China's carbon dioxide emissions from energy under alternative technology assumptions, 2008 – 2050**

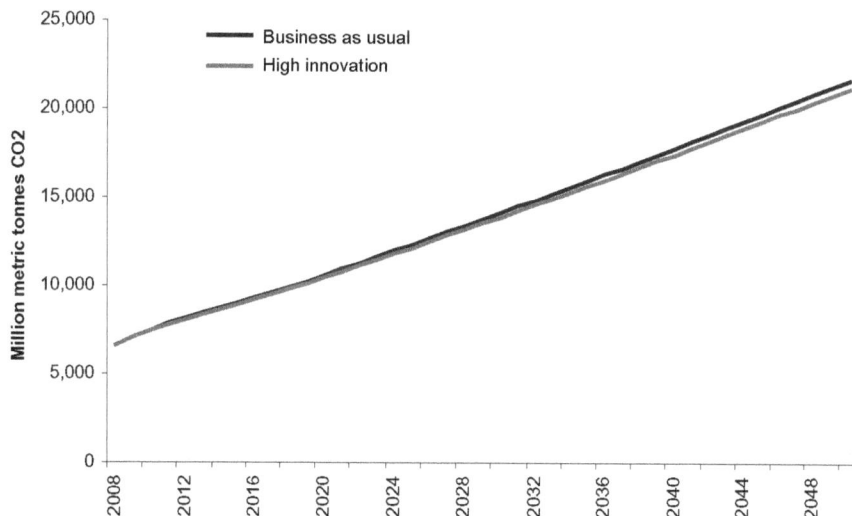

Source: G-cubed model in McKibbin, W. and Wilcoxen, P., 2008. Building on Kyoto: towards a realistic climate change agreement, Paper presented to the Brookings Institution High-Level Workshop on Climate Change, Tokyo, 30 May 2008.

technology as well as to price carbon to encourage substitution away from carbon-intensive inputs is required in a comprehensive approach to tackle the emission of greenhouse gases.

Future research will explore the interaction of alternative technology policies and the cost of carbon abatement under the MWH policy. Combining these approaches offers a potentially important way forward in cementing a global agreement based on economic incentives and technological innovation.

Conclusions

This chapter has summarised recent developments in energy use and carbon dioxide emissions in China. The increase in emissions since 2002 has taken most analysts by surprise and is a significant concern for global policymakers attempting to deal with climate change. As shown in McKibbin and Wilcoxen (2002a), unexpected developments cause the different market-based carbon dioxide reduction mechanisms to create vastly different costs. The international carbon tax and the MWH approach are more economically efficient responses to uncertainty than the cap-and-trade of the Kyoto Protocol.

Because it is very difficult to forecast future energy and emissions paths, concerns about uncertainty could delay or prevent accession by countries (especially developing countries) to a global climate agreement based on rigid targets and timetables. The recent experience of energy use and carbon emissions in China supports the arguments in McKibbin and Wilcoxen (2008) that uncertainty about the economic costs of undertaking binding emission targets is an important problem for a rapidly developing country such as China.

As an alternative, we have outlined the MWH approach, a set of internationally agreed actions that are based on long-term emissions targets and include an explicit compliance mechanism (annual permits) that allows the constraint to be exceeded at a stipulated international price. This approach would reduce emissions but without requiring participating countries to agree to achieve their emissions targets at any cost. Such an approach is not only very consistent with the UN Framework Convention on Climate Change, it is likely to be more viable than the current framework being negotiated under the Kyoto Protocol. China is a pivotal country in the global debate. The more its concerns can be taken into account in the design of a global post-Kyoto system, the more likely it is that the world will begin to take effective action on climate change.

We finish by emphasising the importance of combining a market-based carbon dioxide reduction mechanism with an ambitious program to accelerate the development of green technology. Such a program would probably have a higher chance of success if some important parts of it were based on international collaboration. For example, since China is building a coal-fired power plant each week, there is considerable opportunity to make some of those plants prototypes that could be used to test the scaling up of experimental technologies such as carbon capture and sequestration. On its own, China would hesitate to incur the costs of such experiments because others could quickly learn any useful findings.[24] Clearly, international scientific cooperation paid for by the international community could hasten the progress of a range of new technologies.

Notes

1   The Japanese growth experience since 1870 clearly suggests that the income disparity between China and Western Europe is not independent of Chinese economic policies. In 1870, the average Japanese income was 37 per cent of that of the average Western European income, but after a century of policy-induced convergence of economic institutions in Japan with those in Western Europe and the United States, average Japanese income in 1973 was equal to average income in Western Europe. The growth experiences of South Korea and Taiwan since the early 1960s confirmed that catching-up growth was not unique to Japan.

2   Unless otherwise specified, all dollar figures refer to 1990 dollars.

3   At these growth rates, GDP per capita in 2100 would be $84,400 in Western Europe and $92,800 in China; and GDP per capita in 2150 would be $177,700 in Western Europe and $192,100 in India.

4   Possibly the most authoritative recent statement of this position is IPCC (2007).

5   The first four effects are from page 6, and the fifth is from page 9. On the last effect, the IPPC added that 'while the effects of observed ocean acidification on the marine biosphere are as yet undocumented, the progressive acidification of oceans is expected to have negative impacts on marine shell-forming organisms (eg. corals) and their dependent species'.

6   'As global average temperature increase exceeds about 3.5°C, model projections suggest significant extinctions (40 to 70% of species assessed) around the globe' (IPPC 2007:13–14).

7   The tipping point is defined as when the melting of the Greenland ice-cap becomes irreversible.

8   Raupach is quoted as saying

> …if we manage to bring CO_2 to equilibrium at 450ppm, we would be looking at a temperature rise of 1 to 1.5 degrees above pre-industrial levels, some changes to rainfall patterns, some melting of the Arctic, significant acidification of the oceans through CO_2 rise and so forth. But these are issues which would not cause widespread devastation…If we reach 550ppm, we're getting into 2 to 2.5 degree temperature rise and the amount of climate damage that we would be looking at will in some cases…probably involve crossing thresholds that we can't recover from. If we keep on the present growth projectory then we get there by about 2046 (cited in Beer 2007).

9   The increment was 2.08 ppm in 2002 and 2.54 ppm in 2003; see Kirby (2004). The concentration of atmospheric carbon dioxide is taken to be 380 ppm in 2008.

10 Gulledge (2008:52) has described the proposition that 'future climate change will be smooth and gradual' as a myth: 'The history of climate reveals that climate change occurs in fits and starts, with abrupt and sometimes dramatic changes rather than gradually over time.' Figure 3-1 in Gulledge (2008) makes this point dramatically in the time profile of the number of storms of tropical hurricane force in the North Atlantic in the period 1930–2007.

11 Gulledge (2008:56) points out that

...the models used to project future warnings either omit or do not account for uncertainty in potentially positive feedbacks that could amplify warming (for example, release of greenhouse gases from thawing permafrost, reduced ocean and terrestrial CO_2 removal from the atmosphere), and there is some evidence that such feedbacks may already be occurring in response to the present warming trend.

12 Table 9.2 reports that the United States and China accounted for 35.2 per cent of global carbon dioxide emissions in 2004, and would account for 39.1 per cent in 2010.

13 Strictly speaking, the word 'country' is too narrow. The permits would be valid only within the political jurisdiction of issue. If the relevant jurisdiction is multinational—the European Union, for example—permits could be traded between countries within the broader jurisdiction.

14 This approach is known as a downstream policy because it applies to fuel users. It would also be possible to apply the policy upstream by imposing limits on the carbon embodied in fuels when they are produced—for example, at the mine mouth or well-head.

15 The negotiations, of course, would not be trivial: getting agreement on the annual price would require considerable diplomacy. It is interesting to note that a treaty of this form has a strong built-in incentive for countries to participate in the initial negotiations. Countries that participate will have a role in setting the annual price while those who remain on the sidelines will not. We are indebted to Jonathan Pershing for pointing this out.

16 Countries have different degrees of concern about climate change and different abilities to implement climate policies. A coordinated system of hybrid policies provides participants with the ability to tailor the policy to their own circumstances.

17 A government might prefer a carbon tax if it lacks the institutional and administrative mechanisms needed to operate a permit market.

18 In passing, it is worth noting that anti-competitive behaviour by the incumbents, while unlikely, would have an environmental benefit: it would reduce overall carbon emissions.

19 In contrast, a conventional international permit system could be particularly difficult to enforce because of the links it creates between countries. Restricting sales of permits by non-complying countries, as would be required under the Kyoto Protocol, would harm the interests of compliant countries by raising permit prices. The international links between permit markets therefore provide a strong incentive against enforcement of the agreement.

20 Although long-term permit owners would welcome an increase in the annual price, there is little risk that they would be able to drive prices up on their own. Given that other energy users provide countervailing pressure to keep energy prices low, it is hard to imagine that permit owners would be able to push a government into adopting an inefficiently high price and excessively stringent emissions policy.

21 The exact details of the target are not central to this paper because we will be comparing alternative policies for reaching a single set of targets; however, more rapid cuts in emissions would clearly give different results to those presented here.

22 The excessive amount of long-term permits in the first few years of this policy option means that the global emissions of carbon dioxide in the third policy option exceed the amount of global carbon dioxide emissions in the first and second policy options (whose emissions equal each other's). It is interesting that if China were given an excessive amount of carbon credits in the second policy option, its emissions path and GDP path would still be the same as that shown in Figures 9.7 and 9.8 as long as the extra amount of carbon credits given to China was small and hence had no effect on the world price of carbon credits. As production in China is guided by the world price of carbon credits, it would remain unchanged, and China would just sell off the extra carbon credits and cause the global emissions to be larger than under the first policy option (domestic carbon tax) and the original second policy option (international cap and trade) where the allocated carbon credits were binding from the beginning.

23 A difference arises because the transfer of income across countries with different spending patterns can change carbon dioxide emissions and therefore the price required for an equivalent global target path.

24 This dilemma exists in other forms as well, illustrated by the recent decision of the Virginian regulator of utilities in the United States

> …to turn down an application by the Appalachian Power Company to build a plant that would have captured 90 per cent of its carbon and deposited it nearly two miles underground, at a well that it dug in 2003. The applicant's parent was American Electric Power, one of the nation's largest coal users, and perhaps the most technically able. But the company is a regulated utility and spends money only when it can be reimbursed (cited in Wald 2008).

The Virginia commission said that it was 'neither reasonable nor prudent' for the company to build the plant, and the risks for ratepayers were too great, because costs were uncertain, perhaps double that of a standard coal plant. And in a Catch-22 that plagues the whole effort, the commission said A.E.P. should not build a commercial-scale plant because no one had demonstrated the technology on a commercial scale (Wald 2008).

25 Full details of the model, including a list of equations and parameters, can be found online at www.gcubed.com

26 These issues include: Reaganomics in the 1980s, German unification in the early 1990s, fiscal consolidation in Europe in the mid 1990s, the formation of the North American Free Trade Agreement (NAFTA), the East Asian financial crisis and the productivity boom in the United States.

References

Bagnoli, P., McKibbin, W. and Wilcoxen, P., 1996. 'Future projections and structural change', in N. Nakicenovic, W. Nordhaus, R. Richels and F. Toth (eds), *Climate Change: integrating economics and policy*, CP 96-1, International Institute for Applied Systems Analysis, Vienna:181–206.

Beer, S., 2007. 'Atmospheric CO_2 to reach first danger level by 2028: new research', 22 May 2007. Available from http://www.itwire.com/content/view/12347/1066.

Energy Information Agency (EIA), 2002. *International Energy Outlook*, Department of Energy, Washington, DC.

——, 2007. *International Energy Outlook*, Department of Energy, Washington, DC.

——, various years. *International Energy Outlook*, Department of Energy, Washington, DC.

Garnaut, R., Howes, S., Jotzo, F. and Sheehan, P., 2008. Emissions in the platinum age: the implications of rapid development for climate change mitigation, Paper presented at a seminar at The Australian National University, Canberra.

gcubed.com. Available at http://www.gcubed.com

Gulledge, J., 2008. 'Three plausible scenarios of climate change', in K. Campbell (ed.), *Climatic Cataclysm: the foreign policy and national security implications of climate change*, Brookings Institution Press, Washington, DC.

Intergovernmental Panel on Climate Change (IPPC), 2007. *Climate Change 2007: synthesis report*, Cambridge University Press, Cambridge.

International Monetary Fund (IMF), 2008. *World Economic Outlook*, April, International Monetary Fund, Washington, DC.

Kaya, Y., 1990. Impact of carbon dioxide emission control on GNP growth: interpretation of proposed scenarios, Paper presented to the Intergovernmental Panel on Climate Change Energy and Industry Subgroup, Response Strategies Working Group, Paris.

Kirby, A., 2004. 'Carbon "reaching danger levels"', *BBC News Online*, 13 October. Available from http://news.bbc.co.uk/1/hi/sci/tech/3737160.stm

Kuznets, S., 1966. *Modern Economic Growth: rate, structure, and spread*, Yale University Press, New Haven.

Maddison, A., 2001. *The World Economy: a millennial perspective*, Organisation for Economic Cooperation and Development, Paris.

——, 2007. 'horizontal-file_03-2007.xls', File downloaded from Internet. Available from www.ggdc.net/maddison/Historical_Statistics/horizontal-file_03-2007.xls

McKibbin, W., Pearce, D. and Stegman, A., 2007. 'Long term projections of carbon emissions', *International Journal of Forecasting*, 23:637–53.

McKibbin, W. and Wilcoxen, P., 1998. 'The theoretical and empirical structure of the G-cubed model', *Economic Modelling*, 16(1):123–48.

——, 2002a. *Climate Change Policy After Kyoto: a blueprint for a realistic approach*, The Brookings Institution, Washington, DC.

——, 2002b. 'The role of economics in climate change policy', *Journal of Economic Perspectives*, 16(2):107–30.

——, 2007. 'A credible foundation for long term international cooperation on climate change', in J. Aldy and R. Stavins (eds), *Architectures for Agreement:*

addressing global climate change in the post-Kyoto world, Cambridge University Press, Cambridge:185–208.

——, 2008. Building on Kyoto: towards a realistic climate change agreement, Paper presented to the Brookings Institution High-Level Workshop on Climate Change, Tokyo, 30 May .

Nordhaus, W., 2008. 'Economic analyses of the Kyoto Protocol: is there life after Kyoto?', in E. Zedillo (ed.), *Global Warming: looking beyond Kyoto*, Brookings Institution Press, Washington, DC.

Obstfeld, M. and Rogoff, K., 1996. *Foundations of International Macroeconomics*, MIT Press, Cambridge, Mass.

Wald, M.L., 2008. 'Running in circles over carbon', *New York Times*, 8 June.

Appendix

The G-cubed model

The G-cubed model is an inter-temporal general-equilibrium model of the global economy. The theoretical structure is outlined in McKibbin and Wilcoxen (1998).[25] A number of studies show that the G-cubed modelling approach has been useful in assessing a range of issues across a number of countries since the mid 1980s.[26] Some of the principal features of the model are as follows.

- The model is based on explicit inter-temporal optimisation by the agents (consumers and firms) in each economy (Obstfeld and Rogoff 1996). In contrast with static computable general equilibrium (CGE) models, in the G-cubed model, time and dynamics are of fundamental importance. The G-cubed model is known as a dynamic stochastic general equilibrium (DSGE) model in the macroeconomics literature and as a dynamic inter-temporal general equilibrium (DIGE) model in the CGE literature.

- In order to track the macro-time series, the behaviour of agents is modified to allow for short-run deviations from optimal behaviour due either to myopia or to restrictions on the ability of households and firms to borrow at the risk-free bond rate on government debt. For households and firms, deviations from inter-temporal optimising behaviour take the form of rules of thumb, which are consistent with an optimising agent that does not update predictions based on new information about future events. These rules of thumb are chosen to generate the same steady-state behaviour as optimising agents so that in the long run there is only a single inter-temporal optimising equilibrium of the model. In the short run, real behaviour is assumed to be a weighted average of the optimising and the rule-of-thumb assumptions. Aggregate consumption is therefore a weighted average of consumption based on wealth (current asset valuation and expected future after-tax labour income) and consumption based on current disposable income. Similarly, aggregate investment is a weighted average of investment based on Tobin's 'Q' (a market valuation of the expected future change in the marginal product of capital relative to the cost) and investment based on a backward looking version of 'Q'.

- There is an explicit treatment of the holding of financial assets, including money. Money is introduced into the model through a restriction that households require money to purchase goods.

- The model also allows for short-run nominal wage rigidity (by different degrees in different countries) and therefore allows for significant periods of unemployment depending on the labour-market institutions in each country. This assumption, when taken with the explicit role for money, is what gives the model its 'macroeconomic' characteristics (here again, the model's assumptions differ from the standard market-clearing assumption in most CGE models).

- The model distinguishes between the stickiness of physical capital within sectors and within countries and the flexibility of financial capital, which immediately flows to where expected returns are highest. This important distinction leads to a critical difference between the quantity of physical capital that is available at any time to produce goods and services, and the valuation of that capital as a result of decisions about the allocation of financial capital.

As a result of this structure, the G-cubed model contains rich dynamic behaviour, driven on the one hand by asset accumulation and, on the other, by wage adjustment to a neoclassical steady state. It embodies a wide range of assumptions about individual behaviour and empirical regularities in a general-equilibrium framework. The interdependencies are solved using a computer algorithm that solves for the rational-expectations equilibrium of the global economy. It is important to stress that the term 'general equilibrium' is used to signify that as many interactions as possible are captured, not that all economies are in a full market-clearing equilibrium at each point in time. Although it is assumed that market forces eventually drive the world economy to a neoclassical steady-state growth equilibrium, unemployment does emerge for long periods due to wage stickiness—to an extent that differs between countries due to differences in labour-market institutions.

Table A9.1   **Overview of the G-cubed model (version 80J)**

Regions
  United States
  Japan
  Australia
  Europe
  Rest of the OECD
China
  India
  Oil-exporting developing countries
  Easter Europe and the former Soviet Union
  Other developing countries
Sectors
Energy:
  Electric utilities
  Gas utilities
  Petroleum refining
  Coal-mining
  Crude oil and gas extraction
Non-Energy:
  Mining
  Agriculture, fishing and hunting
  Forestry/wood products
  Durable manufacturing
  Non-durable manufacturing
  Transportation
  Services
  Capital-producing sector

10

The political economy of emissions reduction in China
Are incentives for low carbon growth compatible?

Cai Fang and Du Yang

There is growing, unassailable evidence of severe existing and potential consequences of global climate change and of the relationship between human economic activities and global warming (for example, Stern 2007). The Kyoto Protocol and the Bali climate conference road-map proposed either compulsory or moral requirements for action from all economies—including China, which has the largest population size, the fastest growth rate and the second-largest gross domestic product (GDP) in purchasing power parity (PPP) terms. As research (Thomas 2007) estimates, assuming the ratio of carbon dioxide emissions to GDP remains at the 2001 level, total global emissions will reach as high as 25 billion metric tonnes by 2018. While this will represent an increase in global carbon dioxide emissions of 69 per cent, emissions in China will increase to 9 billion tonnes—a growth of 218 per cent, exceeding all other countries in terms of total emissions.[1]

It is seems to be the case that China's emissions reduction efforts are motivated by international pressure, which the central government passes on to local governments and enterprises. If, however, external pressure is the sole motivating factor for the government to take the issue seriously—and it is not induced endogenously by China's economic growth *per se*—China will face great difficulty reducing emissions because there will be a lack of incentives to carry out emissions reduction strategies. That is, the following questions should be answered before we can be confident about the realisation of the strategic goal. First, does the central government have persistent volition to carry out the policy, in financial and administrative terms? Second, are local governments willing to sacrifice short-term growth for sustainable long-term development? Third, can an enterprise's behaviour in dealing with emissions reduction incentives be compatible with the government's intentions? As has been widely observed, China's rapid economic growth during the transitional phase was stimulated largely by local governments pursuing GDP growth and the resulting enhancement of fiscal revenue. To guarantee an effective implementation of the strategy, any slow-down of economic growth supposedly caused by emissions reduction must be compensated.[2]

Those policy advocates for China's emissions reduction who emphasise the issues of obligation and responsibility but pay no attention to issues relating to capability and incentive, are irrelevant in policy decisions and incomplete in theoretical consistency. The argument of this chapter is that the relative bargaining power and effectiveness of policy advocacy are determined by the dominant priorities and the particular stage of development. While conditions mature, incentives, behaviour and priorities change. The environmental Kuznets curve (EKC), which depicts the relationship between per capita income and environmental quality, is a reflection of such political and economic logic (Deacon 2005). While the EKC is indicative of a simplified correlation of environmental appearance with environmental requirements associated with per capita income, there can be deeper implications behind it. In the case of China, the government's behaviour, characterised so far by its developmental state, which consists of strong development impulses from the central and local governments, responds more sensitively to the demands from potential changes in growth patterns than to the demands from income increases.

In this chapter, we discuss the continuing fundamental changes in China's development stage and the implications for growth patterns. Taking sulphur dioxide emissions as an example, we estimate a Chinese EKC and try to reveal incentive mechanisms and policy focuses for implementing emissions reduction strategies from the empirical findings. The following questions are expected to be answered: 1) can the orientation of growth policy be changed as a result of interaction between central and local governments; 2) are emissions reductions and low carbon growth financially feasible; and 3) can incentives be compatible between the central government, local governments, enterprises and the general population in implementing a strategy of emissions reduction?

Development stage and growth patterns

China's environmental problems are the result of its economic growth pattern, characterised by a reliance on labour and capital inputs rather than on productivity enhancement, although this growth pattern suits a particular development stage. The Chinese stimulation of economic growth is unique in combining central and local governments to form a developmental state, which characterises the outstanding government function in economic development. Consequently, while individual enterprises inherently tend to respond to growth pattern changes—which emissions reduction is the result of—regulations and other government reactions are much more important in guaranteeing the realisation of the strategy. The impending new development stage will help the government shift its policy orientation by making the relevant regulations incentive-compatible among stakeholders.

The political economy of emissions reduction relates to the different reactions of all relevant parties. First, for the central government, decision-making has to do with the recognition of the importance of emissions reduction as required by the change in the development stage. Second, for local governments, incentives to shift growth patterns from input driven to productivity driven and a willingness to sacrifice short-term growth in GDP and fiscal revenue for sustainable growth are more relevant in response to the central government's mandate. Third, for enterprises, which are supposed to care more about profitability than externality, immanent irritants requiring transition from inputs-based to productivity-based expansion are generated by modification of the production factor endowment. Finally, for people, stronger demands for environmental quality come ultimately from the transition from a livelihood dominated by subsistence pressures to comprehensive human development that alters in accordance with development stages. In the final analysis, an environmental strategy can be implemented effectively only by combining the quartet of capacity, responsibility, obligation and incentive.

In his prominent paper on the relationship between economic growth and income inequality, Kuznets (1955) postulates that in the economic growth process, income inequality first increases then declines after reaching a turning point. Environmental economists later applied this inverted U-shape curve to depict a similar relationship between economic growth and environmental quality (Grossman and Krueger 1995). The EKC can also be a useful framework with which to examine whether the Chinese economy possesses inherent momentum to transform its growth pattern to a more environmentally friendly one and, in particular, to understand the political economy behind the transformation.

In more concrete terms, we investigate the ways in which the changes in development stages impact on environmental policy decisions from two aspects. First, the change in development stage requires transformation of growth patterns. The economic growth pattern can be referred to in ways in which factors of production are allocated at micro and macro levels and it can usually be classified by what kinds of sources the economic growth is based on. Thanks to the earlier than expected completion of demographic transition, the entire period of reform and opening-up in China has been characterised by adequate labour force supply and a high savings rate. This demographic dividend resulting from a productive population structure has been realised through marketised resource allocation mechanisms and participation in the global economy. The favourable population factor has provided China's economic growth with a window of opportunity and, therefore, the phenomenon of diminishing return to capital has been deterred (Cai and Wang 2005). In the meantime, economic growth in transitional China has relied heavily on inputs of production factors rather than on productivity improvement. After a short rise in total factor productivity (TFP) and its contribution to growth in the early stages of reform, China's TFP performance has been unsatisfactory since the 1990s (for example, Zheng and Hu 2004; Kuijs and Wang 2005). This was a major attribution of the growth pattern characterised by heavy pollution, high depletion and low efficiency (Kaneko and Managi 2004).

Similar stories were told about the early experiences of the 'Four Tigers' when they created the disputable 'East Asian miracle'. At the time, Krugman (1994) noted that the economic growth in East Asia was fuelled merely by inputs of labour and capital, and the growth of TFP and its share of economic growth were insignificant. Krugman believed the miracle was doubtful; his judgement was based on the neoclassical theory of growth, which assumes diminishing returns to capital due to limited supplies of labour, which was not true in those economies. The facts subsequently showed that once the dual-economy feature of an unlimited labour supply disappeared, those economies had to transform their economic growth patterns from inputs driven to TFP driven—and have since sustained their growth (Bhagwati 1996). After three decades of extraordinary economic growth, fuelled largely by demographic dividends, and as the population ages rapidly and the reservoir of surplus labour in rural areas runs dry, China's labour supply and demand scenario has changed fundamentally since 2004. With the approach of a Lewis turning point, conditions under which economic growth becomes increasingly reliant on productivity improvement rather than on expansion of inputs are maturing (Cai 2008).

In addition, the increase in per capita income induces people's desire for security and quality of life, and their calls for a better environment. A decade ago, the World Bank (1997) estimated that in 1995, financial losses resulting from air and water pollution were worth US$54 billion, accounting for 8 per cent of China's total GDP. Since then—namely, during the period from 1995 to 2006—the real per capita income of urban households increased by 131 per cent, and the real per capita income of rural households increased by 74.8 per cent. As a result of a much faster rate of income growth for the upper group, the richest 20 per cent earned 4.6 times more than the poorest 20 per cent. Per capita income level is the decisive factor in both of the widely used approaches to estimating losses caused by environmental damage—namely, the human-cost and the willingness-to-pay approaches. The upper income group, especially, has stronger bargaining power to have an impact on policy decisions about environmental issues. Therefore, the outstanding performance of income enhancement for Chinese residents must play a role in increasing calls for environmental improvement. Frequent environmental incidents in recent years have shown how quickly and enthusiastically citizens and the press can respond to environmental disasters (Hayward 2005).

The concerns of Chinese residents, scholars, policymakers and, to a lesser extent, enterprises about environmental quality have been translated into the central documents and protocols of the eleventh Five-Year Plan. The documents make repeated calls for transformation of growth patterns and the eleventh Five-Year Plan stipulates restrictive criteria for emissions reductions. As a response partly to those regulations and partly to the increase in prices of raw materials and wages, Chinese manufacturing enterprises successfully improved the efficiency of their usage of intermediate inputs and labour productivity (Kim and Kuijs 2007).

China's environmental Kuznets curve

An effective way of understanding the relationship between economic development and emissions in China is to depict the EKC. As we have mentioned already, economic development and demand for environmental quality as a result of improved living conditions can be represented by per capita GDP. When discussing EKC therefore, we employ provincial panel data to explore the relationship between emissions and per capita GDP and observe when the Kuznets turning point appears. The two main emissions are those of carbon dioxide and sulphur dioxide; however, this chapter describes only the EKC of sulphur dioxide because the data for carbon dioxide emissions are not yet available officially.

Sulphur dioxide is one of the main air pollutants produced by the combustion of sulphur compounds and is of significant environmental concern. Since coal and petroleum, which are the main sources of energy in China, often contain sulphur compounds, their combustion generates sulphur dioxide. In 2005, total sulphur dioxide emissions in China were 25.49 million metric tonnes—the highest in the world. In the eleventh Five-Year Plan, limiting sulphur dioxide emissions is one of the main goals for environmental protection. The plan requires a 10 per cent reduction in sulphur dioxide emissions by 2010—that is, total emissions of sulphur dioxide should not exceed 22.95 metric tonnes.

Data for our empirical analysis were collected from the *China Statistical Yearbook* (NBS, various issues). The variables we chose included sulphur dioxide emissions, per capita GDP, industrialisation levels and population by province from 1991 to 2006. Due to incomplete data for Tibet and Chongqing, we excluded these two provinces from the data sets. Table 10.1 displays the summary of statistics of the variables. Average sulphur dioxide emissions for all observations were 14.02 metric tonnes per 1,000 people; they reached 19.37 metric tonnes per 1,000 people in 2006. Grouping the averages of sulphur dioxide emission levels in the period 1991–2006 by region, we found slightly more emissions generated in the coastal areas than in the central and western areas, but the eastern provinces emitted much less sulphur dioxide (15.2 metric tonnes per 1,000 people) than their central and western counterparts (21.6 metric tonnes per 1,000 people) in 2006 alone. In addition, it is worth noting that the disparities of per capita GDP in 2006 were greater than the time series average.

Table 10.1 **Summary statistics of variables**

Variable	Mean (SD)	Minimum	Maximum	Mean in 2006
All provinces				
SO2 emissions				
(metric tonnes/1,000 people)	14.02 (8.74)	1.48	57.95	19.37 (12.53)
Per capita income (1990 yuan)	4,954 (4,407)	860	34,200	9,743 (7,069)
Population (million)	41.2 (25.3)	45.4	97.2	43.5 (26.6)
Coastal				
SO2 emissions				
(metric tonnes/1,000 people)	14.29 (7.33)	14.8	33.19	15.23 (6.79)
Per capita income (1990=100)	8,516 (5,727)	1,794	34,200	1,712 (766.4)
Population (million)	39.85 (28.6)	67.4	93.1	44.3 (32.9)
Others				
SO2 emissions				
(metric tonnes/1,000 people)	13.88 (9.41)	3.55	57.95	21.55 (14.37)
Per capita income (1990=100)	3,080 (1,499)	860	8,518	5,861 (1,497)
Population (million)	42.0 (23.4)	45.4	97.2	43.0 (23.7)

Note: We define Beijing, Tianjin, Liaoning, Shandong, Jiangsu, Shanghai, Zhejiang, Fujian, Guangdong and Hainan as coastal areas, and the remaining provinces as central and western areas.
Source: National Bureau of Statistics (NBS), various years. *China Statistical Yearbook*, China Statistics Press, Beijing.

We now describe the time trend and regional characteristics in detail. Figure 10.1 shows the changes in China's per capita sulphur dioxide emissions as a whole over time. While there has been a general trend of increases in sulphur dioxide emissions in the past decade, emissions have accelerated since 2002. With rising public attention paid to environmental protection and improved capacity for implementation, the amount of sulphur dioxide limitation has increased more rapidly than emissions. In general, the growing pattern of sulphur dioxide generation can be characterised by a higher rate of removal and bigger increases in emissions in terms of absolute magnitude. This implies that, with enormous economic growth, current efforts of removal and recovery cannot adequately counter emissions.

Figure 10.2 shows the changes in regional emissions of sulphur dioxide over time; the proportion of emissions generated in the central and western provinces of the total enhances over time. The ratio of emissions in 19 interior provinces to that in 10 coastal provinces was 1.57 in 1991, but it increased to 2 in 2006. Given the increasing disparities in economic development between eastern provinces and central and western provinces in the same period, the

Figure 10.1   **Emissions and removals per capita of sulphur dioxide in China, 1991–2006**

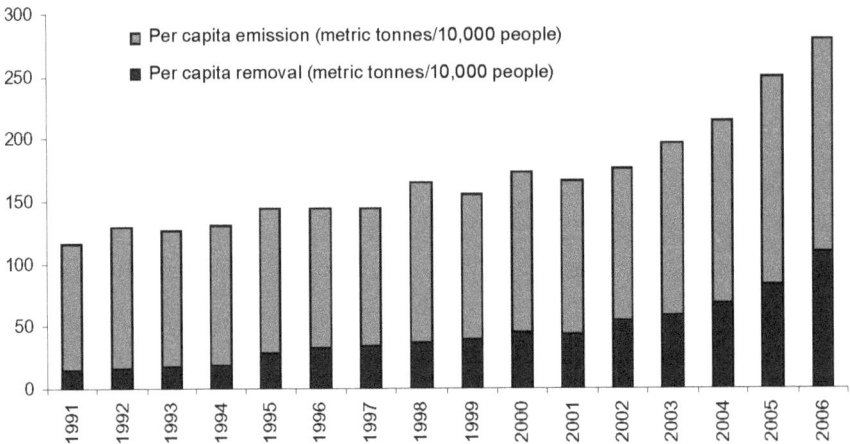

Source: National Bureau of Statistics (NBS), various years. *China Statistical Yearbook*, China Statistics Press, Beijing.

efficiency gains of emissions reduction in costal areas are much higher in real terms, whereas those in the interior provinces are dissatisfied. This rejects any attempt to take emissions in China as a homogenous issue and suggests a methodology that treats the two distinct regions differently while trying to do any meaningful analysis.

The scattered observations in terms of sulphur dioxide emissions at certain times shown in Figure 10.3 further reveal the great heterogeneity between the coastal and interior regions. The horizontal axis represents the levels of per capita GDP in 1990 prices and the vertical axis the levels of per capita sulphur dioxide emissions. At a relatively low level of per capita income for coastal and interior regions, there was no significant difference in emissions between the two regions. As each of the economies grow, the levels of emissions grow, with an even larger divergence in emissions than in economic growth, implying that if there is a sign of EKC, it must be the case that the eastern provinces alone present the path of increase first and then decline in emissions.

To see whether an EKC in terms of a sulphur dioxide emissions pattern exists, we run a regression to examine the relationship between regional economic development and sulphur dioxide emissions based on the following empirical model.

$$S_{i,t} = \alpha_0 + \beta_1 y_{i,t} + \beta_2 y_{i,t}^2 + \beta_3 m_{i,t} + u_i + v_t + \varepsilon_{i,t} \tag{1}$$

in which $S_{i,t}$ is per capita sulphur dioxide emissions of province i in year t, $y_{i,t}$ and $y_{i,t}^2$ are per capita GDP at the 1990 constant price and its square term of province i in year t, and $m_{i,t}$ is the level of industrialisation of province i in year t measured by the ratio of value added in the industrial sector to total GDP. u_i is the province dummy reflecting the persistent provincial difference, such as different patterns of energy consumption, regulation of energy use and environmental protection, preferences for energy consumption and so on. v_t is the year dummy to control the factors that change with time, apart from economic development, such as commodity and energy prices, technology for sulphur dioxide removal and the like. $\varepsilon_{i,t}$ represents randomly disturbing factors apart from time and region. As in other studies on the EKC, the purpose of this estimation is to look at the significance and sign of β_1 and β_2 to decide if they present an inverted-U shape.

Table 10.2 presents the regression results. The three columns list regression results for all 29 provinces, for coastal provinces and for central and western provinces, respectively. The fitness to model varies among regions: that for the model of coastal areas has the highest overall R^2 in three equations, while

Figure 10.2 **Regional composition of sulphur dioxide emissions, 1991 – 2006**

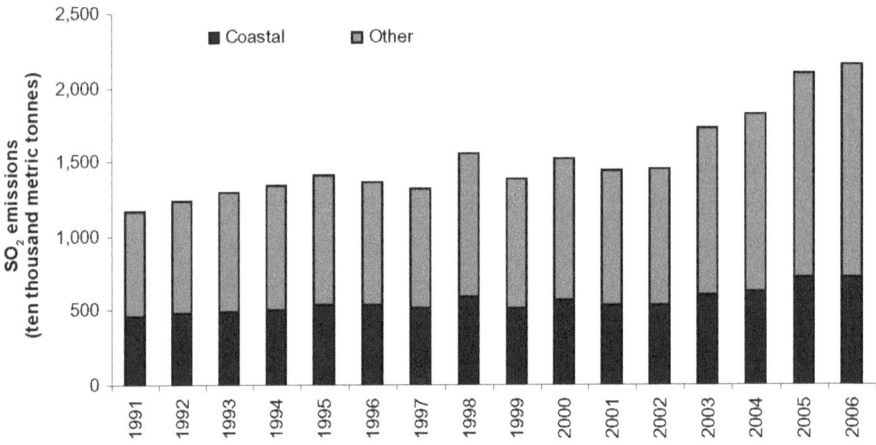

Source: National Bureau of Statistics (NBS), various years. *China Statistical Yearbook*, China Statistics Press, Beijing.

Figure 10.3 **Sulphur dioxide emissions against per capita GDP by province, 1991–2006**

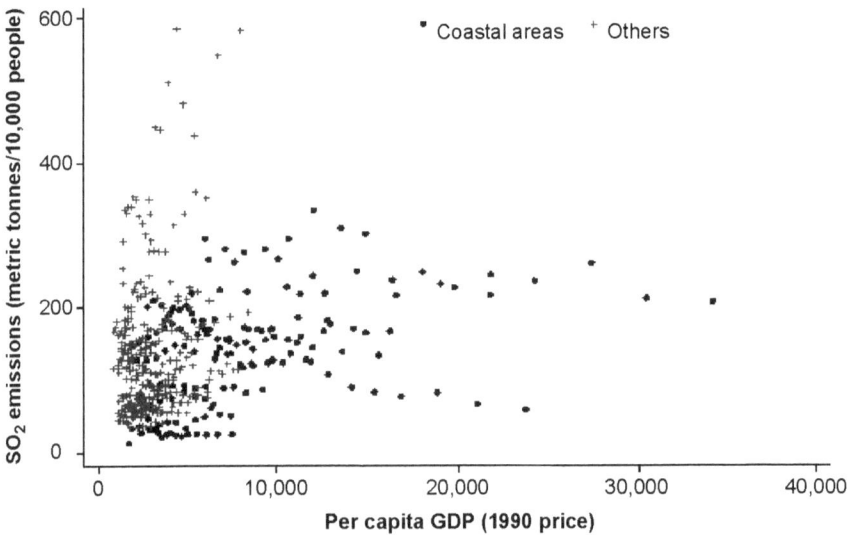

Source: National Bureau of Statistics (NBS), various years. *China Statistical Yearbook*, China Statistics Press, Beijing.

that for other areas is low. The model for coastal areas explains 58 per cent of variations of sulphur dioxide emissions, while that for others explains only 5 per cent.

What concerns us are the signs of the square term of per capita GDP in all three equations, because the negative signs indicate the existence of an EKC, though the coefficients are statistically insignificant for the regression for central and western provinces. For this reason, we can infer that the presence of an inverted-U shape for the pooled sample exists only because the general pattern of sulphur dioxide emissions in coastal areas shows a predictable Kuznets turning point. Based on the regression results in the first column, we can plot a graph that depicts the changing pattern of sulphur dioxide emissions in China as a whole, denoted by 29 provinces (Figure 10.4). Looking at Figures 10.3 and 10.4, the plots of observations for the central and western areas scatter far left of the turning point, while those for the coastal areas stand around the turning point. The huge heterogeneity identifies a need to distinguish between the two regions when observing the EKC in China.

We now use information from the second column in Table 10.2 to predict the EKC for coastal areas, and show the outcomes in Figure 10.5. According to the parameters estimated from the current sample, the turning point appears when per capita GDP reaches RMB18,963 at the 1990 constant price. Beyond this point, emissions are supposed to decrease. With this pattern, provinces that surpassed the turning point in 2007 included Beijing, Tianjin, Shanghai and Zhejing, while Guangdong and Jiangsu were very close to the point in terms of per capita GDP. In other words, in the current circumstances, many provinces in eastern China have already had the capacity and incentives to reduce their sulphur dioxide emissions—namely, to afford low carbon growth.

On the other hand, with accelerating economic growth, the central and western provinces continue their patterns of emissions. In the third column of Table 10.2, we see that coefficients of per capita GDP and its square term are statistically insignificant. If we use a different specification without inclusion of the square term, as shown in the last column of Table 10.2, we see a significant and positive coefficient for the variable of per capita income, which implies that the central and western regions stay at a phase of increasing emissions. As is expected, Figure 10.6 shows that, though diverging, most of the provinces in the region are scrambling in line with monotonously increasing sulphur dioxide emissions. In comparison with the predicted EKC of the eastern provinces, the picture here does not show any sign of an EKC.

Table 10.2   **Economic development and sulphur dioxide emissions: two-way fixed-effect model**

	All provinces	Coastal areas	Interior areas	
GDP per capita in 1990 price	.0033	.010	0.031	.015
	(1.38)	(3.10)	(1.54)	(2.93)
Square term of GDP per capita	-2.09e–07	-2.68e–07	-1.21e–07	–
	(3.35)	(4.06)	(0.8)	
Ratio of secondary industry in GDP	3.77	3.93	4.41	4.60
	(6.26)	(5.99)	(4.63)	(4.99)
Fixed effect, provinces	Yes	Yes	Yes	Yes
Fixed effect, years	Yes	Yes	Yes	Yes
Constant	-44.1	-75.5	-104	-91.86
	(1.67)	(2.23)	(2.60)	(2.48)
R^2				
within	0.47	0.48	0.53	0.52
between	0.06	0.61	0.002	0.0003
overall	0.15	0.58	0.051	0.064
No. of observations	464	160	304	304

Note: t value is in parenthesis.

Figure 10.4   **Environmental Kuznets curve: 29 provinces**

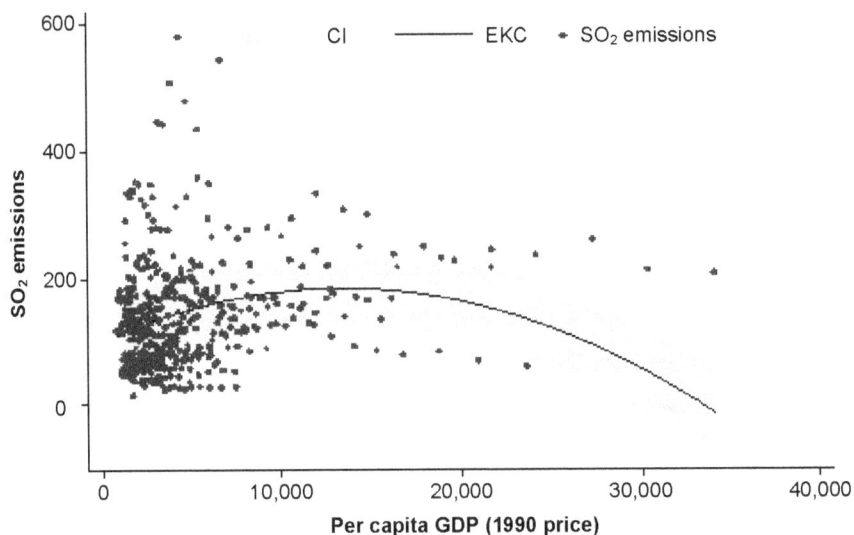

Figure 10.5   **Environmental Kuznets curve: coastal provinces**

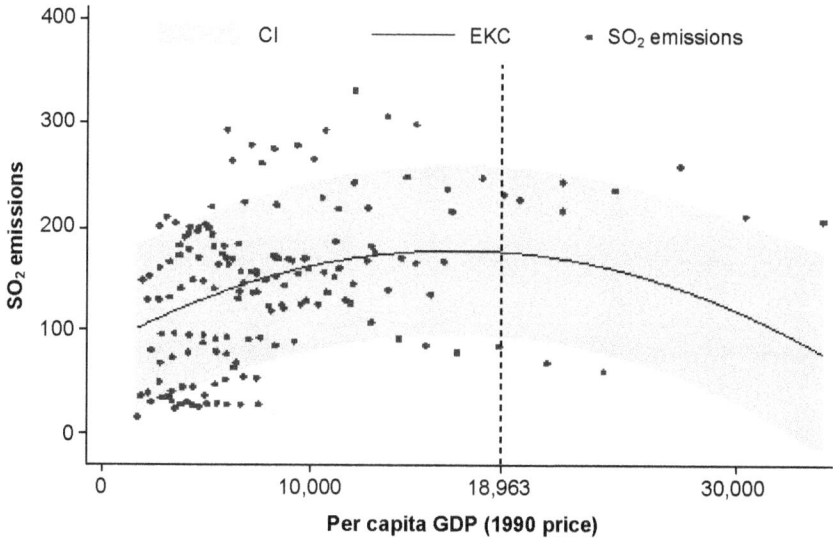

Figure 10.6   **Environmental Kuznets curve: central and western China**

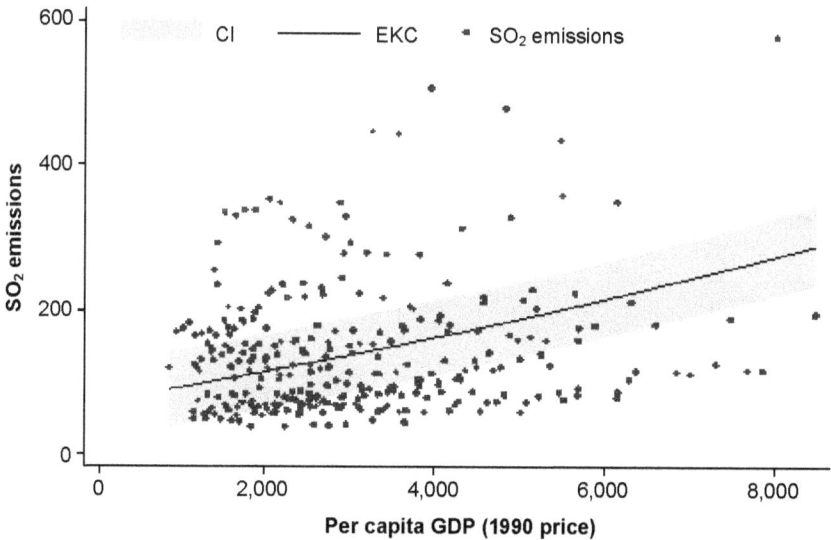

Conclusion and policy implications

Exemplified by sulphur dioxide emissions, the Chinese EKC shows the existence of a relationship between income increases and environmental improvement. There are, however, issues to be considered apart from the general conclusion.

First, while one can expect a future turning point from increases to decline in emissions for China as a whole, most Chinese provinces are still far from reaching that point. The central and western provinces in particular still have a strong desire for economic growth at the cost of the environment in order to catch up with their eastern counterparts. If the previously outlined path is followed, China will have to suffer further environmental degradation before reaching its spontaneous turning point, because the experiences of a spatial transfer of industries show (and the EKC implies) that the latecomers in economic growth tend to receive the transferred industries from their advanced counterparts in accordance not only with their comparative advantage but with their acceptance of environmental degradation—that is, there will be a tendency for the central and western regions to welcome polluting industries transferring from eastern regions. Given the strong desire for growth in the less-developed provinces and the large income gap between Chinese provinces, single incentives such as per capita income are not sufficient to lead those regions to the Kuznets turning point. Genuine changes must rely on the introduction of incentives and regulations based on the need for transformation of growth patterns.

From the regression results, one can see a great heterogeneity of sulphur dioxide emissions among regions, which suggests distinct policy packages for different regions in terms of emissions reduction. For most coastal provinces, which either passed through or are moving towards the Kuznets turning point, the inertial path and intrinsic forces can lead them to reduce emissions spontaneously. As far as the central and western provinces are concerned, it is hard to predict when they will enter the Kuznets turning intervals since the emissions in these areas are accelerating. In this regard, it is essential to enforce regulations to limit their emissions behaviour as total emissions in China are already the highest in the world.

Although we estimated a reasonably fitted EKC and its visible turning point for the eastern regions, observations differ substantially. Even for those observations whose positions are on the right interval of the turning point, they stand at different plots, implying significant heterogeneity among the eastern provinces. In general, the emissions in the east remain high and their decline will be slow.

Studies show that while a general relationship between per capita income and environmental quality has been observed, there are huge differentials across pollutants. Greenhouse gas emissions, while concomitant with those such as sulphur dioxide that are directly harmful to people's health, usually do not follow exactly the same path as other pollutants. As a greenhouse gas with no smell and no immediate harm to health, carbon dioxide emissions have not shown a significant path as the EKC suggests. For instance, the previous empirical studies rarely found an EKC between income levels and carbon dioxide emission patterns. If there are rare cases, they show that the turning point indicating carbon dioxide emissions tending to decline comes much later and requires several times higher income levels than do other pollutants (Webber and Allen 2004).

It is believed widely that because of China's enormous population size, the dominance of manufacturing in its industrial structure and the low efficiency of energy usage, in international rankings, China is positioned high in terms of per capita emissions and low in terms of per GDP emissions. Starting with this feature, there is a tendency for China to converge with rest of the world—that is, its per capita emissions have been found to increase over time, reflecting the development emissions effect, and the per GDP emissions decline, reflecting the progress of technology and improvements in efficiency (Figure 10.7). As is expected, the overall performance of carbon dioxide has not been as good as that of sulphur dioxide.[3]

Even though the EKC is not a sufficient notion for revealing the complete determinants of the rise or fall of emissions, it is useful to justify the existence of the relationship between levels of development and pollution, because it shows that governments, enterprises and people are willing and able to respond positively to changing environmental requirements derived from changes in developmental stages and therefore growth patterns. The predictable EKC and turning point disclose the governmental cognition, determination and policy measures in respect of the environmental issues, and the incentive compatibility between stakeholders—although they by no means imply that the chronic environmental problems can be solved xenogenetically when the time comes. Activities conducted by governments at all levels, such as education and the provision of information, play a role in shortening the time of solving the problem (Deacon and Norman 2004).

Our empirical results show that there is high heterogeneity in sulphur dioxide emissions across provinces, which is also true for the emissions pattern of carbon dioxide; therefore, the policies regarding emissions should be specified regionally. In China's case, given the strong motivations of the central and western provinces to catch up with their eastern counterparts, one of the challenging tasks for the central government is to design a well-functioning

Figure 10.7 Changing trends of carbon dioxide and sulphur dioxide emissions, 1995 – 2006

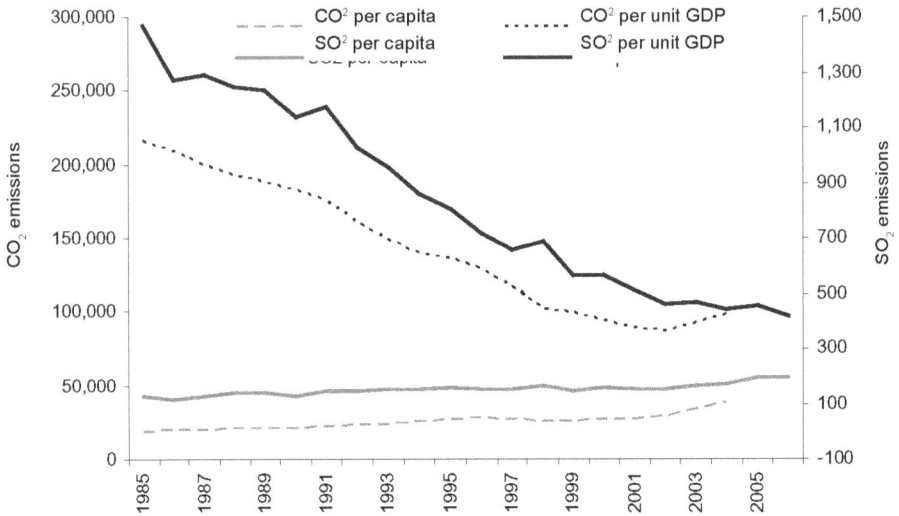

Note: Per capita emissions are measured by tonnes per 10,000 people; per unit GDP emissions are measured by tonnes per 100 million yuan.
Sources: Information on per capita carbon dioxide emissions from World Bank, various years. *World Development Indicators*. World Bank, Washington, DC; information on per GDP carbon dioxide emissions calculated from data from World Bank, various years. *World Development Indicators*. World Bank, Washington, DC, and National Bureau of Statistics (NBS), various years. *China Statistical Yearbook*, China Statistics Press, Beijing; information on sulphur dioxide emissions from National Bureau of Statistics (NBS), various years. *China Statistical Yearbook*, China Statistics Press, Beijing.

mechanism to provide incentives that will translate the implications drawn from changing growth patterns into changes in the function and behaviour of local governments to spur regional economic development. Such policy measures include those that improve transfers between central and local governments and among regions, provide physical and financial incentives for less-developed regions to choose sustainability rather than exploitation of growth potential and to implement emissions reduction policies with incentive compatibility between agents.

The effectiveness of emissions reduction policy lies in the endogenous demands for change in growth patterns and requirements for a better environment. Only when economic development moves to the stage at which economic growth becomes productivity driven can the policy package aimed at significantly reducing greenhouse gas emissions be incentive-compatible with the development motivations of local governments and the behaviour of enterprises. By estimating the EKC, this chapter examined the need for the Chinese economy to implement emissions reduction strategies. The results show that it is not enough to wait for the turning point to come; instead, policy implementation should be strengthened further in order to make incentives compatible between the central and local governments, enterprises and people in a joint effort to reduce emissions and improve the environment.

Notes

1   The figures for China's carbon dioxide emissions are disputable. Apart from official Chinese denial of the figures estimated and published by Western scholars, there is a problem of emissions transfer—that is, in the past two decades, an increasing amount of the polluted, energy-consuming and emission-producing products have been manufactured in China, but consumption has taken place mainly outside China.
2   In his discussion at a workshop, Assar Lindbeck (2008) raised the issue of government incentives for emissions reduction as a sacrifice of growth. His concern was that the central government's imposition of emissions duty on local governments could result in strong resistance from the latter and countermeasures against it, leading to policy failure.
3   Auffhammer and Carson (2008) forecast a likely scenario of a dramatic increase in carbon dioxide emissions in China in the future. As a consequence, China is expected to take the lead in global greenhouse gas emissions.

References

Bhagwati, J.N., 1996. The miracle that did happen: understanding East Asia in comparative perspective, Keynote speech at the Government and Market: The relevance of the Taiwanese performance to development theory and policy conference in honour of Professors Liu and Tsiang, Cornell University, 3 May.

Cai, F., 2008. *Approaching a triumphal span: how far is China towards its Lewisian turning point?*, UNU-WIDER Research Paper 2008/09, Helsinki.

Cai, F. and Wang, D., 2005. 'China's demographic transition: implications for growth', in R. Garnaut and L. Song (eds), *The China Boom and Its Discontents*, Asia Pacific Press and ANU E Press, The Australian National University, Canberra:34–52.

Deacon, R.T., 2005. *Dictatorship, Democracy and the Provision of Public Goods*. Available from http://www.econ.ucsb.edu/~deacon/DictDem11_05X.pdf

Deacon, R.T., 2005. *Dictatorship, Democracy and the Provision of Public Goods*. Available from http://www.econ.ucsb.edu/~deacon/DictDem11_05X.pdf.

Grossman, G. and Krueger, A., 1995. 'Economic growth and environment', *Quarterly Journal of Economics*, 110(2):353–77.

Hayward, S., 2005. 'The China syndrome and the environmental Kuznets curve', *Environmental Policy Outlook AEI Online*. Available from http://www.aei.org/publications/pubID.23617/pub_detail.asp.

Kaneko, S. and Managi, S., 2004. 'Environmental productivity in China', *Economics Bulletin*, 17(2):1–10.

Kim, S.-y. and Kuijs, L., 2007. *Raw material prices, wages, and profitability in China's industry—how was profitability maintained when input prices and wages increase so fast?*, World Bank China Research Paper 8, World Bank, Beijing.

Krugman, P., 1994. 'The myth of Asia's miracle', *Foreign Affairs*, November–December.

Kuijs, L. and Wang, T., 2005. *China's pattern of growth: moving to sustainability and reducing inequality*, World Bank China Research Paper, 2, October.

Kuznets, S., 1955. 'Economic growth and income inequality', *American Economic Review*, 45(1):1–28.

Lindbeck, A., 2008. Workshop presentation to China Economics of Climate Change—Towards a Low-Carbon Economy, Inception Meeting, St Brännbo, Stockholm, 14–15 February.

National Bureau of Statistics (NBS), various years. *China Statistical Yearbook*, China Statistics Press, Beijing

Stern, N., 2007. *The Economics of Climate Change: the Stern review*, Cambridge University Press, Cambridge.

Thomas, M., 2007. *Climate Change and the Stern Review: an overview and comment from Future in Our Hands Network*. Available from http://www.climatecooperation.org/index.php?title=Stern_Review/Mike_Thomas_2

Webber, D.J. and Allen, D.O., 2004. *Environmental Kuznets curves: mess or meaning?*, School of Economics Discussion Papers, (0406), University of the West of England. Available from http://ideas.repec.org/s/uwe/wpaper.html

World Bank, 1997. *Clear Water, Blue Skies: China's environment in the new century*, World Bank, Washington, DC.

World Bank, various years. *World Development Indicators*. World Bank, Washington, DC.

Zheng, J. and Hu, A., 2004. *An empirical analysis on inter-provincial productivity growth during the reform period*, Center for China Situation Studies Working Paper, (1):2–26.

Acknowledgements

The authors thank Wang Meiyan and Zhang Binbin for their excellent assistance in data processing.

11

The environmental consequences of foreign direct investment in China

Qun Bao, Yuanyuan Chen and Ligang Song

China has been the largest recipient of foreign direct investment (FDI) in the developing world since 1990, and there has been a profound change in China's foreign investment policy in the past decade. It is acknowledged generally that FDI has played a significant role in promoting China's rapid economic growth through meeting the gaps of capital shortage, pushing technology spill-over towards local firms and improving the degree of China's economic openness (Cheung and Lin 2004; Yao 2006). Foreign direct investment has, however, also had some negative impacts on China's economy, which have increasingly aroused concern. Among them, the environmental consequences of the entry of foreign firms have attracted much attention: FDI could be a contributing factor to the serious environmental damage that has accompanied China's rapid economic growth. Questions can be asked about whether further increases in FDI will intensify pressure for environmental protection in China. While foreign investments in pollution-intensive industries, including mining, have accounted for a large share of China's total FDI,[1] does foreign investment cause more pollution emissions and therefore environmental damage, as is the fear of most people in China?

One of the most popular viewpoints about the effect of economic openness on environmental quality is the 'pollution haven hypothesis' (PHH), which supposes that developing countries in general have comparative advantages in polluting sectors due to their relatively lax environmental regulation. As a result, by relocating pollution-intensive industries from richer countries to poor countries through either international trade or FDI, multinational firms can achieve lower production costs in the host country. A corollary is the 'race to the bottom' phenomenon—that is, in order to attract more foreign investment, developing countries might deliberately undervalue their environmental damage, and lower their environmental standards. A number of empirical studies have examined the relationship between foreign investment and local environmental pollution in host countries; however, their estimation results are generally mixed. Two possible reasons could explain why tests on the pollution haven hypotheses have so far not been able to provide conclusive results. The first is that econometric results could suffer from omitted variables and model specification as well as from the pollution-emission indicators the authors have chosen. Second, while foreign firms might cause increased pollution emissions in host countries, they also help to reduce local pollution emissions by adopting cleaner technology and through the productivity spill-over effect. As a result, the effect of foreign investment on local environmental quality could be ambiguous, which implies that the relationship between FDI and pollution emissions could be non-linear.

Since a linear relationship has generally been assumed in most previous studies, few researchers have examined the possible non-linear impact of foreign investment on local environmental pollution. In this chapter, we attempt to check empirically whether there is such a non-linear effect of FDI on pollution in host countries, by using China's panel data of provinces. The remainder of this chapter is organised as follows: the next section provides a review of the literature and the motivation for our research. Section three introduces the model specification and data descriptions. Section four presents our empirical results, and section five provides the conclusion.

Literature review

A number of econometric studies have examined whether differences in environmental standards among host countries can act as determinants of the location of foreign investment, and three approaches have been used (Dean et al. 2004).[2] The first approach is a cross-country data study. For example, in their study on the location choice of multinational firms across Eastern Europe and the former Soviet Union, Smarzynska and Wei (2001) consider the roles played by corruption and environmental regulation, and their estimation results do not support the idea that lower environmental standards lead to increased foreign investment. The second approach is an inter-state data study, which investigates mainly whether environmental stringency affects the location decisions of firms across the United States (for example, Levinson 1996; Keller and Levinson 2002; List et al. 2004; Henderson and Millimet 2007). The third approach is an inter-industry data study. For example, Eskeland and Harrison (2003) examine the location decisions of foreign firms across different industries within countries, including Mexico and Morocco. Similar to Smarzynska and Wei (2001), the authors find that abatement costs are not a significant variable in determining the location of foreign firms among manufacturing industries within one country. Additionally, Eskeland and Harrison (2003) found that foreign ownership was related significantly to lower pollution intensity.

What should be mentioned is that empirical work designed to test the pollution haven hypothesis has so far not been able to provide conclusive results,[3] which could be explained by two factors. First, the econometric results could suffer from the problems of omitted variables, model specification and measurement errors (Letchumanan and Kodama 2000). As well as these problems, the estimated effects of foreign investment on environmental quality in a host country are likely to be influenced by the pollutant indicator proxies adopted. Second, it has also been proposed that foreign investment could help improve local environmental quality in host countries. For example, since multinational firms apply a universal environmental standard, they tend to diffuse their clean technology among their local counterparts in the host country, which generates the 'pollution halo hypothesis' (Birdsall and Wheeler 1993; Chudnovsky and Lopez 1999). Additionally, foreign investment could cause technological spill-overs to the local firms, crowding out inefficient local firms and improving the efficiency of energy and resource usage, which helps to decrease local pollution emissions (Wayne and Shadbegian 2002; Liang 2006).

The impacts of the entry of foreign firms on China's environmental quality and pollution emissions have also been studied. For example, by using the data from about 1,000 industrial firms in three Chinese provinces, Wang and Jin (2002) explore the differences in the pollution control performance of industries with different types of ownership. They find that foreign-owned and collectively owned enterprises have the best environmental performance in terms of water pollution discharge intensity, while state-owned enterprises and privately owned enterprises in China are the worst performers. To explain their findings, Wang and Jin suggest that foreign-owned enterprises might adopt more environmentally friendly technology in their production than other firms. Similarly, based on data from China's 260 cities, Liang (2006) also shows that the entry of foreign investment doesn't harm environmental quality in China, since a negative relationship is found between FDI and air pollution. Specifically, a 1 per cent increase in FDI will decrease sulphur dioxide emission intensity by 0.6–0.7 per cent, according to Liang (2006). A simultaneous estimation technique has been applied in He (2002) to decompose the impact of FDI on sulphur dioxide emissions into the scale, technique and composition effects. By using China's city-level panel data from 1993–99, He (2002) reached a similar conclusion: that the entry of FDI helped reduce sulphur dioxide emissions. Additionally, others considered the role of industrial composition and sources of foreign investment. For example, Dean et al. (2004) find that the environmental consequence of FDI varies with its sources. While FDI originating from Hong Kong, Taiwan and Macao is attracted to provinces with weak environmental controls, FDI from non-Chinese sources is attracted to areas with higher levels of skilled labour and high pollution levies.

Table 11.1 Selected literature on the relationship between FDI and pollution

Authors	Sample data	Pollutant proxies	Conclusions
Eskeland and Harrison (2003)	US outward investment in four countries, including Mexico and Morocco	Various proxies, including air and water pollution	+ (compared with local firms)
Smarzynska and Wei (2001)	Foreign investments across 25 transitional economies	Inventory of toxins released	–
Wang and Jin (2002)	Data from more than 1,000 firms in China	SO2, suspended particulate matter	+ (compared with state-owned enterprises and private enterprises)
Dean et al. (2004)	2,886 manufacturing joint-venture projects during 1993–96 in China	Water pollution	Foreign investment sources do matter
He (2002)	China's city-level panel data from 1993–99	SO2	+
Liang (2006)	Data from China's 260 cities	SO2	+

Note: The '+' sign denotes that foreign investments helped improve local environmental quality in host countries.
Source: Authors' own summary.

Since a linear relationship between FDI and pollution emissions has generally been assumed in these studies, few have considered a non-linear possibility—as far as we know. Could the environmental impact of FDI be non-linear? There are two main motivations for our research. First, the logic of the non-linear FDI–pollution nexus is very similar to the well-known environmental Kuznets curve (EKC) hypothesis, which states that the relationship between economic growth and pollution emissions could follow an inverted-U curve— namely, environmental degradation worsens in the early stages of growth, but eventually reaches a peak and starts declining as incomes exceed a certain threshold.[4] The logic of an EKC relationship is intuitively appealing. In the first stage of industrialisation, pollution grows rapidly because high priority is given to increasing material output, and people are more interested in increasing jobs and incomes than having clean air and water (Dasgupta et al. 2002). The rapid growth results inevitably in greater use of natural resources and increased emissions of pollutants, which in turn put more pressure on the environment. People are too poor to pay for abatement and disregard the environmental consequences of growth. In the late stage of industrialisation, as incomes rise, people value the environment more, the regulatory institutions become more effective and, as a consequence, pollution levels decline. The EKC hypothesis therefore posits a well-defined relationship between the level of economic activity and environmental pressure (defined as the level of concentration of pollution or flow of emissions, depletion of resources and so on). Such an inverted-U curve relationship could naturally extend to the FDI–pollution nexus: in the initial stage of FDI utilisation, the entry of multinational firms leads to a higher scale of production activities in host countries. Meanwhile, people in the poor host country have low demands for a clean environment, as the pollution damage is trivial, so fewer resources will be allocated to abate environmental pollution. As a result, foreign investment causes pollution emissions in host countries to rise; however, as the income levels of the host country's residents increase steadily with the accumulation of foreign investment, people become less tolerant of declining environmental quality, propelling the government to implement more stringent environmental regulations, and, therefore, more pollution abatement endeavours will be undertaken. Meanwhile, as foreign firms adopt more environmentally friendly techniques, the entry of FDI can cause a spill-over effect of clean technology to local firms, which eventually affects the industrial structure in host countries. As a result, once a certain threshold has been passed, foreign investment will be beneficial to local environmental quality.[5]

Second, concerning the relationship between FDI and pollution in China, it is hard to provide strong evidence to support the linear assumption. Although it has been found (Wang and Jin 2002; He 2002; Liang 2006) that foreign investment generally helps to reduce China's pollution emissions, the environmental effect of FDI varies significantly between different regions in China, as shown in Table 11.2. While the eastern region has accounted for a much larger share of FDI than other regions in the past due to its geographical location and its infrastructure facility advantage, provinces in the eastern region have not performed much better than those in other regions in terms of environmental protection. For example, Guangdong ranks the highest in terms of attracting FDI, with the largest average share of 27.5 per cent. Its pollution behaviour is, however, very similar to Hubei, Hunan and Guizhou, all of which are in central or western China. Emission levels of four pollution indicators in Guangdong, including 'water', 'cod' (chemical oxygen demand), 'smoke' and 'solid', have a mean value similar to those in Hubei; however, the mean value of FDI in Hubei is less than one-tenth of that in Guangdong. Additionally, even within the eastern region, the performance of individual provinces varies. For instance, during 1992–2004, the emission levels of three pollutants, excluding cod, showed an increasing trend in Guangdong, which implies that it emits more pollution to the local environment than other eastern provinces. In comparison, Shanghai behaves much better than Guangdong, as its emission levels of the four pollutants all decline, except solid. Such casual observation demonstrates that, in terms of local environmental quality, while some provinces benefit from the entry of foreign firms, others lose. In other words, the potential relationship between FDI and the local environment could be non-linear rather than the linear one assumed previously.

Model specification and data description

To examine the non-linear impact of foreign investment on China's pollution emissions, we use China's panel data from 29 provinces during 1992–2004. Tibet is excluded from the sample due to the unavailability of data for certain indicators. Chongqing became a municipality only in 1997, so we integrate its data into that for Sichuan Province as a whole. Our estimation model is set as follows.

$$P_{it} = c_i + \gamma_t + \alpha_1 fdi_{it} + \alpha_2 fdi_{it}^2 + \alpha_3 scale_{it} + \alpha_4 comp_{it} + \alpha_5 tech_{it} + \beta CV_{it} + \varepsilon_{it} \quad (1)$$

in which P_{it} denotes various pollutant emissions in province i for year t; fdi_{it} denotes FDI in province i for year t; c_i and γ_t denote regional and time-specific

Table 11.2   **FDI and pollution emissions among different provinces**

	FDI US$100 million	cod (10 million kg)	smoke (10 million kg)	gas (10 million kg)	water (100 billion kg)	solid (100 billion kg)
Guangdong	113.76	29.51 (-12)	23.63 (2.4)	72.96 (61.51)	12.89 (2.27)	0.17 (0.08)
Jiangsu	61.73	35.01 (-13.43)	46.03 (-14)	112.92 (-1.46)	22.38 (3.86)	0.31 (0.22)
Shanghai	37.77	10.02 (-15.84)	10.24 (-9.56)	38.82 (-16.41)	9.29 (-8.06)	0.13 (0.06)
Shandong	34.73	68.79 (-28.03)	55.13 (-6.6)	174.73 (-71.28)	10.42 (4.22)	0.53 (0.39)
Fujian	34.25	14.67 (-9.81)	8.24 (3.00)	18.55 (16.93)	6.79 (5.18)	0.19 (0.26)
Liaoning	22.44	29.66 (-23.82)	56.54 (-33.1)	81.38 (-42.48)	11.96 (-6.12)	0.75 (0.08)
Zhejiang	20.86	29.12 (-6.55)	20.27 (2.80)	56.80 (27.71)	12.45 (4.86)	0.13 (0.13)
Beijing	16.21	4.79 (-9.07)	7.65 (-7.7)	22.16 (-24.38)	2.90 (-2.70)	0.11 (0.04)
Tianjin	15.70	6.29 (-5.74)	8.12 (-0.40)	22.18 (-2.88)	2.04 (0.15)	0.05 (0.03)
Hubei	10.72	27.77 (-19.8)	24.86 (4.40)	49.38 (11.95)	11.95 (4.71)	0.24 (0.13)
Hebei	7.76	36.18 (11.78)	54.12 (8.60)	109.27 (28.56)	9.53 (3.73)	0.77 (1.11)
Hunan	7.47	31.43 (-12.4)	33.97 (17.5)	57.67 (21.86)	13.41 (-5.52)	0.22 (0.12)
Hainan	6.27	3.30 (-5.32)	1.39 (-0.10)	2.21 (-0.26)	0.76 (-0.34)	0.01 (-0.00)
Jiangxi	6.11	11.95 (-13.11)	21.72 (-8.50)	31.86 (15.23)	5.51 (-1.97)	0.44 (0.31)
Guangxi	6.10	56.85 (27.14)	36.82 (25.00)	66.91 (28.02)	9.37 (2.89)	0.20 (0.19)
Henan	5.02	45.06 (-5.76)	52.33 (27.70)	71.50 (53.55)	10.10 (2.23)	0.33 (0.29)
Sichuan	4.82	49.32 (7.29)	73.23 (20.50)	153.40 (6.25)	17.90 (0.32)	0.49 (0.31)
Heilongjiang	4.14	19.56 (-15.23)	43.63 (-11.9)	26.41 (-2.48)	6.04 (-3.22)	0.31 (-0.06)
Anhui	3.69	22.15 (-25.8)	23.74 (-3.5)	38.74 (1.82)	7.30 (-3.32)	0.29 (0.12)
Jilin	3.16	20.69 (-14.13)	29.63 (-3.3)	22.32 (-1.87)	3.94 (-1.18)	0.16 (0.05)
Shaanxi	3.03	11.03 (6.68)	35.21 (-2.9)	63.57 (3.79)	3.40 (-0.14)	0.23 (0.22)
Shanxi	1.82	16.98 (-21.67)	64.72 (49.4)	94.27 (35.05)	3.76 (-0.91)	0.61 (0.62)
Neimenggu	1.26	17.04 (5.24)	40.40 (-48.40)	66.78 (29.61)	2.46 (-0.41)	0.27 (0.26)
Yunan	1.06	19.56 (-7.35)	17.93 (-1.20)	31.31 (15.07)	3.97 (-0.56)	0.26 (0.22)
Xinjiang	0.75	12.46 (5.83)	12.80 (0.10)	23.01 (7.95)	1.73 (-0.09)	0.07 (0.07)
Guizhou	0.69	5.26 (-2.99)	27.37 (-5.70)	66.04 (-11.50)	2.40 (-1.28)	0.22 (0.32)
Gansu	0.61	5.06 (-2.40)	14.57 (-1.30)	36.21 (8.35)	2.98 (-1.92)	0.15 (0.09)
Qinghai	0.52	0.40 (-0.34)	4.60 (2.10)	2.87 (4.30)	0.45 (-0.18)	0.03 (0.02)
Ningxia	0.25	6.57 (2.33)	9.93 (-2.40)	20.63 (4.23)	0.92 (0.16)	0.04 (0.02)

Note: We list the mean value of each indicator during 1992–2004, and the values in parentheses are the emission levels in 2004 minus those in 1992 for each pollutant—measuring the change in pollution emissions. water is industrial polluted water; solid is industrial solid wastes; cod is chemical oxygen demand in industrial water pollution; gas is sulphur dioxide emissions; smoke is industrial smoke.
Source: Authors' own calculations, using pollution emissions data from various issues of State Environmental Protection Administration (SEPA), various years. China Environment Yearbook, China Environment Yearbook Press; and FDI data from National Bureau of Statistics (NBS), various years. China Statistical Yearbook, China Statistics Press, Beijing.

effects respectively; ε_{it} is a normally distributed error term. What should be mentioned here is that we include fdi_{it} and fdi_{it}^2 in our estimation model: if the coefficient of fdi_{it}^2 is estimated as zero, $\alpha_2 = 0$, then only a linear relationship exists between foreign investment and pollution emissions. If $\alpha_2 < 0$, it implies that there is an inverted-U curve relationship between fdi and pollution emissions, while a positive α_2 indicates a U-shape relationship. In the meantime, we can further calculate the turning-point along the inverted-U curve by taking the first-order differentiation of the empirical model as:

$$fdi^* = -\frac{\alpha_1}{2\alpha_2} \qquad (2)$$

Similar to Grossman and Krueger (1995) and Antweiler et al. (2001), the indicators *scale*, *comp* and *tech* measure the scale effect, composition effect and technique effect respectively, while *CV* represents other variables that could affect the level of pollution emissions.

We choose the indicators and provide our data description as follows.

Five pollutant indicators are chosen in this study to measure different pollution emissions—two water pollutants (industrial polluted water emissions and chemical oxygen demand in industrial water pollution), two air pollutants (industrial smoke emissions and sulphur dioxide emissions) and one solid pollutant (industrial solid wastes)—from 1992 to 2004 in 29 of China's provinces and municipalities. Table 11.3 lists the description of all the pollutant variables and our data sources. We chose five different pollutant indicators because we could not compare only the impacts of foreign investment on different

Table 11.3   Five indicators of pollution emissions in China

Pollutant	Unit	Symbol
1 Industrial polluted water emissions	100 billion kg	*water*
2 Chemical oxygen demand in industrial water pollution	10 million kg	*cod*
3 Sulphur dioxide emissions	10 million kg	*gas*
4 Industrial smoke emissions	10 million kg	*smoke*
5 Industrial solid wastes	100 billion kg	*solid*

Sources: State Environmental Protection Administration (SEPA), 1993–2005. *China Environment Yearbook*, China Environment Yearbook Press, Beijing.

pollutants; rather, we needed to also ensure the robustness of our estimated inverted-U curve. Our original pollutant data from 1992–97 were collected by dividing all industries into 18 sectors, while the data for 1998–2004 were calculated based on 43 industrial sectors, due to changes in environmental pollution measurement methods in China.

It was Grossman and Krueger (1995) who first introduced three effects on environmental quality change. The first is the scale effect, in which a larger economic scale means more production activity and therefore higher requirements for natural resources, resulting in more pollution emissions. The second is the composition effect, which reflects the impact of adjustments in industrial composition and changes in factor-input combinations on environmental pollution. The last is the technique effect, in which technological progress and the use of environmentally friendly technology helps to decrease pollution emissions.

Like Copeland and Taylor (2003), our empirical model also considers the three effects on pollution emissions. The scale effect is measured using regional gross domestic product (gdp_{it}), which is the real gross domestic product (GDP) after removing the effect of inflation. The composition effect is identified as the combination of factor inputs—namely, physical capital per capita, which is the ratio of physical capital (k_{it}) to labour inputs (l_{it}). It is found that capital-intensive industries can generally cause more pollution emissions than other industries; therefore, the coefficient of physical capital per capita is expected to be positive. Zhang et al. (2004) estimated the physical capital stock for China's 30 provinces by using a long period of sample data from 1952 to 2000; here we follow their method in obtaining the capital stock data. The role of human capital accumulation is also considered. As in Barro and Lee (2001), we use the average educational attainment to measure the level of human capital stock among different regions, which is the ratio of total educational attainment to the total population. Specifically, the educational attainment is specified as six years for primary school graduates, and nine, 12 and 16 years for junior middle school graduates, senior middle school graduates and university graduates respectively. The unit of human capital stock (hc) is therefore educational years per capita, and we expect its coefficient to be negative, as higher human capital accumulation attracts cleaner industries to invest. Finally, similar to Antweiler et al. (2001), we use the lagged term of China's gross national product (GNP) per capita to capture the role of the technique effect, since it is acknowledged widely that a higher GNP per capita reflects a higher level of technology use. Due to data availability, here we use GDP per capita as a proxy for the technique effect.

A number of pollution control variables are also included. The first is population density ($density_{it}$), which is the population per square kilometre; and, as Antweiler et al. (2001) point out, its coefficient could be negative since a higher population density causes greater marginal pollution damage. The second is environmentally related research and development expenditure (rd_{it}), which is used to measure pollution abatement endeavours, and the ratio of rd_{it} to GDP is used. Another indicator is the number of environmentally related institutions ($agency_{it}$). We expect that the coefficients of rd_{it} and $agency_{it}$ will be negative, to reflect the role of environmental regulation and pollution abatement efforts in reducing industrial emissions.

The data for the five pollutant emissions, as well as the other two control variables, rd_{it} and $agency_{it}$, are collected from various issues of the *China Environment Yearbook* (SEPA, various years). Data for other indicators were collected from the *China Statistical Yearbook* (NBS various years). Except for rd_{it}, the logarithm value of all other variables is used. For our panel data estimation method, the usual Hausman test was employed to choose fixed-effect (FE) or random-effect (RE) models. If the value of the Hausman test statistic is larger than the critical value, the null hypothesis can be rejected and fixed effects are preferred to random effects. The White cross-section method is derived by treating the panel regression as a multi-variance regression (with an equation for each cross-section), and by computing White-type robust standard errors for the system of equations. This estimator is robust for cross-equation (contemporaneous) correlations as well as different error variances in each cross-section (Wooldridge 2002). The basic statistical information is shown in Table 11.4.[6] It can be seen that significant regional disparity exists in terms not only of various pollution emissions, but in attracting foreign investment. We therefore estimate the inverted-U cure not only for our national sample data, but for controlling different regions in the next section.

Estimation results

The total sample estimation result

We first estimate the impact of foreign investment on China's pollution emissions for the total sample of 29 provinces and the results are shown in Table 11.5.

Among the five pollutant indicators we have chosen, the inverted-U curve relationship between FDI and pollution emissions is generally supported. The coefficients of fdi_{it}^2 are estimated to be significantly negative, while the coefficients of fdi_{it} are positive. Such results demonstrate that foreign investment has a non-linear impact on environmental pollution, rather than the linear one assumed in most studies. We can further calculate the threshold

Table 11.4   **Basic statistical information on pollution and FDI**

	Total sample	Coastal region	Non-coastal region
water	7.21 (5.57)	9.24 (5.91)	5.78 (4.85)
cod	22.29 (18.81)	27.02 (22.45)	18.96 (14.91)
gas	56.38 (43.28)	64.91 (50.18)	50.36 (36.58)
smoke	29.62 (22.04)	27.35 (21.64)	31.22 (22.21)
solid	0.27 (0.22)	0.28 (0.27)	0.26 (0.17)
FDI (US$100 million)	14.92 (23.91)	31.47 (30.55)	3.24 (2.91)

Note: We list the mean value for each indictor, and the values in parentheses are the standard deviations.
Source: Authors' own calculations.

Table 11.5   **Total sample estimation results**

	water	*cod*	*gas*	*smoke*	*solid*
fdi_{it}	0.538***	0.511**	0.192***	0.319***	0.098**
	(6.983)	(4.575)	(2.691)	(3.539)	(2.484)
fdi_{it}^2	-0.025***	-0.025***	-0.009**	-0.018***	-0.005***
	(-7.340)	(-5.248)	(-2.481)	(-3.742)	(-2.297)
gdp_{it}	0.450***	0.816***	0.538***	-0.819***	0.408***
	(9.261)	(12.46)	(3.366)	(-4.059)	(4.426)
$(k/l)_{it}$	-0.792***	-0.367***	0.023	0.826***	0.221***
	(-16.01)	(-6.169)	(0.214)	(6.129)	(3.911)
hc_{it}	-0.011	-0.061*	0.042	0.001	0.079***
	(-0.352)	(-1.823)	(0.908)	(0.016)	(3.172)
$tech_{it}$	0.509***	-0.233***	-0.295***	0.004	-0.148***
	(11.96)	(-3.622)	(-4.323)	(0.051)	(-3.471)
$density_{it}$	0.277***	0.157***	-2.257***	-2.007***	-0.881***
	(9.009)	(4.248)	(-4.221)	(-2.973)	(-2.608)
rd_{it}	-0.087**	-0.406*	-0.250***	-0.224**	0.003
	(-2.521)	(-2.984)	(-3.072)	(-2.186)	(0.059)
$agency_{it}$	0.194***	0.054	0.061	-0.059	-0.011
	(4.177)	(0.984)	(0.826)	(-0.643)	(-0.262)
$adj-R^2$	0.876	0.731	0.958	0.923	0.964
χ^2	216.58	253.05	188.74	207.60	192.23
Hausman	31.57	33.97	40.45	38.86	44.29
observations	377	377	377	377	377

*** statistically significant at 1 per cent level
** statistically significant at 5 per cent level
* statistically significant at 10 per cent level
Note: *t*-statistic values are in parentheses; χ^2–*statistic* is used to test whether the specific cross-section and period effects are both significant at the same time; and *Hausman–test* is used to specify the estimation panel model.
Source: Authors' own estimations.

value of foreign investment according to the estimation results. Let us take industrial polluted water emissions as an example. Table 11.5 shows that the coefficients of fdi_{it}^2 and fdi_{it} are -0.025 and 0.538 respectively, which implies that the turning-point along its inverted-U curve is fdi^*=7.68, meaning that if the real foreign investment for a certain region is lower than US$768 million, increases in foreign investment will cause local pollution emissions to rise. On the other hand, if the real foreign investment is larger than this threshold value, environmental pollution will fall as new foreign firms enter the area. Comparing the real foreign investment for China's 29 provinces with such threshold value fdi^*, we can reach a rough conclusion about the impact of foreign investment on local industrial polluted water emissions. The mean value of foreign investment for 18 provinces is lower than US$768 million, among which the foreign investment in Hunan Province is US$747 million, which is very near the turning-point along the inverted-U curve. It demonstrates that foreign investment generally causes local pollution emissions in the 18 provinces to go up, and therefore has a negative effect on local environmental quality. For the remaining 11 provinces, industrial polluted water emissions will decline with the entry of foreign investors; foreign investment in Hubei (US$776 million) is a little larger than the threshold value.

We can further observe that even though an inverted-U curve is generally estimated, the shape of inverted-U curves for different pollutant indicators varies significantly, especially for their turning-points (fdi^*).[7] Let us compare the two water pollutants, *water* and *cod*. For the estimation result for *cod*, the coefficients of fdi_{it}^2 and fdi_{it} are -0.025 and 0.511 respectively, and we can calculate the turning-point along its inverted-U curve, which is fdi^*=US$334 million—much lower than the threshold value in the industrial polluted water emission estimation. For the other three indicators, we obtain their threshold values as US$669 million (*gas*), US$122 million (*smoke*) and US$2.683 billion (*solid*). The threshold values for different pollutant indicators determine how foreign investment affects local environmental pollution among different regions. For example, since the turning-point for the estimation of industrial solid wastes is as high as US$2.683 billion, only five provinces have a larger mean value of foreign investment than the threshold value: Guangdong (US$11.376 billion), Jiangsu (US$6.173 billion), Shanghai (US$3.778 billion), Shandong (US$3.473 billion) and Fujian (US$3.425 billion). These five provinces are in coastal eastern China. In comparison, as the turning-point for industrial smoke emissions is only US$122 million, the value of foreign investment for most provinces is larger, implying that although an inverted-U curve relationship has been estimated, most provinces have been in the right part along the inverted-U curve. Table 11.6 compares our estimated threshold value of FDI for five pollutants with the mean of real utilised FDI among 29 provinces. It is easy to see that five provinces—Guangdong, Jiangsu, Shanghai, Shandong and Fujian—are located in the right part of the inverted-U curve for all five pollutants, while seven provinces from the western region are still located in the left part of the curve, which implies that more foreign investment in those provinces could cause pollution emissions to rise.

The effects of other pollution control variables are also shown in Table 11.5. According to Antweiler et al. (2003), scale effect and composition effect can cause pollution emissions to rise, while the technique effect leads pollution emissions to fall. Our estimations, however, demonstrate that it is not straightforward to obtain a clear-cut conclusion concerning the three effects, since the estimated coefficients of the three effects vary significantly for different pollutant indicators (Table 11.7). For instance, since capital-intensive goods can cause a higher level of pollution emissions, the coefficient of physical capital per capita is supposed to be positive, which implies that the increase in the capital/labour ratio will cause more pollution emissions. According to Table 11.5, however, while the coefficients of *gas*, *smoke* and *solid* are estimated to be positive, they are significantly negative for the other two indicators, *water* and *cod*. For the other control variables, environmentally related research and development expenditure shows a significantly negative effect on pollution

---

Table 11.6　**The threshold effect of FDI for five pollutants** (US$ billion)

---

solid (2.683)
Guangdong (11.376), Jiangsu (6.173), Shanghai (3.778), Shandong (3.473), Fujian (3.425)
water (0.768)
Hunan (0.747)
gas (0.669)
Hainan (0.627), Jiangxi (0.612), Guangxi (0.610), Henan (0.503), Sichuan (0.482), Heilongjiang (0.414), Anhui (0.369)
cod (0.334)
Jilin (0.316), Shaanxi (0.303), Shanxi (0.182)
smoke (0.122)
Neimenggu (0.121), Yunan (0.107), Xinjiang (0.075), Guizhou (0.070), Gansu (0.061), Qinghai (0.053), Ningxia (0.026)

Note: The numbers in parentheses refer to our calculated threshold values of FDI for different pollutants, or the mean of utilised FDI among different provinces during the sample period.
Source: Authors' own calculations.

emissions, which verifies the role of the adoption of environmentally friendly new technology and technological progress. The effects are, however, estimated to be ambiguous for the two indicators—namely, the number of environmentally related institutions and population density.

Sub-sample regression results

As mentioned already, since significant regional development disparity exists among different regions in China, we further estimate the impacts of foreign investment on local pollution emissions by dividing our total sample into two sub-samples: the coastal eastern region, and the inland region. By doing this, we can compare not only the effects of foreign investment on local environmental pollution for different regions, we can examine whether the inverted-U curve still holds robustly for our sub-sample estimation.

The estimation results for the coastal eastern regions are shown in Table 11.8. First, similar to our total sample estimation results, the coefficient of fdi_{it}^2 is estimated to be significantly negative, which means that the inverted-U curve relationship between FDI and pollution emissions holds true if considering only the coastal eastern regions in the model estimation. Meanwhile, the three effects (scale, composition and technique) are also estimated to be ambiguous. Take the composition effect measured by the ratio of physical capital to labour as an example. Although its coefficient is negative for the five pollutant indicators, it doesn't show statistically significant estimates for industrial smoke pollution and solid wastes. Furthermore, it is found that the negative composition effect is larger for *cod* than for industrial polluted water emissions and *gas*. Finally, the role of the other control variables also differs from that in the total sample estimation. Take environmentally related research and development expenditure as an example. Similar to the results reported in Table 11.5, the coefficient of *rd* is significantly negative, as reported in Table 11.8. This finding supports the role of technological progress in reducing pollution emissions. It is also found, however, that the estimated role of research and development expenditure is generally larger than that obtained using the total sample. Since most research and development activities are taking place in the coastal eastern regions, we can conclude that research and development activities play a more significant role in improving environmental quality in coastal regions than in inland regions.

We also estimate whether the effects of foreign investment on local pollution emissions for the 17 inland provinces are non-linear, and the results are shown in Table 11.9. As for the relationship between foreign investment and pollution emissions, the inverted-U curve still holds true for two indicators: industrial

Table 11.7 Estimation results of the three effects

	water	cod	gas	smoke	solid
Scale effect	+***	+***	+***	+***	+***
Composition effect	–***	+***	+	+***	+***
Technique effect	+***	–***	–***	+	–***

Note: The composition effect is measured as the ratio of physical capital to labour. ***, **, * mean that the estimated co-efficient of various effects is statistically significant at 1, 5 and 10 per cent level respectively.
Source: Authors' own summary, based on the estimation results reported in Table 11.5.

Table 11.8 Estimation results for coastal eastern regions

	water	cod	gas	smoke	solid
fdi_{it}	1.264**	0.694***	0.717**	0.998***	0.326*
	(2.054)	(2.678)	(2.492)	(3.562)	(1.719)
fdi_{it}^2	-0.054**	-0.037***	-0.033***	-0.052***	-0.016*
	(-2.209)	(-3.695)	(-2.681)	(-4.471)	(-1.670)
gdp_{it}	-0.306	0.613	0.398*	-0.050	0.831***
	(-1.174)	(3.896)	(1.924)	(-0.204)	(7.516)
$(k/l)_{it}$	-0.097**	-0.174**	-0.093***	-0.032	-0.003
	(-1.997)	(-2.401)	(-2.749)	(-0.926)	(0.828)
hc_{it}	0.029	-0.142*	-0.005	0.056	0.103***
	(0.314)	(-1.782)	(-0.089)	(0.786)	(2.926)
$tech_{it}$	0.267**	-0.512**	-0.165*	0.279*	-0.148***
	(2.144)	(-3.718)	(-1.843)	(1.947)	(-2.884)
$density_{it}$	-6.289***	0.385***	-1.242***	-3.770***	-2.052***
	(-7.204)	(3.363)	(-2.947)	(-5.313)	(-5.280)
rd_{it}	-0.313**	-0.504**	-0.270***	-0.493***	-0.075
	(-2.032)	(-2.076)	(-2.801)	(-5.402)	(-0.883)
$agency_{it}$	0.454***	0.401***	0.201**	0.102	-0.007
	(3.092)	(2.675)	(1.993)	(1.381)	(-0.138)
χ^2	201.78	168.24	159.91	179.57	213.07
Hausman	36.76	27.40	31.19	40.77	35.54
$adj - R^2$	0.936	0.782	0.932	0.959	0.921
observations	156	156	156	156	156

*** statistically significant at 1 per cent level
** statistically significant at 5 per cent level
* statistically significant at 10 per cent level
Note: t-statistic values are in parentheses; χ^2 is used to test whether the specific cross-section and period effects are both significant at the same time; and *Hausman* is used to specify the estimation panel model.
Source: Authors' own estimations.

polluted water emissions and industrial smoke emissions. The estimation results demonstrate, however, a U-curve for the other three indicators. This finding could imply that in the initial stage, foreign investment helps to reduce environmental pollution and, once certain turning-points are passed, further increases in foreign investment cause a deterioration in local environmental quality. Furthermore, the role of other variables also varies in comparison with the results reported in Tables 5 and 8. Again, take the effect of environmentally related research and development expenditure as an example. It is found that more environmentally related research and development expenditure significantly decreases the level of pollution emissions; however, its coefficient is estimated to be insignificant for the inland provinces, while it is positive for the pollutant *cod*.

Comparisons of estimation results among different regions

The regional disparity on the FDI–pollution nexus can be determined by comparing the estimation results for different provinces, and the results are shown in Table 11.10.

By comparing the estimation results for the total sample and the coastal eastern regions, we find that although an inverted-U curve relationship is generally estimated to hold for all five pollutant indicators, the shape and especially the levels of the turning-point of different inverted-U curves for the two samples vary significantly. Generally speaking, although the threshold value of the industrial solid wastes curve is obviously larger in the total sample (US$2.683 billion) than in the coastal regions (US$1.598 billion), the threshold values in coastal regions are larger than in the total sample for the other four pollutant indicators. According to Bao et al. (2007), the threshold value of foreign investment is possibly determined by the following factors: the first is the industrial structure of foreign investment—that is, the sector into which foreign investment flows. Specifically, if more foreign firms enter the pollution-intensive sectors in the host country, the threshold value of foreign investment is expected to be higher; in other words, there is a positive relationship between the pollution intensity of foreign investment and the threshold value of fdi^*. The second factor is how FDI affects the marginal pollution damage in host countries, which depends essentially on the income levels of local residents. Generally speaking, a higher income level leads to a higher requirement for a clean environment, and hence the marginal damage caused by pollution emissions will be accordingly more severe. The third factor is the contribution of foreign investment to local economic development. Generally, with the accumulation of foreign investment, the marginal contribution of foreign firms to the local economy will fall accordingly. Therefore, to compare the threshold

Table 11.9   **Estimation results for inland provinces**

	water	cod	gas	smoke	solid
fdi_{it}	1.170***	-0.156*	-0.085*	0.247**	-0.087**
	(8.085)	(-1.633)	(-1.701)	(2.120)	(-1.949)
fdi_{it}^2	-0.056***	0.009*	0.008**	-0.012**	0.006**
	(-7.356)	(1.654)	(2.251)	(-1.980)	(2.331)
gdp_{it}	-0.014	-0.912***	0.874***	-0.056	0.410***
	(-0.150)	(-3.843)	(3.307)	(-0.576)	(5.619)
$(k/l)_{it}$	-0.160**	0.140***	-0.061**	-0.067	0.055***
	(-2.064)	(2.936)	(-2.127)	(-1.011)	(3.343)
hc_{it}	-0.073	-0.097	0.033	0.303***	-0.009
	(-1.321)	(-0.974)	(0.329)	(5.067)	(-0.266)
$tech_{it}$	-0.038	0.136	-0.505***	-0.499***	-0.124***
	(-0.605)	(1.364)	(-4.444)	(-9.239)	(-3.184)
$density_{it}$	0.253***	4.389***	-1.728**	-0.009	3.895***
	(5.934)	(2.534)	(-2.409)	(-0.215)	(5.591)
rd_{it}	-0.248	0.004	-0.080	-0.165	-0.010
	(-1.286)	(0.027)	(-0.744)	(-0.791)	(-0.139)
$agency_{it}$	0.742***	0.077	0.072	0.737***	-0.064
	(6.776)	(0.516)	(0.689)	(7.259)	(-0.943)
χ^2	239.03	188.37	173.69	214.06	226.51
Hausman	34.65	37.21	29.06	32.78	37.93
$adj-R^2$	0.781	0.901	0.908	0.916	0.928
observations	221	221	221	221	221

*** statistically significant at 1 per cent level
** statistically significant at 5 per cent level
* statistically significant at 10 per cent level
Note: t-statistic values are in parentheses; χ^2 is used to test whether the specific cross-section and period effects are both significant at the same time; and *Hausman* is used to specify the estimation panel model.
Source: Authors' own estimations.

values of foreign investments among different regions, we need to know how foreign investments affect local environmental pollution through the above three effects. Since the income level in coastal regions is generally much higher than in the inland provinces, it holds true that the marginal pollution damage must be larger for coastal provinces. Combining the other two effects, however, our estimation results show that the threshold values for coastal provinces are unexpectedly higher. Due to data limitation, we cannot measure the industrial composition effect of foreign investment without the industrial location data for foreign investment for different provinces, which could be a topic for future research.

Table 11.10 **Comparisons of the FDI–pollution nexus among different regions** (US$ million)

	water	cod	gas	smoke	solid
Total sample	Inverted-U	Inverted-U	Inverted-U	Inverted-U	Inverted-U
	(768)	(334)	(669)	(122)	(2,683)
Coastal regions	Inverted-U	Inverted-U	Inverted-U	Inverted-U	Inverted-U
	(1,211)	(3,966)	(1,105)	(181)	(1,598)
Inland regions	Inverted-U	U curve	U curve	Inverted-U	U curve
	(344)	(58)	(20)	(295)	(14)

Note: The figures in parentheses are the estimated threshold values for the turning-point for different shapes of the estimated curves.
Source: Authors' own summary.

We can also compare the estimation results for inland provinces with those obtained using the total sample. Specifically, although it is found that the threshold value for the inverted-U curve among inland provinces is generally smaller, most inland provinces are still located along the left side of the inverted U-curve—that is, foreign investment in those regions is still causing increased pollution emissions. Take industrial polluted water emissions as an example: the threshold value for the inland provinces is estimated to be US$344 million—much smaller than the value for our total sample (US$1.211 billion). We find, however, that 10 of 17 inland provinces have a mean value of foreign investment lower than the threshold value. This suggests that a significant regional disparity exists in terms of their ability to attract foreign investment. While the mean value of foreign investment for the coastal provinces is US$3.147 billion, the mean value for the 17 inland provinces is much smaller (US$324 million). Since most inland provinces are still in the initial stage of attracting foreign investment, a positive relationship generally occurs between foreign investment and local pollution emissions, suggesting that foreign investment has a negative effect on local environmental quality in those inland regions.

Conclusions

In this chapter, we investigate the possible non-linear impact of foreign investment on local pollution emissions in China. The hypothesis that an inverted-U curve relationship exists between foreign investment and pollution is supported empirically using the panel data for 29 Chinese provinces. The fact that an inverted U-curve exists between foreign investment and pollution emissions suggests that foreign investment helps, in general, to reduce pollution emissions in China. The empirical results also show that there are significant regional disparities with respect to the impact of foreign investment on local pollution levels. More specifically, those threshold values for most of the pollutants are much higher in some of the most developed coastal regions than in those relatively less developed inland regions in China. This implies that for those more developed regions, more foreign investment inflows will contribute to further reduction of pollution levels, while further increases in foreign investment in inland provinces will continue to a worsening of their environment.

These findings do not, however, imply that the only effective way to alleviate environmental protection pressure in the early stages of development—especially in those less developed inland regions—is to attract more foreign investment, until a certain threshold value along the inverted-U curve is passed. In other words, it is misleading to conclude that the environmental damage caused by the entry of foreign firms in the initial stage can be remedied naturally by the continual accumulation of foreign investment. This is mainly because, as the empirical results show, most Chinese provinces have a long way to go in attracting foreign investment before they reach the turning-point along the inverted-U curve, as the real value of foreign investment in these regions is much lower than the estimated threshold value—namely, their FDI–pollution relationship is still located along the left side of the inverted-U curve. In other words, they have to continue to pay a significantly high cost of environmental degradation as a result of attracting foreign investment and economic growth. Given that China has set a target of reducing overall pollution emissions, whether these inland provinces can achieve their objectives in reducing emissions along the path of growth and development will be crucial for China to meet its targets. These inland provinces will need financial and technological support to enable them to comply with toughened government emissions regulations, similar to the way developing countries are treated in fulfilling their obligations to reduce emissions. Firms in the more developed regions in China and foreign investment can help inland provinces to fulfil their obligations.

It is also in the interest of inland regions to reduce emissions by improving environmental standards, even though pollution abatement endeavours will involve additional costs, which usually increase with pollution emissions. The costs associated with pollution damage, however, could be much higher, especially in cases where environmental damage cannot be remedied in the short term. Land erosion, deforestation, radiation pollution, the loss of species diversity and even the extinction of certain species are a few examples. These considerations require that governments regulate and monitor environmental pollution even in the early phase of industrialisation when environmental qualities are most compromised. Furthermore, the shape and threshold value of the turning-point of the inverted-U curve depend essentially on the industrial composition of foreign investment (Bao et al. 2007). Therefore, in order to avoid the trap of a race to the bottom, more attention should be paid to the industrial structure of foreign investment rather than to the scale of those investments. Finally, since the role of pollution control variables is incorporated in this study, the results suggest that the combination of complementary policies should also be considered in order for China to better deal with the trade-off between economic growth and environmental pollution—by implementing policies aimed at alleviating environmental pressures, such as increases in environmentally related research and development and provision of technical support.

Notes

1 For example, in 2006, the share of FDI in secondary industry was as large as 63.59 per cent; it was 2.03 per cent for electricity, gas and water production, 1.09 per cent for the building industry and 0.73 per cent for the mining industry. All four industries are commonly regarded as being pollution intensive.

2 For a review of the literature on empirical studies of the pollution haven hypothesis, see Dean et al. (2004) and Copeland and Taylor (2004).

3 For example, in their empirical studies of the relationship between foreign investment and environmental regulation and pollution emissions, the estimation results remain puzzling. While some authors—such as List and Co (2000), Keller and Levinson (2002) and Fredriksson et al. (2003)—empirically support the pollution haven hypothesis, others deny the effect of the stringency of environmental policies on the location choice of foreign firms, or the PHH exists only under certain conditions (for example, Levinson 1996; List et al. 2004; Henderson and Millimet 2007). A review of the literature on the PHH tests can be found in Copeland and Taylor (2004).

4 It was Grossman and Krueger (1991) who first proposed the existence of such an inverted-U curve in their study of the environmental consequences of the North American Free Trade Agreement (NAFTA). For a review of the literature on EKC studies, see Stern (1998) and Dinda (2004).

5 In one of our recent working papers (Bao et al. 2007), by introducing international capital flows into the original framework of Copeland and Taylor (2003), we provide mathematical proof of such an inverted-U curve relationship between foreign investment and pollution emissions in host countries. Additionally, as long as the environment is regarded as a normal good, this inverted-U shape conclusion is very robust—whether foreign firms invest in pollution-intensive or clean sectors.

6 As usual, the coastal region includes 12 provinces and municipalities, such as Beijing, Tianjin and Hebei. The remaining 17 provinces all belong to the inland regions, while Chongqing is combined with Sichuan's provincial data.

7 Some authors, such as Wang and Jin (2002) and Liang (2006), empirically find a negative effect of foreign investment on pollution emissions in China. Similarly, we also estimate the linear

effect of foreign investment on the five pollutants, and our estimation results are consistent with those authors' findings—except that the coefficient of foreign investment is insignificant for industrial solid wastes; it is significantly negative for the other four pollutants: -0.018 (*water*), -0.049 (*cod*), -0.168 (*gas*), -0.008 (*smoke*).

References

Antweiler, W., Copeland, S. and Taylor, M., 2001. 'Is free trade good for the environment?', *American Economic Review*, 91(4):877–908.

Bao, Q., Chen, Y. and Song, L., 2007. *Foreign direct investment and pollution in host countries: a theoretical approach*, Working Paper, Department of International Trade and Economics, Nankai University, Tianjin.

Barro. R. and Lee, J.-W., 2001. 'International data on educational attainment: updates and implications', *Oxford Economic Papers*, 53(3):541–63.

Birdsall, N. and Wheeler, D., 1993. 'Trade policy and industrial pollution in Latin America: where are the pollution havens', *Journal of Environment & Development*, 2(1):137–47.

Cheung, K.-y. and Lin, P., 2004. 'Spillover effects of FDI on innovation in China', *China Economic Review*, 15:25–44.

Chudnovsky, D. and Lopez, A., 1999. TNCs and the diffusion of environmentally friendly technologies to developing countries, Mimeo., Copenhagen Business School Cross Border Environmental Project.

Copeland, B. and Taylor, S., 2003. *Trade and the Environment*, Princeton University Press, Princeton, NJ.

——, 2004. 'Trade, growth and the environment', *Journal of Economic Literature*, 42(1):7–71.

Dasgupta, S., Laplante, B., Wang, H. and Wheeler, D. 2002. 'Confronting the Environmental Kuznets Curve', *Journal of Economic Perspectives*, 16(1): 147–168.

Dean, M., Lovely, E. and Wang, H., 2004. Foreign direct investment and pollution haven, evaluating the evidence from China, Mimeo., US International Trade Commission, Washington, DC.

Eskeland, G. and Harrison, E., 2003. 'Moving to greener pastures? Multinationals and the pollution haven hypothesis', *Journal of Development Economics*, 70:1–23.

Fredriksson, P., List, J. and Millimet, D., 2003. 'Bureaucratic corruption, environmental policy and inbound US FDI: theory and evidence', *Journal of Public Economics*, 87:1407–30.

Grossman, G. and Krueger, A., 1991. *Environmental impacts of a North American Free Trade Agreement*, NBER Working Papers 3914, Cambridge, Mass.

He, J., 2002. *Pollution haven hypothesis and environmental impacts of foreign direct investment: the case of industrial emission of sulfur dioxide (SO₂) in China*, Working Paper, University of Auvergne, France.

Henderson, D. and Millimet, D., 2007. 'Pollution abatement costs and foreign direct investment inflows to US states: a nonparametric reassessment', *Review of Economics and Statistics*, 89:178–83.

Keller, W. and Levinson, A., 2002. 'Pollution abatement costs and foreign direct investment inflows to the US states', *Review of Economics and Statistics*, 84:691–703.

Letchumanan, R. and Kodama, F., 2000. 'Reconciling the conflict between the "pollution-haven" hypothesis and an emerging trajectory of international technology transfer', *Research Policy*, 29:59–79.

Levinson, A., 1996. 'Environmental regulations and manufacturers' location choices', *Journal of Public Economics*, 62:5–29.

Liang, F., 2006. Does foreign direct investment harm the host country's environment?, Mimeo., Hass School of Business, University of California, Berkeley.

List, J. and Co, C., 2000. 'Environmental regulations on foreign direct investment', *Journal of Environmental Economics and Management*, 40:1–20.

List, J., McHone, W. and Millimet, D., 2004. 'Effects of environmental regulation on foreign and domestic plant births: is there a home field advantage?', *Journal of Urban Economics*, 56:303–26.

National Bureau of Statistics (NBS), various years. *China Statistical Yearbook*, China Statistics Press, Beijing.

Smarzynska, B. and Wei, S.-J., 2001. *Pollution havens and foreign direct investment: dirty secret or popular myth?*, NBER Working Paper 8465.

State Environmental Protection Administration (SEPA), various years. *China Environment Yearbook*, China Environment Yearbook Press, Beijing.

Stern, D.I., 1998, 'Progress on the environmental Kuznets curve?', *Environment and Development Economics*, 3:173–96.

Wang, H. and Jin, Y., 2002. *Industrial ownership and environmental performance: evidence from China*, World Bank Policy Research Working Paper 2936.

Wayne, B. and Shadbegian, R., 2002. *When do firms shift production across states to avoid environmental regulation?*, NBER Working Papers 8705.

Wooldridge, J., 2002. *Econometric Analysis of Cross Section and Panel Data*, The MIT Press, Cambridge, Mass.

Yao, S., 2006. 'On economic growth, FDI and exports in China', *Applied Economics*, 38:339–51.

Zhang, J., Wu, G.Y. and Zhang, J.P., 2004. 'Estimation of China's provincial physical capital stock: 1952–2000', *Economic Research*, 10:5–44.

12

The impact of global warming on Chinese wheat productivity

Liangzhi You, Mark W. Rosegrant, Cheng Fang and Stanley Wood

The adoption of modern varieties and the increased use of irrigation and fertilisers during the 'Green Revolution' dramatically increased crop yields all over the world (Evenson and Gollin 2003b; Rosegrant and Cline 2003). The Green Revolution enabled food production in developing countries to keep pace with population growth (Conway and Toenniessen 1999). Crop yield growth has slowed since the 1990s (Evenson and Gollin 2003b; Rosegrant and Cline 2003), but continued crop yield increases are required to feed the world in the twenty-first century (Rosegrant and Cline 2003; Cassman 1999) given the continuing decline in the amount of land suitable for grain production due to urbanisation and industrialisation. Food security, in particular in developing countries, remains a challenge. This challenge is made worse by the adverse effects of predicted climate change in most food-insecure developing countries (Rosenzweig and Parry 1994).

Given the large body of research that has been done to quantify the contributions of crop productivity (Evenson and Gollin 2003a; Evenson and Gollin 2003b), we know factors such as modern varieties, increasing input use and better farm management contribute greatly to crop yield growth. Our knowledge on the impact of climate on crop productivity, however, remains quite uncertain. While many researchers have evaluated the possible impact of global warming on crop yields using mainly indirect crop-simulation models (for example, Rosenzweig and Parry 1994; Brown and Rosenberg 1997; Reilly et al. 2003), there have been relatively few direct assessments of the impact of observed climate change on past crop yields and growth, except for a few studies (Nichalls 1997; Carter and Zhang 1998; Naylor et al. 2002; Lobell and Asner 2003; Peng et al. 2004). In a recent study, Peng et al. (2004) reported that rice yields declined with higher night-time temperatures. Lobell and Asner (2003) showed that corn and soybean yields in the United States could drop by as much as 17 per cent for each degree of increase in the growing-season temperature. Although climate is the major uncontrollable factor that influences crop development, it is difficult to separate this influence from other factors such as the increased use of modern inputs and intensified crop management, which were introduced during the Green Revolution. In fact, one major concern with the above-mentioned studies is the simplification of approximating such non-climate contributions as a linear trend (Gu 2003; Godden et al. 1998).

In this paper, we use crop-specific panel data to investigate the climate contribution to Chinese wheat-yield growth. We find that global warming has a significantly negative impact on wheat yields in China, but the magnitude of impact is less than that reported by previous studies in other regions.

Data and method

We use time series and cross-section data from 1979 to 2000 for 22 major wheat-producing provinces in China and the corresponding climate data such as temperature, rainfall and solar radiation during this period. Wheat input and output data are from China's *State Statistical Yearbook* and the *Rural Statistical Yearbook* (NBS 1979–2002a, 1979–2002b), and *China Agricultural Cost and Return Yearbook* (State Price Bureau 1979–2002). Climate data are from the Climatic Research Unit at the University of East Anglia. The data-set used is CRU TS 2.0 (Mitchell et al. 2004). The provincial climate parameters are calculated by averaging all the values of those pixels within the provinces. China grows winter wheat and spring wheat. The majority of wheat production in China, about 80–90 per cent, is winter wheat. Winter wheat is grown throughout most of eastern and southern China, while spring wheat is grown in northeast and western China. Winter and spring wheat are grown in northern China. The growing season for wheat varies from province to province. The annual climate data are monthly averages during the wheat-growing seasons, taking account of the changing growing seasons by province.

The analytical challenge is to separate the non-climate effects on crop yields from the climate change effects. We hypothesise the crop yield as a function of crop inputs, technology, management, land quality and climate factors. The initial explanatory variables for the yield equation include inputs such as land, labour, chemical fertiliser, seeds, pesticide, machinery, irrigation and other physical inputs; regional production specialisation; climate variables such as temperature, precipitation and solar radiation; a set of regional dummy variables; and two institutional change dummy variables. In this study, the labour input is measured in terms of working days from the survey data. The previous study (Stavis 1991) found the marginal return to labour input was negligible due to the huge labour surplus in agriculture in China. Our own estimation confirms this finding: labour and draft animals have a negative sign for the wheat-yield equation, indicating that the impact of these two variables on yield was negligible. The inputs of labour and draft animals are therefore not included in the model. The physical inputs are measured in expenses per unit of harvested area, and are selected based on the sign and level of statistical significance. We included chemical fertiliser, seeds, pesticide and machinery individually, and combined the rest of the inputs into an aggregated category of 'other inputs'. The regional production specialisation variable is represented by the share of wheat in the total crop area in that province. This variable is created to reflect the other factors such as soil quality and other regional government supports to wheat production. It is expected that the regions with a high share of crop production have more suitable land and a better environment for wheat production and therefore higher wheat yields than other regions. Admittedly, this variable could potentially be an endogenous variable, as the trade-off between the amount of area to sow with a grain crop and the amount of area to sow with a cash crop depends on trade-offs that involve yields and relative productivity and profitability. The Hausman–Wu procedure (Wu 1973; Hausman 1978) was used to test the exogeneity of the share of area planted with wheat. Predicted wheat areas are not significant in the test equation, indicating that it is exogenous for the yield equation. A set of regional dummy variables is used to represent time-persistent, regional differences in social, economic and natural endowments not accounted for by the other variables. During our study period (1979–2000), China undertook major policy reforms: the Household Responsibility System in the early 1980s and the new development in agricultural policy in the late 1990s. We used time-specific dummy variables to reflect these two major policy changes. Finally, a time trend is used to represent the factor due to technological change during this period.

Finally, a Cobb-Douglas form of wheat-yield function is specified as follows

$$\ln Yield_{it} = (\alpha_0 + \alpha_1 t) + \sum_j \beta_j \ln X_{jit} + \gamma \ln S_{it} + w \ln Climate_{it}$$
$$+ \sum_{r=2}^{7} \delta_\gamma D_\gamma + \sum_{I=1}^{2} r_I D_I + \varepsilon_{it} \tag{1}$$

in which ln is natural log, $t = 1, 2...22$ denotes observations from the years from 1979 to 2000. $Yield_{it}$ refers to the wheat yield for Chinese province i at time t (the time trend from 1979–2000); X represents the conventional inputs per hectare of sown wheat area including seeds, fertiliser, pesticide, machinery and other inputs such as irrigation, manure and animal power; S denotes the share of area under wheat in the total sown area, reflecting the regional specialisation (including land quality) in wheat production; $Climate$ is the climate variables including temperature, rainfall and solar radiation during the wheat-growing season. We approximate solar radiation with cloud cover expressed in percentages. Therefore, the higher the cloud cover, the weaker is the sun's radiation. We include a set of regional dummy variables, D_r, to represent time-persistent, regional differences in social, economic and natural endowments not accounted for by other variables.[1] Time-specific dummy variables, D_l, capture the effects of two major policy reforms in agriculture from 1979 to 1985, and from 1995 to 2000. $\alpha, \beta, \gamma, w, \delta$ and r are parameters to be estimated and ε is the error term.

Estimation and results

We first perform an augmented Dickey-Fuller unit root test to test the stationarity of dependent and independent variables. No problems are found. The model is estimated by an SAS package. Since the ordinary linear square (OLS) estimation has auto-correlation problems, we also estimate Equation 1 using an auto-regressive error model with a one-year lag (AR1). The constant variance error (no heteroscedasticity) assumptions are examined by plots between the predicted values and residuals using the AR1 estimation. The plot (not reported here) shows that the assumptions for Equation 1 are reasonably held. We also examined another plot between the predicted value and the time trend and found no auto-correlation problem. Another potential problem could be an omitted-variable bias, where some temperature-related variables (such as diseases or pests) that affect wheat yield have been left out of Equation 1. We perform the Ramsey (1969) regression specification error test (RESET) for omitted variables. The test is passed (P > 28 per cent). The assumptions of normal distribution for errors, outliers and linearity are also diagnosed and these assumptions are found to still hold. In addition, we estimated the equation with fixed effects and random effects but found little difference.

The estimated results are reported in Table 12.1. The OLS estimates for all parameters for physical inputs are significant at the 10 per cent level or below with the expected signs.

Table 12.1   **Estimated wheat-yield function in China, 1979–2000**

Explanatory variables	OLS	AR1
Constant	7.534 (32.12)***	7.482 (33.22)***
Ln fertiliser	0.127 (1.60)***	0.136 (4.47)***
Ln seeds	0.180 (4.64)***	0.153 (4.19)***
Ln pesticide	0.056 (4.71)***	0.051 (4.66)***
Ln machinery	0.024 (1.95)**	0.027 (2.29)**
Ln other inputs	0.043 (1.60)*	0.042 (1.76)*
Ln share of wheat	0.065 (2.32)**	0.057 (2.41)**
Ln temperature	−0.269 (−10.01)***	−0.268 (−11.97)***
Ln precipitation	−0.043 (−1.34)	−0.039 (−1.26)
Ln cloud cover	0.083 (0.96)	0.067 (0.78)
Time	0.021 (4.96)***	0.021 (4.15)***
Regional dummy (Northeast)	−0.141 (−2.29)**	−0.193 (−3.44)***
Regional dummy (North)	−0.113 (−0.29)	−0.120 (−0.35)
Regional dummy (Northwest)	−0.414 (−9.88)***	−0.407 (−9.47)***
Regional dummy (Central)	−0.119 (−2.49)***	−0.107 (−2.63)***
Regional dummy (Southeast)	−0.011 (−0.27)	−0.015 (−0.43)
Regional dummy (Southwest)	−0.387 (−7.74)***	−0.403 (−9.16)***
Institutional dummy (1979–85)	0.051 (1.40)	0.048 (1.03)
Institutional dummy (1995–2000)	−0.093 (−2.54)***	−0.098 (−2.11)*
Degree of freedom	462	461
Adjusted R2	0.801	0.835

* 0.10 level of statistical significance
** 0.05 level of statistical significance
*** 0.01 level of statistical significance
Note: The dependent variable is Ln (wheat yield). Numbers in parentheses are t-values.

The AR1 estimates differ slightly from OLS with some improvements, and all parameters are still significant at or below the 10 per cent level. We will therefore refer only to the AR1 results in the rest of the paper. As expected, the regional specialisation is correlated positively with wheat productivity. The regional dummies in northeast, northwest, central and southwest China are statistically significant. While the institutional dummy between 1979 and 1985 has a positive sign—meaning that the policy reforms during this period did contribute to wheat productivity growth—it is not significant. On the other hand, the changes in agricultural policy after 1995 had a negative impact on wheat productivity, which was measurable at the 10 per cent level of statistical significance. We find no significant relationships between wheat yield and rainfall or solar radiation.

The temperature has, however, a significantly negative effect on wheat yield. Because we use a double-log functional form, the estimated coefficients are elasticities in the above equation. The coefficient for temperature, –0.27, means a 1 per cent increase in the growing-season temperature could reduce wheat yield by 0.27 per cent.

Since our major focus is to measure the contribution of growing-season temperatures on wheat yields, it is convenient to treat other terms in Equation 1 as 'residual' effects. By subtracting the non-climate terms from the wheat yield, we single out the wheat-yield change due to climate change. We define $Yield^{Climate}$ as

$$\ln Yield^{Climate} = \ln Yield_{it} - (\alpha_0 + \alpha_1 t) - \sum_{j=1}^{5} \beta_j \ln X_{jit} - \gamma \ln S_{it}$$
$$- \sum_{r=2}^{7} \delta_\gamma D_\gamma - \sum_{I=1}^{2} r_I D_I \qquad (2)$$

Figure 12.1 shows the relationship between this net wheat-yield change and the relative change in wheat growing-season temperatures. The downward slope of the trend line shows clearly the negative impact of rising temperatures on wheat yields in China.

Figure 12.1 **Correlation between growing-season temperatures and wheat-yield change due to climate change**

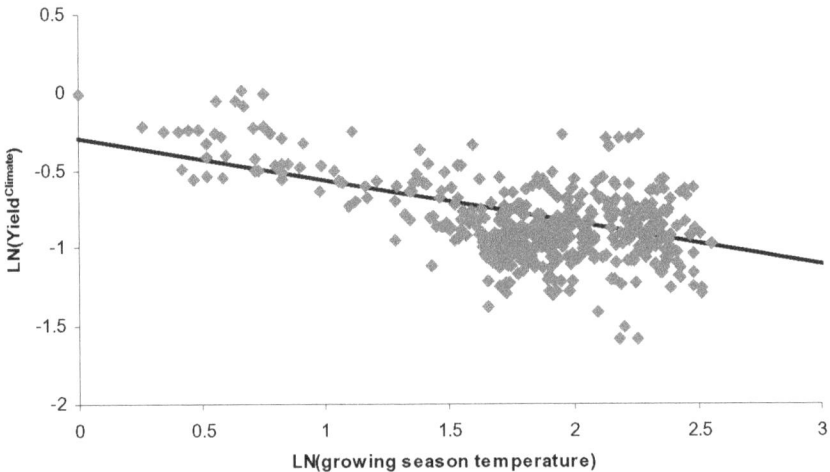

Note: The slope for the regression line is –0.268, R^2=0.84, n=461.

Across wheat-growing provinces in China, the growing-season temperatures vary from 5ºC to 18ºC. A one-degree increase in temperature is therefore equivalent to 5.6–20 per cent of relative change. Since our result shows that a 1 per cent increase in growing-season temperature could reduce wheat yield by 0.27 per cent, this means a 1.5 to 5.4 per cent decline in wheat yield for each 1ºC increase in temperature in China. This estimated effect of temperature on wheat yield is smaller than that found in the previous three studies of: rice in the Philippines (Peng et al. 2004), wheat in Australia (Nichalls 1997) and corn and soybean in the United States (Lobell and Asner 2003). Table 12.2 shows a comparison of these studies. The reasons for this difference are twofold: it could reflect the non-linear effect of physical inputs and crop management on crop yields (Gu 2003; Godden et al. 1998), or it could imply that the temperature effect on crop yields varies from one region to another, or from crop to crop.

Table 12.2   **Comparison: impact of 1ºC increase in growing-season temperature**

Study	Crop	Location	Impact (%)
Nichalls (1997)	Wheat	Australia	+30~+50
Lobell and Asner (2003)	Corn, soybean	United States	−17
Peng et al. (2004)	Rice	Philippines	−10
Our study	Wheat	China	−2~−5

To assess the relative contribution of rising growing-season temperatures on the wheat yield, we take the first derivative of Equation 1 with respect to t (Lin 1992; Fan and Pardey 1997).

$$\frac{\partial \ln Yield_{it}}{\partial t} = \alpha_1 + \sum_j \beta_j \frac{\partial \ln X_{jit}}{\partial t} + \gamma \frac{\partial \ln S_{it}}{\partial t} + w \frac{\partial \ln Climate_{it}}{\partial t} +$$
$$\sum_{r=2}^{7} \delta_\gamma \frac{\partial D_r}{\partial t} + \sum_{r=1}^{2} r_I \frac{\partial D_I}{\partial t} + \frac{\partial \varepsilon_{it}}{\partial t} \tag{3}$$

Table 12.3 reports the growth accounting based on the estimate of the wheat-yield function in column 1 of Table 12.1. The total wheat-yield growth from 1979 to 2000 was 85.41 per cent. From the accounting in Table 12.3, it

Table 12.3   Accounting for wheat-yield growth

Explanatory variable	Estimated coefficient (1)	1979–2000 Change in explanatory variable (2)	Contribution to growth (%) (3)=(1)X(2)
Inputs			64.25 (75.23)
Chemical fertiliser	0.136	255.00	34.68 (40.60)
Pesticide	0.051	220.33	11.13 (13.03)
Machinery	0.027	324.62	8.70 (10.19)
Seeds	0.153	64.39	9.85 (11.53)
Other inputs	0.043	−2.43	−0.10 (−0.12)
Specialisation	0.057	−7.80	0.44 (−0.52)
Temperature	−0.268	7.57	−2.03 (−2.37)
Residual*			23.63 (27.67)
Total growth			85.41 (100)

* an accounting residual derived by netting out the effects of inputs, specialisation and temperature. Here it reflects mainly the impact of agricultural research and development and institutional change.
Note: The estimated coefficients are taken from Table 12.1, and the change in explanatory variable refers to percentage growth of that variable from 1979–81 to 1998–2000 (three-year averages are taken to avoid atypical years). The numbers in parentheses are the percentage shares of contribution to total wheat-yield growth, with total yield growth set at 100.

appears that 75.23 per cent of this yield growth comes from the increased use of physical inputs. Rising temperatures attributed to 2.37 per cent of the decline in wheat yield. This negative contribution is relatively small compared with that of physical inputs, which underlines the necessity of including physical inputs in the regression analysis of crop yield–climate interactions[2]

Conclusion

Since the introduction of rural reforms in China in the late 1970s, agricultural production of and productivity for wheat have increased significantly. While the majority of wheat productivity increase is due to the increased use of physical inputs and institutional change, the gradual increase in growing-season temperatures in the past few decades has had a measurable effect on wheat productivity. In this paper, we have evaluated the impacts of climate and non-climate factors on wheat-yield growth in China, and found that a 1 per cent increase in wheat growing-season temperatures reduced the yield by about 0.3 per cent. The rising temperatures from 1979 to 2000 cut wheat-yield growth by 2.4 per cent. There is a deficiency in the current literature about how to measure the influence of climate on productivity. Authors fail frequently to distinguish between climate factors and the influence of modern inputs and management practices on productivity. We emphasise the necessity of including such major influencing factors as physical inputs into crop yield–climate functions in order to have an accurate estimation of climate impact on crop yields. With so much uncertainty about the potential impacts of climate change, it is essential to first evaluate the impacts past climate changes have had on agricultural productivity. Our study demonstrates a clear need to synthesise climate and crop-specific management and inputs data in order to investigate the impact of climate change.

In China, providing enough food to feed more than 1.3 billion people is always a challenge. There is increasing concern about the impacts of climate change on Chinese food security. Our study shows that climate change does have a measurable negative impact on wheat productivity. This negative impact will probably become worse with accelerating climate change in the future. Our study demonstrates the need to consider climate change and its effects on crop productivity in order to meet the food-security goals in China as well as in other developing countries. There is also a need to extend such studies to other regions—in particular, to food-insecure countries where climate change will have the most severe adverse impacts on crop productivity.

Notes

1   The seven regions in China are: Northeast (Heilongjiang, Liaoning, Jilin), North (Beijing, Tianjin, Hebei, Henan, Shandong, Shanxi, Shaanxi), Northwest (Nei Menggu, Ningxia, Xinjiang, Tibet, Qinghai, Gansu), Central (Jiangxi, Hunan, Hubei), Southeast (Shanghai, Jiangsu, Zhejiang, Anhui), Southwest (Sichuan, Guizhou, Yunnan), South (Guangxi, Fujian, Hainan, Guangdong).

2   Simple de-trending of wheat yields and temperature while ignoring the physical inputs finds no significant relationship between wheat yield and temperature ($R^2 < 0.001$).

References

Brown, R.A. and Rosenberg, N.J., 1997. 'Sensitivity of crop yield and water use to change in a range of climatic factors and CO_2 concentrations: a simulation study applying EPIC to the central USA', *Agricultural and Forest Meteorology*, 83:171–203.

Carter, C. and Zhang, B., 1998. 'Weather factor and variability in China's grain supply', *Journal of Comparative Economics*, 26:529–43.

Cassman, K.G., 1999. 'Ecological intensification of cereal production systems: yield potential, soil quality, and precision agriculture', *Proceedings of the National Academies of Science USA*, 96:5952–9.

Climatic Research Unit, n.d. Climate data, Climatic Research Unit, School of Environmental Sciences, University of East Anglia, UK. Available from www.cru.uea.ac.uk.

Conway, G. and Toenniessen, G., 1999. 'Feeding the world in the twenty-first century', *Nature*, 402:C55–8.

Evenson, R.E. and Gollin, D., 2003a. *Crop Variety Improvement and Its Effect on Productivity: the impact of international agricultural research*, CAB International, Wallingford, UK.

——, 2003b. 'Assessing the impact of the Green Revolution, 1960 to 2000', *Science*, 300:758–62.

Fan, S. and Pardey, P., 1997. 'Research, productivity, and output growth in Chinese agriculture', *Journal of Development Economics*, 53:115–37.

Godden, D., Batterham, R. and Drynan, R., 1998. 'Comment on "Climate change and Australian wheat yield"', *Nature*, 391:447.

Gu, L., 2003. 'Comment on "Climate and management contributions to recent trends in US agricultural yields"', *Science*, 300:1505b.

Hausman, J., 1978. 'Specification tests in econometrics', *Econmetrica*, 46:1251–71.

Lin, J.Y., 1992. 'Rural reforms and agricultural growth in China', *American Economic Review*, 82:34–51.

Lobell, D. and Asner, G., 2003. 'Climate and management contributions to recent trends in US agricultural yields', *Science*, 299:1032.

Mitchell, T.D., Carter, T.R. Jones, P.D., Hulme, M. and New, M., 2004. A comprehensive set of climate scenarios for Europe and the globe (unpublished).

National Bureau of Statistics (NBS), 1979–2002a. *Rural Statistical Yearbook, 1979–2002*, China Statistics Press, Beijing.

National Bureau of Statistics (NBS), 1979–2002b. *State Statistical Yearbook, 1979–2002*, China Statistics Press, Beijing.

Naylor, R., Falcon, W., Wada, N. and Rochberg, D., 2002. 'Using El Niño–southern oscillation climate data to improve food policy planning in Indonesia', *Bulletin of Indonesian Economic Studies*, 38:75–88.

Nichalls, N., 1997. 'Increased Australian wheat yield due to recent climate trends', *Nature*, 387:484–5.

Peng, S., Huang, J., Sheehy, J.E., Laza, R.C., Visperas, R.M., Zhong, X., Centeno, G.S., Khush, G.S. and Cassman, K.G., 2004. 'Rice yields decline with higher night temperature from global warming', *Proceedings of National Academies of Science USA*, 101:9971–5.

Ramsey, J.B., 1969. 'Tests for specification error in classical linear least squares regression analysis', *Journal of the Royal Statistical Society*, B31:250–71.

Reilly, J., Tubiello, F., McCarl, B., Abler, D., Darwin, R., Fuglie, K., Hollinger, S., Izaurralde, C., Jagtap, S., Jones, J., Mearns, L., Ojima, D., Paul, E., Paustian, K., Riha, S., Rosenberg, N. and Rosenzweig, C., 2003. 'US agriculture and climate change: new results', *Climatic Change*, 57:43–69.

Rosegrant, M.W. and Cline, S.A., 2003. 'Global food security: challenge and policies', *Science*, 302:1917–20.

Rosenzweig, C. and Parry, M., 1994. 'Potential impact of climate change on world food supply', *Nature*, 367:133–8.

State Price Bureau, 1979–2002. *China Agricultural Cost and Return Yearbook, 1979–2002*, State Price Bureau, Beijing.

Stavis, B., 1991. 'Market reforms and changes in crop productivity: insight from China', *Pacific Affairs*, 64:371–83.

Wu, D., 1973. 'Alternative tests of independence between stochastic regressors and disturbances', *Econometrica*, 41:733–50.

13

Understanding the water crisis in northern China

How do farmers and the government respond?

Jinxia Wang, Jikun Huang, Scott Rozelle, Qiuqiong Huang and Lijuan Zhang

Increasing evidence indicates that China is facing serious water shortages, especially in the north of the country. These shortages are a result not only of falling water supplies, but of rising water demand. There is evidence of falling supplies of surface-water resources and the related closure of rivers. Because of climate change and human activity, in the past two decades, run-off in some major river basins in northern China has declined significantly, resulting in the decrease of available surface-water resources. For example, run-off in the Hai River Basin has decreased by 41 per cent (Ministry of Water Resources 2007). Run-off in other river basins has also decreased—from 9–15 per cent in the Liao, Yellow and Huai River Basins. Because of declining surface-water supplies and increasing competition for water among regions, the water in some river basins (such as those of the Yellow and Hai Rivers) in northern China cannot flow to the lower regions (Wang and Huang 2004).

With declining surface-water resources, farmers in northern China have begun to explore ground-water resources and ground water has become the dominant source of water for irrigation in northern China. In the early 1950s, ground-water irrigation was almost non-existent in northern China (Wang et al. 2007a); in the 1970s, it rose to 30 per cent of the total irrigation water. After the economic reforms in the late 1970s, ground-water irrigation continued to expand, reaching 58 per cent in 1995. In 2004, most irrigation in northern China came from ground-water resources, and the share of ground-water irrigated areas increased to nearly 70 per cent.

Unfortunately, the development of the use of ground water has resulted in an overdraft of these resources and many environmental problems. According to a comprehensive survey completed by the Ministry of Water Resources in 1996, the overdraft of ground water was one of China's most serious resource problems (Ministry of Water Resources and Nanjing Water Institute 2004). In the late 1990s, the annual rate of overdraft exceeded 9 billion cubic metres. More than one-third of the volume of overdraft is from deep wells, many of which might not be renewable. Ground-water overdraft also causes many environmental problems. The most obvious effect of ground-water overdraft is the falling water table. For example, based on our research in Hebei Province, the shallow water table has dropped about 1 metre per annum (Wang and Huang 2004); the deep water table has also dropped quickly as a result. The drop rate of the deep water table was more than 2 metres per annum. Overdraft of ground-water resources can also cause other environmental problems, such as land subsidence, the intrusion of seawater into freshwater aquifers, desertification and the depletion of stream flows previously supplied by natural ground-water discharge (Wang et al. 2007a).

The rapidly growing industrial sector, an expanding farming sector and an increasingly wealthy urban population also compete for China's limited water resources. Between 1949 and 2004, total water use in China increased by 430 per cent, which was similar to the global average increase of 400 per cent, but greater than the average for developing countries (Wang et al. 2005a). Rising industrial and urban growth rates have caused China's water allocations to be directed increasingly to non-agricultural uses. From 1949 to 2004, the share of water use for irrigation declined from 97 per cent to 65 per cent of total water use (Ministry of Water Resources 2004). At the same time, the share of industrial water use increased from 2 to 22 per cent; the share of domestic water use increased from 1 to 13 per cent.

Faced with the decline of water availability and increased water demand, many people think China has a water crisis—at least, this is the perception of some scholars and policymakers within and outside China. For example, even back in 1999, Chinese Prime Minister Wen Jiabao warned of the dire water situation in China and of looming water shortages (McAlister 2005). Senior officials from the Ministry of Water Resources pointed out that China was fighting for every drop of water, and the water crisis was threatening national grain production. Brown (2000) predicted that falling water tables in China might soon raise food prices everywhere. Nankivell (2004) demonstrated that China was now at a point where critical decisions must be made to resolve water issues. Although some other observers have made more moderate predictions, they also suggest that many agricultural producers will have to forgo irrigation (Crook and Diao 2000).

Despite many existing discussions and dire predictions, some researchers (including ourselves) are not sure whether China is facing a water crisis. The major problem is that most discussions about water are based on observations of producers or users in a single location, not on large-scale field-level data. It is difficult to judge, based only on these observations, the seriousness of water shortages and it is also difficult to conclude whether China is facing a water crisis. In addition, relying only on some observations does not allow us an overall picture of water-shortage issues, especially regional differences.

The overall goal of our research is to establish the facts about whether there is a water crisis in China, especially in the north of the country. In order to realise our overall goal, we will pursue the following objectives. First, we will identify the status and trend of water shortages. Second, we want to understand the response of government and its effectiveness in addressing the water crisis. Third, we want to understand the responses of farmers and whether their role is helping or hurting.

This chapter is organised as follows. In the next section, based on our large field survey in northern China, we provide some facts about the water crisis by measuring several indicators relevant to water shortages. In the following section, we discuss the response of government to the water crisis, including ground-water policy and irrigation management reform of surface-water resources. In section four, we will further discuss farmers' responses to increasing water scarcity, focusing on the following issues: the digging and privatisation of tube-wells, developing ground-water markets, behavioural change to increasing water charges and the adoption of water-saving technologies.

Data

Our analysis is based on the data we collected as part of two recent surveys designed specifically to address irrigation practices and agricultural water management. The first, the China Water Institutions and Management (CWIM) survey, was conducted in September 2004. Enumerators conducted surveys of community leaders, ground-water managers, surface-water irrigation managers and households in 80 villages in Hebei, Henan and Ningxia Provinces. The villages were chosen according to geographic location (which, in the Hai River Basin, often correlated with water-scarcity levels). In Hebei, villages were chosen from counties near the coast, near the mountains and in the central region between the mountains and the coast. In Henan and Ningxia Provinces, villages were chosen from counties bordering the Yellow River and from counties in irrigation districts varying distances from the Yellow River. The 2004 CWIM survey was the second round of a panel survey, the first phase of which was conducted in 2001.

We conducted a second survey, the North China Water Resource (NCWR) survey, in December 2004 and January 2005. This survey of village leaders from 400 villages in Inner Mongolia, Hebei, Henan, Liaoning, Shaanxi and Shanxi Provinces used an extended version of the community-level village instruments of the CWIM survey. Using a stratified random sampling strategy for the purpose of generating a sample representative of northern China, we first sorted counties in each of our regionally representative sample provinces into one of four water-scarcity categories: very scarce, somewhat scarce, normal and mountain/desert. We randomly selected two townships within each county and four villages within each township. Combining the CWIM and NCWR surveys, we visited a total of six provinces, 60 counties, 126 townships and 448 villages.

The scope of the surveys was quite broad. Each of the survey questionnaires included more than 10 sections. Among the sections, there were those that focused on the nature of rural China's water resources, the common types of wells and pumping technology. There were several sections that examined the most important water problems, government water policies and regulations and a number of institutional responses (for example, tube-well privatisation). Although sections of the survey asked about surface and ground-water resources, we will focus mostly on those villages that have ground-water resources (in some cases, whether they are using them or not). The survey collected data on many variables for two years: 2004 and 1995. By weighting our descriptive and multivariate analysis with a set of population weights, we are able to generate point estimates for all of northern China.

Facts about water shortages

The water-shortage situation can be gauged by examining some relevant indicators, in addition to the assessment of farmers, who are directly influenced by water. In this section, we will examine the water situation in northern China from the following three aspects. First, we will examine the overall situation of water shortages through farmers' judgements based on their intuition. Second, we will examine this issue by checking the status and changes of the supply reliability of surface and ground-water resources. Finally, we will analyse changes in the water table over time.

Farmers' judgements of water shortages

During our field survey, enumerators asked village leaders to characterise the nature of water resources in their village in 1995 and 2004. The leaders chose one of three answers: water is not short (at least currently); water is short but the shortage is not severe; and water is short and frequently constrains agricultural production (or water scarcity is very severe). Based on the responses of village leaders, we can make a preliminary judgement of the situation and trend of water shortages in northern China.

Based on farmers' judgements, most villages in northern China are facing water shortages, and the outlook is not positive. Survey results show that in 2005, 70 per cent of villages in our samples reported that they were facing water shortages—at least, farmers in these villages felt there were problems with water availability (Table 13.1). Only 30 per cent of villages did not have water-supply problems. In addition, among those villages where farmers reported water shortages, 16 per cent of villages were facing severe water shortages.

More importantly, water shortages have become more serious in the past decade. Survey results show that from 1995 to 2004, the degree of water shortage continued to increase, and the share of villages short of water increased by 5 per cent. The share of villages with severe water shortages also increased, by 2 per cent. Therefore, based on the farmers' judgements, water shortages were very serious. Though not all villages are facing such problems, most are. In some villages, the water shortage is so serious it has become one of the important constraining factors on agricultural production.

Changes in surface and ground-water supply reliability

The reliability of the water supply is another indicator that can measure the degree of water shortages in rural areas. If the surface-water supply is reliable, water will not be a constraining factor on agricultural production in the short term.

Survey results show that surface-water reliability is very low in northern China. Presently, most villages in northern China do not have a reliable surface-water supply (Table 13.2). During the period 2001–04, 61 per cent of villages did not have reliable surface-water resources.

In the past 10 years, water reliability has declined remarkably. From 1991 to 1995, surface-water supplies in most villages were very reliable; 64 per cent of villages reported that they could access a reliable water supply at that time (Table 13.2). From 2001 to 2004, however, the share of villages with a reliable supply of surface water declined to 39 per cent—a decrease of 25 per cent.

At the same time, the share of villages without a reliable supply of water also increased significantly—from 36 to 61 per cent. The decline of surface-water reliability represents an increase in the degree of water shortage.

The reliability of ground-water supplies has also declined (Table 13.2). Unlike surface-water resources, ground water has always been considered a reliable water resource. Ground water has, however, become a less reliable resource for farmers. For example, from 1991 to 1995, 91 per cent of villages reported that their ground-water supplies were very reliable, and only 9 per cent indicated that there were problems with the reliability of their ground-water supply. From 2001 to 2004, however, villages with a reliable ground-water supply declined by 5 per cent. The villages without a reliable ground-water supply increased—from 9 to 15 per cent. The reliability of supply has therefore worsened—for surface and ground-water resources—which indicates that there is an increasing shortage of water in northern China.

Table 13.1 **Farmers' judgements of water shortages in villages in northern China**

Water-shortage situation	Share of sample villages (%)	
	1995	2004
No shortage	35	30
Shortage	65	70
When short, severly short	14	16

Source: Authors' survey in 2004 (NCWR survey data set).

Table 13.2 **Water-supply reliability in villages in northern China**

Water-supply reliability	Share of sample villages (%)			
	Surface-water reliability		Ground-water reliability	
	1991–95	2001–04	1991–95	2001–04
Reliable	64	39	91	85
Not reliable	36	61	9	15
Total	100	100	100	100

Source: Authors' survey in 2004 (NCWR survey data set).

Decline of the water table

Field surveys reveal that not all regions in northern China are experiencing falling water tables (Wang et al. 2007a). According to our data, there was no fall in the water table in 25–33 per cent of the villages in northern China using ground water in 1995 and 2004.[1] In 8.5–16 per cent of villages (one-third to one-half of villages reporting no fall in the water table), respondents told the enumerators that the water table was higher in 2004 than in 1995. In another 10–17 per cent of villages, the average annual fall in the water table was less than 0.25 metres. In other words, in more than one-third to one-half of northern China's villages using ground water in the past decade, ground-water resources have shown little or no decline since the mid 1990s.

Although, based on our data, most villages are in, or are nearly in, balance, we are not arguing that ground-water problems do not exist. In fact, there are still a large number of villages in which the water table is falling. Before classifying these villages as inappropriate ground-water resource exploiters (although some of them might be), it is important to remember that a village's water resources might not be overexploited, even if the water table is falling. Even under the most rationally planned ground-water utilisation strategy, therefore, there will be a share of villages in China in which we can expect the water table to be falling. In addition, if we follow the Ministry of Water Resources' definition of serious overdraft, only 10 per cent of villages using ground water in the past decade had water tables falling at a rate greater than 1.5 metres per annum. Of course, such a rate of decline is not just serious; it is a crisis.

In summary, then, the point we want to make is that in many places—indeed, in most places in northern China—it is possible that water resources are not being misused. We do not, however, want to minimise the problems that are occurring in some places. There are a large number of rural areas in which the water table appears to be falling at a dangerously fast pace. Where the resource is being misused, steps will eventually be required to protect the long-term value and use of the resource. It is, however, important to realise that many of the required measures (discussed in the next section) will have a number of associated costs in their adoption, affecting productivity and perhaps reducing incomes. Because measures to counter overdraft are not needed in all villages, leaders should not take a 'one-size-fits-all' approach; doing so could inflict unnecessary costs on producers in communities where overdraft conditions do not exist.

Government responses

Faced with increasing water shortages, the Chinese government has taken many steps to improve the management of surface and ground-water resources. It has responded not only by issuing many laws, regulations and policies, it has tried to encourage local regions to reform their water management. In this section, we will focus on two major responses that the government has made: its ground-water policies, and reform of the management of surface-water resources for irrigation. We will also try to examine the effectiveness of the implementation of these responses in resolving water-shortage problems in northern China.

Issuing ground-water policies

Government officials in China have put some effort into issuing laws, regulations and policies to manage ground-water resources, although they have been limited (Wang et al. 2007b). For example, according to China's National Water Law, which was revised in 2002, all property rights to ground-water resources belong to the State. This means that the right to use, sell and/or charge for water rests ultimately with the government. The law does not allow ground-water extraction if pumping is going to be harmful to the long-term sustainability of ground-water use. Beyond the formal laws, a number of policy measures (such as regulations controlling the right to drill tube-wells, the spacing of tube-wells and the collection of water-resource fees) have been set up, in part to rationally manage use of the nation's ground-water resources. Compared with regulations concerning other issues, however, such as flood control, the construction of water-related infrastructure projects and surface-water management initiatives, the number of regulations relevant to ground-water management is very small. More importantly, at the national level, there is not one water regulation that is focused specifically on ground-water management issues.

Even more important than the lack of official laws and policy measures for ground-water management has been the insufficient effort put into implementing existing laws (Wang et al. 2007a). Certainly, part of the problem is a history of neglect. In fact, at the ministerial level, the division of ground-water management is still relatively small. There are far fewer officials working in this division than in other divisions, such as flood control, surface-water system management and water transfer. Moreover, unlike the case of surface-water management (Lohmar et al. 2003), there has been no effort to bring management of aquifers that span jurisdictional boundaries under the ultimate control of an authority covering government and private entities that use water extracted from different parts of the aquifer. According to Negri (1989), without a single body controlling the entire resource, it becomes difficult to implement policies that attempt to manage the resource in a manner that is sustainable, or optimal, in the long term.

Whether due to lack of personnel or other implementation-related difficulties, few regulations have had any affect within China's villages (Wang et al. 2007b). For example, according to our survey data, less than 10 per cent of well owners obtained a permit before drilling, despite the nearly universal regulation requiring a permit. Only 5 per cent of village respondents believed that well-drilling decisions required consideration of the spacing between wells. Even more tellingly, water-extraction charges were not charged in any village, and there were no quantity limits put on well owners. In fact, in most villages in China, ground-water resources are all but completely unregulated. This does not mean, however, that policy and governance do not have an impact, at least indirectly, on agricultural ground-water use.

Reforming management of surface-water resources for irrigation

Since the early 1990s, faced with increasing water shortages, Chinese leaders have begun to consider community-level irrigation management reform as a key part of their strategy to combat China's water problems. According to our survey in Ningxia Province, since the early 1990s and especially after 1995, reform has successively established Water User Associations (WUAs) and contracting systems in place of collective management (Figure 13.1) (Wang et al. 2007c).[2] The share of communities that manage water by collective. declined from 91 per cent in 1990 to 23 per cent in 2004. Contracting has developed more rapidly than WUAs. By 2004, 57 per cent of villages managed their water under contract and 19 per cent managed theirs through WUAs. Our survey in six provinces also found a similar reform trend for irrigation management (Figure 13.2) (Huang et al. 2007). From 1995 to 2004, collective management declined from 90 to 73 per cent, while at the same time, WUAs increased from 3 to 10 per cent and contracting management increased from 5 to 13 per cent.

Although reforming the institution in name might be important, the nature of incentives within the institution might be more important (Wang et al. 2005c).[3] According to our data, not all reformed management institutions (WUAs or contracting management) can establish incentive mechanisms. For example, in 2001, on average, leaders in only 41 per cent of villages offered WUA and contracting (or non-collective) managers with incentives that could be expected to induce them to save water in order to earn excess profit. In the remaining villages, although there was a nominal shift in the institution type (that is, leaders claimed that they were implementing WUAs or contracting), in fact, from an incentive point of view, the WUA and contracting managers were operating without imposed incentives. In these villages, managers are similar to leaders in a collectively managed village in that they do not have a financial incentive to save water.

Research results show that nominally reforming irrigation management has no significant impact on water use; however, when irrigation managers face effective incentives, they are able to reduce water use (Wang et al. 2005c). Our results, based either on descriptive statistics or econometric analysis, show that there is no significant relationship between water use and nominal irrigation management reform. We found, however, that in villages that provided water managers with strong incentives, water use fell sharply. The incentives must also have improved the efficiency of the irrigation systems since the output of major crops, such as rice and maize, did not fall, and rural incomes and poverty remained unchanged statistically. Although our study needs to be undertaken in other areas before the results can be generalised to the rest of China, at least in the sample sites that provided their managers with incentives, water management reform has been a win-win policy in resolving water shortage problems.

Farmers' responses

Although the government response has not been very effective in resolving water shortages, this does not mean that farmers will not respond on their own. In fact, farmers have already begun to act. In this subsection, we will examine farmers' responses to increasing water shortages. These include digging tube-wells, privatisation of tube-wells, the emergence of ground-water markets, behavioural changes to increasing water prices and the adoption of water-saving technologies.

Digging tube-wells. The most obvious response by farmers to increasing water shortages is to dig tube-wells. According to national statistics, the installation of tube-wells began in the late 1950s (Wang et al. 2007a). Since the mid 1960s, the installation and expansion of tube-wells across China has been nothing less than phenomenal. In 1965, there were only 150,000 tube-wells in all of China (Shi 2000). Since then, the number has grown steadily. By the late 1970s, there were more than 2.3 million tube-wells. After stagnating during the early 1980s, a period when the area of land under irrigation—especially that serviced by surface water—fell, the number of tube-wells continued to rise. By 1997, there were more than 3.5 million tube-wells; by 2003, the number rose to 4.7 million.

Figure 13.1 **Evolution of irrigation management in Ningxia Province, 1990–2004**

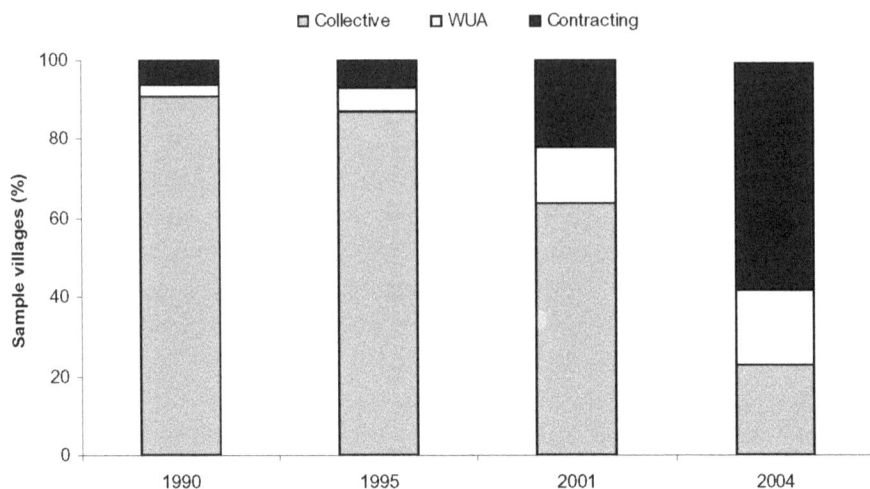

Source: Authors' survey in 2001 and 2004 (CWIM survey data set).

Figure 13.2 **Changes in water management institutions, 1995–2004**

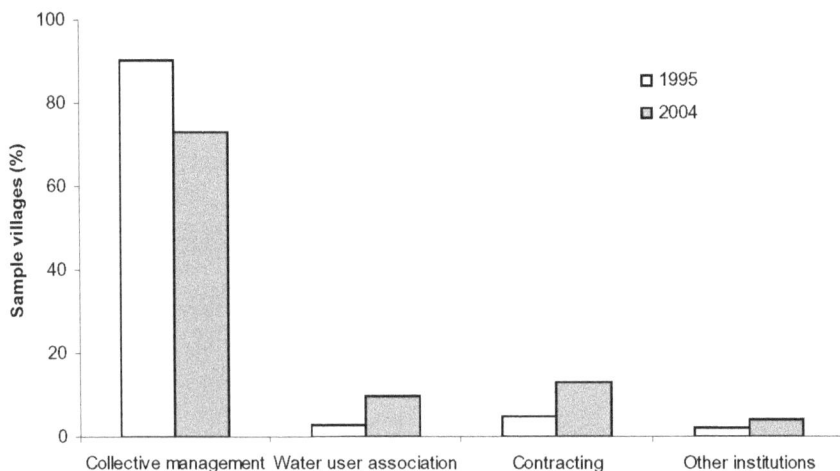

Note: 'Other institutions' includes the four types of mixed institutions: 1) Water User Association combined with collective management; 2) Water User Association combined with contracting; 3) contracting combined with collective management; 4) Water User Association, contracting and collective management.
Source: Authors' surveys in 2001 and 2004 (CWIM and NCWR surveys data sets).

While the growth of the use of tube-wells reported by the official statistical system is impressive, we have reason to believe that the numbers are significantly understated (Wang et al. 2007a). According to the NCWR survey, on average, each village in northern China had 35 wells in 1995. When extrapolated regionally, this means that there were more than 3.5 million tube-wells in the 14 provinces in northern China by 1995. According to our data, the rate of increase in the number of wells has also grown rapidly. By 2004, the average village in northern China had 70 wells, suggesting that the rise in tube-well construction since the mid 1990s has risen even faster than indicated in the official statistics. We estimate that, by 2004, there were more than 7.6 million tube-wells in northern China. At least in our sample villages, the number of tube-wells grew by more than 12 per cent annually between 1995 and 2004. According to our data, a significant share of the new wells are located in areas that are making allowances for the expansion of cropping area, increased intensity of cropping and rising yields. While the rise in tube-wells will not necessarily result in increased consumption of water in all areas, in some cases it will. The digging of new tube-wells is therefore one reason for increasing water shortages.

Privatisation of tube-wells. Faced with increasing water shortages, farmers respond not only by digging new tube-wells that can sustain their production, they demand changes to institutions that can improve their water management. Among all the institutional responses, the privatisation of tube-wells is perhaps the most prominent response by farmers in more than two decades. According to our survey, since the early 1980s, the ownership of tube-wells in northern China began to shift sharply. For example, in Hebei Province, collective ownership of tube-wells diminished from 93 per cent in the early 1980s to 56 per cent in the late 1990s (Wang et al. 2006). At the same time, the share of private tube-wells in the total increased from 7 to 64 per cent. Data from the NCWR survey largely support these findings (Wang et al. 2006). In 1995, collective ownership accounted for 58 per cent of tube-wells in the average ground-water-using village. From 1995 to 2004, however, collective ownership of tube-wells diminished and accounted for only 30 per cent of wells in 2004. In contrast, during the same period, the share of private tube-wells in the total increased from 42 to 70 per cent.

Our findings also demonstrate that the privatisation of tube-wells has promoted the adjustment of cropping patterns while having no adverse impacts on crop yields; more importantly, privatisation has not accelerated the drop of water tables. Econometric results show that, after privatisation, farmers expanded the area of land sown with water-sensitive and high-value crops, such as wheat and non-cotton cash crops (which are mainly horticultural crops)

(Wang et al. 2006). It is perhaps because of the rising demand for horticultural crops that some individuals have become interested in investing in tube-wells. Such results are consistent with the hypothesis that when tube-well ownership shifts from collective to private and water is managed more efficiently (as shown in Wang and Huang 2002), producers are able to cultivate relatively high-value crops, which in some cases demands greater attention from tube-well owners. In addition, our research indicates that the shift from collective to private tube-well management does not accelerate the drop in water tables. With the expansion of private ownership of tube-wells, however, and ground-water markets, water table levels will continue to decline. How farmers choose to respond can, therefore, contribute to the water crisis. Effective management of ground-water resources is needed urgently.

The growing importance of ground-water markets. In response to demand for water in an environment increasingly dominated by private and privatised wells, ground-water markets have begun to emerge in recent years as a way for many producers in rural China to access water (Zhang et al. 2008a). While this is new in China, it appears to be following a pattern similar to that observed in parts of South Asia (Shah 1993). In the 1980s, ground-water markets were all but non-existent in China and, even by the mid 1990s, according to the NCWR survey data, only a small share of villages (21 per cent) had ground-water markets. By 2004, however, tube-well operators in 44 per cent of villages were selling water. Across all villages, about 15 per cent of private tube-well owners sold water. Although ground-water markets exist in less than half of northern China's villages, the numbers are still significant: we estimate that farmers in more than 100,000 villages are accessing water through ground-water markets. Moreover, in villages that have them, these markets play an important role in transferring large volumes of water to a large share of households.

Our research further indicates that ground-water markets have not only improved the equity of water use, they have improved the efficiency of water use in northern China. According to the survey data, ground-water markets have provided poorer farmers with opportunities to access water and have therefore reduced potential income gaps (Zhang et al. 2008b). Specifically, households in the sample that buy water from ground-water markets are poorer than water-selling households (Figure 13.3). In addition, when farmers buy water from ground-water markets, they use less water than those who have their own tube-wells or those who use collective wells (Zhang et al. 2008b). Crop yields, however, do not fall because of this. In addition, our results show that ground-water markets in the North China Plain do not have a negative effect on incomes.

Figure 13.3 **Differences of per capita cropping income and total income between water-selling households and water-buying households, 2004** (yuan)

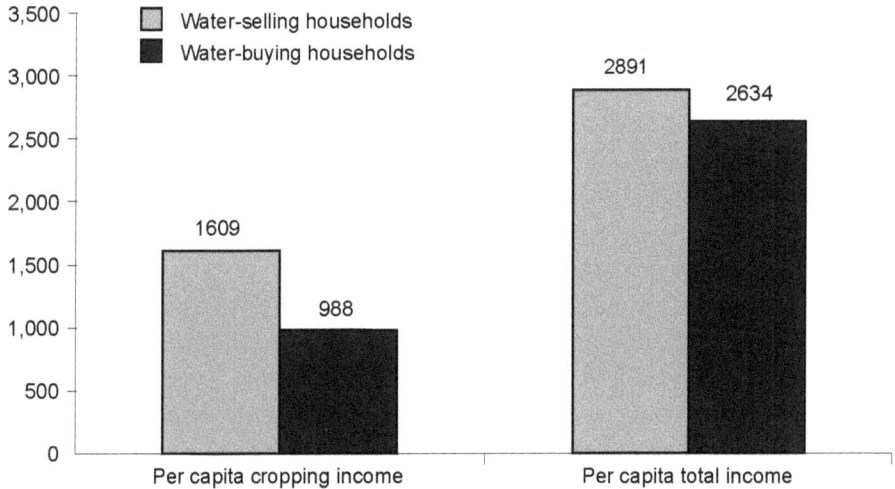

Source: Authors' surveys in 2001 and 2004 (CWIM survey data set).

In summary, ground-water markets help farmers access water (that is, they do not need to invest in their own well) and, when they do so, their water use is reduced, but crop yields and income are not negatively influenced. With increasing water scarcity, increasing water-use efficiency is a very important issue and has been addressed by policymakers. The emergence of water markets is therefore an effective way to provide irrigation services.

Farmers' responses to increasing water prices. As the water table falls, prices in the ground-water market increase. For instance, according to the CWIM survey data, when the level of the water table in our Hebei villages declined from 4.4 metres below the surface to 77.5 metres, the price of ground water for wheat-producing households increased from 0.08 to 0.56 yuan per cubic metre (Wang et al. 2007b). Since pumping costs rise as the water table falls, this indicates that, to some extent, ground-water prices reflect the scarcity value of water resources. Across our CWIM survey sample of Hebei wheat producers, we estimated that pumping costs rose by 0.005 yuan per cubic metre of ground water extracted for each additional metre of pumping depth.

More importantly, as ground-water prices rise and water becomes scarce, farmers respond by reducing their water use and changing their cropping patterns. Analysis of the behaviour of Hebei wheat farmers in the CWIM survey data set indicates that, when ground-water prices increased from 0.08 to 0.56 yuan/cubic metre, water use per hectare decreased from 6,433 cubic metres to 2,154 cubic metres (Wang et al. 2007b). In addition, we found that, as water tables fell, there was an increase in the share of cash crops (such as horticultural crops, cotton and peanuts). That is, when the depth of the water table increases from 4.7 to 79 metres below the surface, the area's share of cash crops increases from 13 to 41 per cent. Our results therefore imply that as the water table falls, water becomes increasingly scarce and the costs of acquisition rise, farmers consider not only how much water to use, but how much value can be produced by that water use.

Despite potential resource conservation benefits, the rise of water prices (or costs to farmers for acquisition of self-pumped water) will have an inevitable negative impact on farmers (Wang et al. 2007b). In other work that we have undertaken using the same data set, we estimate that doubling the price of ground water in Hebei Province causes 75 per cent of wheat-producing farmers to lose money on cropping activities and has a negative effect on agricultural output (Huang et al. 2006). Given the government's interest in maintaining rural incomes, any use of a pricing policy must be accompanied by complementary policies that can offset the negative effect of price increases.

Adopting water-saving technologies. Another possible response to water shortages is the adoption of water-saving technologies (Blanke et al. 2007). The NCWRS survey covered three sets of water-saving technologies: traditional, household based and community based. Traditional technologies—such as border, furrow and levelling technologies—are agronomy based, highly divisible and have generally been used by farmers in China even before the establishment of the communist State. Household-based technologies—such as surface piping, plastic film, conservation tillage and drought-resistant plant varieties—are highly divisible, require low fixed costs and little collective action. Community-based technologies—such as ground pipes and lined canals—require not only collective action for adoption and maintenance, they require high fixed costs.

Results show that although water-saving technologies for agriculture have been emphasised by many policymakers and researchers, the adoption rate is still very low. Despite a relatively high initial level of adoption (35 per cent in the 1950s), in 2004, only 52 per cent of villages adopted traditional water-saving technologies (Table 13.3). Household-based and community-based water-saving technologies have been adopted mainly since the 1980s. In 2004, half of all villages adopted household-based water-saving technologies and only 24 per cent of villages adopted community-based water-saving technologies. Comparing the adoption rate with the share of sown areas, the adoption rate of these three kinds of water-saving technologies is even lower. For example, the share of sown areas adopting community-based water-saving technologies was only 7 per cent; even traditional water-saving technologies reached only 25 per cent. The low adoption rate of water-saving technologies also indicates that in China, there is still much policy space in which to promote such technology uptake. How to effectively promote the adoption of water-saving technologies is one of the important challenges for policymakers.

The increase in the rate of adoption of household-based water-saving technologies is much higher than for the other two types of technologies. Since the 1980s, the share of villages adopting household-based water-saving technologies increased 669 per cent (Table 13.3). Although not very low, the increase in the rate of adoption of community-based water-saving technologies was only 64 per cent of household-based water-saving technologies. During the reform period, the adoption of traditional technologies grew slowly compared with the other two technologies—only 16 per cent. The low increase in the rate of adoption of traditional water-saving technologies was possibly due to their already high adoption rate in the pre-reform and early reform eras. The high increase in the rate of adoption of household-based water-saving technologies implies that, faced with increasing water scarcity, farmers are more responsive than communities in general and government in China.

Based on these descriptive contours, it is unclear what is driving the path of adoption of community-based technologies; however, work by Blanke et al. (2006) suggests that it is likely that there are two sets of forces that are both encouraging and holding back adoption (Wang et al. 2007b). On the one hand, the increased scarcity of water resources is almost certainly pushing up demand for community-based technologies. On the other hand, the predominance of household farming in China (Rozelle and Swinnen 2004) and the weakening of the collective's financial resources and management authority (Lin 1991) has made it more difficult to gather the resources and coordinate the effort needed to adopt technologies that have high fixed costs and that involve many households in the community. In contrast, household-based technologies might be adopted more widely because of their relatively low fixed costs, divisibility and minimal coordination requirements.

Table 13.3 **Adoption rate of water-saving technologies over time in northern China's villages**

	Traditional water-saving technologies	Community-based water-saving technologies	Household-based water-saving technologies
Adoption rate (%)			
Share of villages			
1950	35	0	4
1980	45	5	7
1995	49	15	27
2004	52	24	50
Share of sown areas			
1995	20	3	8
2004	25	7	21
Increase in rate of adoption (%)			
Share of villages			
(2004/1980)	16	433	669
Share of sown areas			
(2004/1995)	25	133	163

Source: Authors' survey in 2004 (NCWR survey data set).

Conclusion

The primary goal of this chapter was to sketch a general picture of China's water-shortage situation, and to establish the facts, especially for northern China. Our findings show that, based on farmers' judgements, 70 per cent of villages are facing increasing water shortages. Among the villages short of water, 16 per cent find their agricultural production severely constrained by the shortage. Supplies of surface and ground water have become less reliable than before. In more than one-third to one-half of China's villages using ground water in the past decade, ground-water resources have shown little or no decline since the mid 1990s. Though most villages are in, or are nearly in, balance, there is still a large number of villages in which the water table is falling. If we follow the Ministry of Water Resources' definition of serious water overdraft, 10 per cent of villages using ground water in the past decade have water tables that are falling at a rate greater than 1.5 metres per annum.

There is, therefore, a water crisis in northern China; however, the crisis does not affect all areas. There are many parts of China in which water resources have not deteriorated, but half of northern China is suffering from rapidly falling water tables. Policies to address water-shortage problems should therefore be targeted carefully.

Faced with a water crisis, the Chinese government has begun to make a number of policy responses, but implementation has not been very effective. It has responded to the issue with many laws, regulations and policies, and has also tried to encourage local regions to reform water management. Whether due to a lack of personnel or other implementation-related difficulties, few regulations have had any effect within China's villages. Although there has been some progress in the reform of surface-water irrigation management in northern China, much of this has been nominal. Nominal reform cannot realise the policy goal of increasing water-use efficiency. Only those reforms that provide managers with incentives can reduce crop water use. Unfortunately, only a small percentage of reform can establish effective incentive mechanisms. In the future, the water crisis will continue to grow, especially as competition among water users increases, and if there is no effective implementation of water policies and management reform.

Where water is becoming scarce, farmers and community leaders have been responding. The most obvious response from farmers is the digging of new tube-wells. In addition, farmers have taken control of most of the well and pump assets; and farmers are increasingly taking on responsibility for transferring water from those who have wells to those who demand water. Farmers also are increasingly figuring out ways to conserve this scarce resource. Farmers do not, however, always respond in a manner that conserves water. Why? The major reason is that farmers do not always face good incentives. Our research shows that when they are given good incentives, they do save water. The government therefore cannot ignore the response of farmers; in fact, it needs to use this responsiveness to reduce the adverse effects of the water crisis and encourage conservation.

Finally, we think most of the blame for the water crisis should be put on the government, because its response has been largely ineffective. It has not created the institutions and infrastructure that will provide the incentives to make farmers save water. We believe a sustainable environment needs to be built on effective water pricing and water rights policies; to make these work, a huge commitment is needed to set up the institutions and infrastructure to implement them. Although this is a huge job, we believe it will be more effective and much cheaper than the South to North Water Transfer Project.

Notes

1 In our survey, we asked village leaders about the average ground-water depth during the year and the 'static' ground-water level. We told village leaders that the static level of the water table was the level that existed at a time immediately before the irrigation season (for example, in the North China Plain this would be about March). According to our respondents, there were differences in the statistics for the changes in the water table when using average or static ground-water levels. According to our data, the static level produced numbers that suggested there were fewer villages in which the water table was falling.

2 'Collective management' implies that the village leadership takes responsibility directly through the village committee for water allocation, canal operation and maintenance and fee collection; WUAs are theoretically a farmer-based, participatory organisational system that is set up to manage the village's irrigation water; 'contracting' is a system in which the village leadership establishes a contract with an individual to manage the village's water.

3 When managers have partial or full claim on the earnings of the water management activities (for example, on the value of the water saved by water management reform), we say that they face strong incentives (or that the manager is managing 'with incentives'). If the income from their water management duties is not linked to water savings, they are said to manage 'without incentives'.

References

Blanke, A., Rozelle, S., Lohmar, B., Wang, J. and Huang, J., 2007. 'Water saving technology and saving water in China', *Agricultural Water Management*, 87:139–50.

Brown, L.R., 2000. *Falling Water Tables in China May Soon Raise Prices Everywhere*. Available from http://www.qmw.ac.uk/~ugte133/courses/environs/cuttings/water/china.pdf.

Crook F. and Diao, X., 2000. 'Water pressure in China: growth strains resources', *Agricultural Outlook*, January–February, Economic Research Service, US Department of Agriculture:25–9.

Huang, Q., Rozelle, S., Howitt, R., Wang, J. and Huang, J., 2006. *Irrigation water pricing policy in China*, Working Paper, Center for Chinese Agricultural Policy, Chinese Academy of Sciences, Beijing.

Huang, Q., Rozelle, S., Wang, J. and Huang, J., 2007. *Water user associations and the evolution and determinants of management reform: a representative look at northern China*, Working Paper, Center for Chinese Agricultural Policy, Chinese Academy of Sciences.

Lin, J., 1991. 'Prohibitions of factor market exchanges and technological choice in Chinese agriculture', *Journal of Development Study*, 27(4):1–15.

Lohmar, B., Wang, J., Rozelle, S., Dawe, D. and Huang, J., 2003. *China's agricultural water policy reforms: increasing investment, resolving conflicts, and revising incentives*, Agriculture Information Bulletin Number 782, US Department of Agriculture, Economic Research Service, Washington DC.

McAlister, J., 2005. *China's water crisis*, Deutsche Bank China Expert Series. Available from http://www.cbiz.cn/download/aquabio.pdf.

Ministry of Water Resources and Nanjing Water Institute, 2004. *Groundwater Exploitation and Utilization in the Early 21st Century*, China Water Resources and Hydropower Publishing House, Beijing.

Ministry of Water Resources, 2004. *Water Resources Statistical Yearbook*, Ministry of Water Resources, Beijing.

Ministry of Water Resources 2007. *Water Resources Bulletin*, Ministry of Water Resources, Beijing.

Nankivell, N., 2004. China's mounting water crisis and the implications for the Chinese Communist Party, Paper presented at the twelfth annual CANCAPS Conference, Quebec City, 3–5 December 2004. Available from http://www.cancaps.ca/conf2004/nankivell.pdf

Negri, D. H., 1989. 'The common property aquifer as a differential game', *Water Resources Research,* 25(1):9–15.

Rozelle, S. and Swinnen, J., 2004. 'Success and failure of reforms: insights from transition agriculture', *Journal of Economic Literature*, 42(2):404–56.

Shah, T., 1993. *Irrigation Services Markets for Groundwater and Irrigation Development: political economy and practical policy*, Oxford University Press, Bombay.

Shi, Y., 2000. Groundwater development in China, Paper presented at the Second World Forum, The Hague, 17–22 March.

Wang, J. and Huang, J., 2004. 'Water problems in the Fuyang River Basin', *Natural Resources Transactions*, 19(4):424–9.

Wang, J. and Huang, J. and Rozelle, S., 2002. 'Groundwater management and tubewell technical efficiency', *Journal of Water Sciences Advances*, 13(2):259–63.

Wang, J., Huang, J. and Rozelle, S., 2005a. 'Evolution of tubewell ownership and production in the North China Plain', *Australian Journal of Agricultural and Resource Economics*, 49(2):177–95.

Wang, J., Huang, J., Blanke, A., Huang, Q. and Rozelle, S., 2007a. 'The development, challenges and management of groundwater in rural China', in M. Giordano and K.G. Villholth (eds), *The Agricultural Groundwater Revolution: opportunities and threats to development*, Comprehensive Assessment of Water Management in Agriculture Series, Cromwell Press, Trowbridge:37–62.

Wang, J., Huang, J., Huang, Q. and Rozelle, S., 2005b. 'Privatization of tubewells in north China: determinants and impacts on irrigated area, productivity and the water table', *Hydrogeology Journal*, 14:275–85.

Wang, J., Huang, J., Rozelle, S., Huang, Q. and Blanke, A., 2007b. 'Agriculture and groundwater development in northern China: trends, institutional responses, and policy options', *Water Policy*, 9(2007)(S1):61–74.

Wang, J., Huang, J., Xu, Z., Rozelle, S., Hussain, I. and Biltonen, E., 2007c. 'Irrigation management reforms in the Yellow River Basin: implications for water saving and poverty', *Irrigation and Drainage Journal*, 56:247–59.

Wang, J., Xu, Z., Huang, J. and Rozelle, S., 2005c. 'Incentives in water management reform: assessing the effect on water use, productivity and poverty in the Yellow River Basin', *Environment and Development Economics*, 10:769–99.

——, 2006. 'Incentives to managers and participation of farmers: which matters for water management reform in China?', *Agricultural Economics*, (34)2006:315–30.

Zhang, L., Wang, J., Huang, J. and Rozelle, S., 2008a. 'Development of groundwater markets in China: a glimpse into progress', *World Development*, doi:10.1016/j.worlddev.2007.04.012.

Zhang, L., Wang, J., Huang, J., Rozelle, S. and Huang, Q., 2008b. *Irrigation service markets for groundwater in the North China Plain: impact on irrigation water use, crop yields and farmer income*, Working Paper, Center for Chinese Agricultural Policy, Chinese Academy of Sciences, Beijing.

Acknowledgments

The authors acknowledge financial support from the National Natural Sciences Foundation (70733004) in China, the Knowledge Innovation Program of the Chinese Academy of Sciences (KSCX2-YW-N-039), the International Water Management Institute, the Food and Agriculture Organization of the United Nations and the Comprehensive Assessment of Water Management in Agriculture.

14

The impact of air pollution on mortality in Shanghai

Health and Mortality Transition in Shanghai Project Research Team

Climate change, air pollution and improving public health are some of the most serious challenges facing humankind at the beginning of the twenty-first century. While considerable effort has been made to understand the impact of changing climatic conditions and air quality on population health, our knowledge of their relationship is still rather limited, and this is especially the case in developing countries, where detailed and reliable data for environmental conditions and population health are often difficult to find.

To further improve our knowledge of declining mortality rates in the past half-century, the major characteristics of current mortality patterns and their future changes in China, researchers from The Australian National University, Fudan University, Shanghai Municipality Centre for Diseases Control and Prevention and the University of Cambridge have been conducting a detailed study of health and mortality transition in Shanghai. This project, which has been funded by a research grant from the Wellcome Trust, aims to achieve the following objectives: to provide detailed information about changes in causes of death and mortality levels in Shanghai in recent decades, to gain a better understanding of changing mortality patterns and their relationship with major socioeconomic and environmental factors, and to further examine the process of China's epidemiological transition and some proposed mortality patterns and their changes.

The project has collected and digitalised detailed population and mortality data collected from seven urban districts in Shanghai from the late 1950s to the beginning of the twenty-first century. In addition, it has gathered detailed information on weather for the past four decades and air-quality data for recent years. These data are used in this study to explore the relationship between air pollution, weather and mortality in Shanghai.

Shanghai, like many large cities in developing countries, has been experiencing a rapid transition. Fast economic growth and urban expansion have led to considerable changes in people's living environments, which in turn could greatly influence improvements in population health and socioeconomic development. The impact of changing air quality and climate on population health is an issue that has attracted great attention in recent years. Compared with some other Chinese cities, Shanghai's air quality has been reasonably good; however, in comparison with industrialised countries and the air-quality guidelines recommended by the World Health Organization (WHO), there is still a long way to go to control air pollution in Shanghai (Brajer and Mead 2004; Kan et al. 2007; WHO 2005). For example, out of 3,149 days between 1998 and 2007, when the data were available, the level of particulate matter less than 10 micrometres (μm) in aerodynamic diameter (PM_{10}) reached the standard recommended by the WHO air-quality guidelines on only 29.1 per cent of days; the levels of nitrogen dioxide and sulphur dioxide reached their standards on 50 per cent and 10 per cent of the days, respectively. The situation was worse in cool seasons when the proportion of days when these pollutants were lower than or reached the level recommended by the WHO guidelines fell to 23.9 per cent (for PM_{10}), 39.2 per cent (for nitrogen dioxide) and 6.8 per cent (for sulphur dioxide). The proportion of days when the levels of these three pollutants all reached the WHO guidelines was even lower: about 5 per cent.

An increasing number of studies exploring the link between air pollution and population health have been conducted in recent years, and they have shown that air pollution has considerable effects on morbidity and mortality in most of the populations being investigated (Kenney and Ozkaynak 1991; Xu et al. 1994, 2000; Ostro et al. 1996; Touloumi et al. 1997; Lee et al. 1999; Wong et al. 2001; Kan et al. 2003, 2007). Despite this, there is still considerable uncertainty about the nature and the strength of such effects, which is partly attributable to the following facts. First, the number of studies, particularly those conducted in developing countries, is still limited. Second, most available single-city studies use different statistical models. Because real time-series data such as air quality and mortality often have greater complexity than many commonly used models allow, conclusions drawn from these studies tend to be sensitive to model specification (Erbas and Hyndman 2005).

To improve our knowledge about the relationship between air quality and population health, especially in the context of rapid urban development in developing countries, this study systematically analyses the impact of air pollution on mortality in Shanghai. On the basis of this analysis, it also compares the impact of air pollution on daily mortality between Shanghai and Hong Kong—two major cities in Asia.

Data and methods

Mortality data collected by Shanghai Municipality Centre for Diseases Control and Prevention are used in this study. With financial support provided by the Wellcome Trust, the Health and Mortality Transition in Shanghai Project has digitalised nearly 300,000 death certificates issued between 1956 and 2001 in seven urban districts: Nanshi, Luwan, Xuhui, Changning, Zhabei, Hongkou and Yangpu. These death certificates record the following information for each deceased: name, sex, date of birth, date of death, race, personal ID number, home address, educational level, occupation, marital status, age at death, the principle cause of death and place of death (although there are variations in the amount of information included over time). They also record the level of the hospital (provincial, county or township) that certified the death, major methods of diagnosis (autopsy, biopsy, laboratory test, clinical diagnosis or inference on the basis of available evidence) and other useful information. In this study, deaths are classified into groups according to the International Classification of Diseases Revision Nine (ICD-9). Since information on the date of death is available, daily mortality patterns can be examined.

The meteorological data for Shanghai between 1956 and 2001 are extracted from a data set made available by China's Meteorological Bureau. They provide information on daily average air temperature, relative humidity and atmospheric pressure for most years. The collection of air-quality data started late in Shanghai. Detailed information of levels of major pollutants such as PM_{10}, sulphur dioxide and nitrogen dioxide, which can be accessed directly from the web site of the Shanghai Environmental Protection Bureau, became available only from the late 1990s. For this reason, our analysis of the relationship between air quality and mortality has to be restricted to the years 2000 and 2001 when detailed data for mortality, weather and air quality were all available.

Since one of the objectives of this chapter is to compare the impact of air pollution on daily mortality between Shanghai and Hong Kong, we have chosen to use the method used in Wong et al. (2001). In the analysis, Poisson regression was used with daily mortality counts as the dependent variable. To obtain the core model for the mortality outcomes in which we are interested, smoothed (through the use of the Loess smoothing function) terms for trends on days, seasonality, temperature, humidity and dummy variables for days of the weeks were fitted as the independent variables. Like Wong et al., we considered lag effects of temperature and humidity, but other factors such as the effects of holidays and influenza epidemics (measured by a large number of hospital admissions for influenza) were not included in our estimation. A detailed discussion of the method can be found in Wong et al. (2001).

The analysis was conducted in the following steps. After obtaining expected mortality counts (ξ) from the core model, Poisson regression was fitted on pollutant concentrations in the form of

$$\log[E(Y_t)] - \log(\xi_t) = \alpha + \beta x_\tau \tag{1}$$

to obtain the log relative risk (β) estimated with offset on log (ξ). In examining the impact of each pollutant on mortality, we estimated the effect of pollutant concentration for the current and previous days, and identified the best lagged day by 'Akaike's information criterion' (AIC). (We also examined the effect of the average level of pollutant concentration on the current and previous days; these results correlate highly with those estimated from the pollutant level for a single day and are not reported here.) The effect of each pollutant was then adjusted for auto-correlation using general least square regression and for the co-pollutant, which was determined using linear regression.

We then further estimated the effect of pollutant concentration on mortality in warm and cold seasons and the effect of the interaction between the season and the pollutant being examined. To make the division of seasons consistent with those used in Wong et al. (2001), we included the six months from October to March in the cold season, and those from April to September in the warm season. Because Shanghai's climatic conditions differ from those in Hong Kong, it might be more appropriate to divide the seasons in different ways; this has been examined in our analysis, but the results are not reported here. When estimating the effect of a single pollutant, we used the same model as that described above but with two additional independent variables: a dummy variable for the season (1 for the warm season and 0 for the cold season) and a variable representing the interaction between the season and the pollutant. We then add two more variables—the concentration of the co-pollutant identified earlier and the interaction of the co-pollutant and the season—into the model to estimate the effect of the major pollutant adjusting for the effect of the co-pollutant. In the analysis, we have further examined the exposure–response relationship between the concentration level of the pollutant and mortality outcomes in warm and cold seasons, although detailed results will not be discussed in this chapter. All analyses were conducted using the statistical software *R*.

Results

Table 14.1 presents summarised statistics of mortality counts, weather and major pollutants for the study period, by the two seasons specified earlier. While this seasonal division might not be the best way to group months, it has been adopted to make the analysis consistent with that used in the Hong Kong study. Alternative ways of dividing seasons have been considered in the data analysis, but their influence on the results is relatively small.

Table 14.1 **Summary statistics of mortality outcome, air pollution levels and meteorological measures by season**

	No. (days)	Mean	Standard deviation	Minimum	P_{10}	Median	P_{90}	Maximum
Mortality counts								
Non-accidental								
Cool	365	110.2	18.5	62	86	109	134	164
Warm	366	85.3	11.2	58	72	84	101	119
Cardiovascular disease								
Cool	365	40.1	9.3	16	28	39	53	65
Warm	366	28.6	5.9	14	21	28.5	36	50
Respiratory disease								
Cool	365	13.9	5.5	3	7	13	21	32
Warm	366	8.5	3.4	2	4	8	13	22
Air pollution concentrations ($\mu g/m^3$)								
Nitrogen dioxide								
Cool	365	61.9	36.1	20	32	48	100	222
Warm	366	43.6	29.5	13	20	34	82	190
Sulphur dioxide								
Cool	365	44.8	16.2	12	24	45	67	98
Warm	366	37.6	13.2	10	20	38	55	74
PM_{10}								
Cool	365	73.9	45.7	9	33	62	131	358
Warm	366	65.1	30.5	16	38	59.5	95	338
Meteorological measurements:								
Temperature (ºC)								
Cool	365	10.6	5.9	-2.1	3.6	9.8	20.1	24.4
Warm	366	23.8	5.1	9.8	16.3	24.2	30	33.2
Humidity (%)								
Cool	365	74.5	12.9	36	57	76	92	99
Warm	366	77.5	10.9	46	62	79	90	98

Sources: Authors' calculations.

Shanghai is located on China's east coast and has a subtropical monsoon climate. As shown in Table 14.1, Shanghai's average air temperature was 23.8 degrees Celsius (ºC) in the warm months and 10.6 ºC in the cold months in 2000 and 2001—about 3 ºC and 8 ºC lower than the average temperatures recorded in Hong Kong in the two respective seasons over the period from 1995 to 1997. Shanghai has a greater seasonal difference in air temperature than Hong Kong. In Shanghai, the average relative humidity for the cold months was close to that for the warm months. These levels and their seasonal difference are similar to those observed in Hong Kong.

Statistics for pollutant concentrations are presented in the middle panel of Table 14.1. In comparison with warm seasons, in cold seasons, higher levels of concentration of all three listed pollutants were found. The average levels of air pollution in the cold and warm seasons are 61.9 and 43.6 micrograms (μg) per cubic metre for nitrogen dioxide, 44.8 and 37.6 μg/cubic metre for sulphur dioxide and 73.9 and 65.1 μg/cubic metre for PM_{10}, respectively. The comparison between these results and those recorded in Hong Kong between 1995 and 1997 suggests that, in Shanghai, the level of nitrogen dioxide was slightly lower than in Hong Kong, but the level of sulphur dioxide was much higher, and their seasonal difference was greater than that observed in Hong Kong. In Shanghai, the level of PM_{10} was also markedly higher than in Hong Kong, especially in the warm months. Another notable difference found between the two cities is that Shanghai's maximum levels of nitrogen dioxide and PM_{10} are much higher than those recorded in Hong Kong.

In Shanghai, a larger number of non-accidental deaths took place in cold seasons than in warm seasons; this is particularly true for deaths caused by cardiovascular and respiratory diseases. The ratios of cold-month to warm-month deaths for the three mortality groups specified in Table 14.1 were 1.29 (non-accidental deaths), 1.40 (cardiovascular disease) and 1.64 (respiratory disease). These ratios were higher than those recorded in Hong Kong between 1995 and 1997, which were 1.17 (non-accidental deaths), 1.34 (cardiovascular disease) and 1.15 (respiratory disease).

Table 14.2 presents statistical results showing the effect of listed pollutants on the number of all non-accidental deaths and the number of deaths due to cardiovascular and respiratory diseases. These pollutants and the three types of deaths are listed in the first column of the table. The second column of the table presents the lag day, which is the day on which the level of the given pollutant has the strongest association with the number of deaths recorded on the day of observation. We compared the lag effects for up to three days, and the best single lagged day was identified by the AIC.

Table 14.2　Relative risk (RR) and 95 per cent confidence interval (CI) of the best single lagged-day effects by linear extrapolation for a tenth–ninetieth percentile change in pollutant concentration, 2000–01 (whole year)

Cause of mortality	Lag day	Unadjusted RR (95% CI)	P-value	Auto-correlation adjusted RR (95% CI)	P-value	Co-pollutant*	Adjusted for co-pollutant RR (95% CI)	P-value
Nitrogen dioxide								
Non-accidental	0	1.0002 (0.9999–1.0004)	0.127	1.0003 (1.0000–1.0006)	0.042	SO_2	1.0000 (0.9997–1.0003)	0.513
Cardiovascular disease	0	1.0002 (0.9998–1.0005)	0.209	1.0002 (0.9998–1.0006)	0.194	SO_2	1.0002 (0.9997–1.0006)	0.304
Respiratory disease	2	1.0000 (0.9993–1.0006)	0.550	1.0000 (0.9992–1.0007)	0.520	SO_2	0.9990 (0.9982–0.9999)	0.960
Sulphur dioxide								
Non-accidental	1	1.0012 (1.0007–1.0017)	0.000	1.0010 (1.0005–1.0016)	0.001	NO_2	1.0011 (1.0003–1.0019)	0.012
Cardiovascular disease	1	1.0009 (1.0002–1.0017)	0.022	1.0009 (1.0000–1.0017)	0.040	NO_2	1.0004 (0.9992–1.0017)	0.276
Respiratory disease	2	1.0028 (1.0014–1.0043)	0.001	1.0027 (1.0012–1.0043)	0.002	NO_2	1.0043 (1.0020–1.0066)	0.001
PM_{10}								
Non-accidental	0	1.0002 (1.0000–1.0004)	0.080	1.0002 (1.0000–1.0004)	0.096	SO_2	0.9998 (0.9996–1.0001)	0.836
Cardiovascular disease	0	1.0000 (0.9997–1.0003)	0.416	1.0000 (0.9997–1.0004)	0.415	SO_2	0.9999 (0.9995–1.0003)	0.670
Respiratory disease	2	1.0006 (1.0001–1.0012)	0.036	1.0007 (1.0001–1.0013)	0.029	SO_2	1.0005 (0.9997–1.0013)	0.143

* the co-pollutant that produced the least significant effect in the pollutant after adjustment

The effect of the given pollutant on mortality outcomes, in the form of relative risks (RRs) and their 95 per cent confidence interval (CI), is shown in column three and its significant level is shown in column four. These results show that the concentration of sulphur dioxide on the best single lagged day is significantly associated with all three mortality outcomes on the day of observation, and the levels of PM_{10} are significantly related to the deaths caused by respiratory diseases. These results are not, however, adjusted for auto-correlation and co-pollutants. After adjusting for auto-correlation (using the general least square method made available by the statistical software R), some changes were observed in the results—shown in columns five and six of the table. These changes are generally small, indicating that the influence of auto-correlation on the results is moderate or weak. We then further adjusted the model to take into account the impact of co-pollutants (listed in column seven), which were identified through the use of linear regression. These results are presented in columns eight and nine. After the model has been adjusted for auto-correlation and co-pollutants, the effects of nitrogen dioxide and PM_{10} on mortality outcomes are statistically insignificant. The concentration of sulphur dioxide is, however, still associated significantly with the number of all non-accidental deaths and those caused by respiratory diseases. These results are broadly similar to those in Wong et al. (2001).

Table 14.3 presents statistical results for the effects of specified pollutants on mortality outcomes in the warm and cold seasons, which have been estimated using a model similar to that described above. To estimate the impact of a single pollutant without taking into account the effect of the co-pollutant, two additional variables—a dummy season variable and the interaction between the given pollutant and the season—were added into the model. When the impact of the co-pollutant was adjusted, two further variables—the co-pollutant and the interaction of the co-pollutant and the season—were also included in the model. These results show that, in the warm season, the effect of all listed pollutants on mortality outcomes is not statistically significant, regardless of whether the effect of the co-pollutant is adjusted. In the cold season, however, the concentration of nitrogen dioxide and sulphur dioxide is related significantly to all mortality outcomes if the effect of the co-pollutant is not adjusted. After the impact of the co-pollutant was taken into account, most of these significant associations remained except that between the concentration of nitrogen dioxide (after being adjusted for the co-pollutant sulphur dioxide) and the number of deaths caused by respiratory diseases. There is no significant association between the concentration of PM_{10} and specified mortality outcomes, regardless of whether the model was adjusted

Table 14.3 Relative risk (RR) and 95 per cent confidence interval (CI) of the best single lagged-day effects by linear extrapolation for a tenth–ninetieth percentile change in pollutant concentration, 2000–01 (without[a] and with[b] adjustment for a co-pollutant)

Cause of mortality	Co-pollutant	Warm season		Cool season		Between-season
		RR (95% CI)	P-value	RR (95% CI)	P-value	P-value
Nitrogen dioxide						
Non-accidental	-	0.9997 (0.9989–1.0005)	0.736	1.0007 (1.0003–1.0010)	0.001	0.009
	SO₂	0.9993 (0.9983–1.0003)	0.877	1.0004 (1.0000–1.0009)	0.038	0.087
Cardiovascular disease	-	0.9998 (0.9986–1.0011)	0.584	1.0006 (1.0001–1.0011)	0.031	0.147
	SO₂	0.9997 (0.9983–1.0012)	0.612	1.0006 (1.0000–1.0013)	0.045	0.458
Respiratory disease	-	0.9993 (0.9970–1.0015)	0.706	1.0014 (1.0005–1.0024)	0.008	0.048
	SO₂	0.9990 (0.9963–1.0017)	0.724	1.0008 (0.9996–1.0020)	0.131	0.064
Sulphur dioxide						
Non-accidental	-	1.0004 (0.9988–1.0021)	0.335	1.0015 (1.0008–1.0022)	0.000	0.110
	NO₂	1.0009 (0.9989–1.0029)	0.229	1.0016 (1.0008–1.0025)	0.001	0.097
Cardiovascular disease	-	1.0001 (0.9976–1.0027)	0.465	1.0015 (1.0004–1.0025)	0.015	0.187
	NO₂	1.0011 (0.9981–1.0042)	0.269	1.0015 (1.0002–1.0028)	0.031	0.156
Respiratory disease	-	1.0026 (0.9978–1.0075)	0.190	1.0025 (1.0004–1.0046)	0.026	0.959
	NO₂	1.0043 (0.9985–1.0100)	0.113	1.0026 (1.0002–1.0051)	0.040	0.980
PM₁₀						
Non-accidental	-	1.0002 (0.9996–1.0009)	0.267	1.0002 (0.9999–1.0005)	0.125	0.901
	SO₂	1.0002 (0.9994–1.0010)	0.341	0.9998 (0.9995–1.0002)	0.809	0.177
Cardiovascular disease	-	1.0004 (0.9994–1.0014)	0.260	1.0003 (0.9999–1.0007)	0.086	0.943
	SO₂	1.0007 (0.9995–1.0018)	0.174	0.9999 (0.9994–1.0004)	0.683	0.044
Respiratory disease	-	1.0015 (0.9996–1.0034)	0.094	1.0005 (0.9997–1.0012)	0.145	0.230
	SO₂	1.0012 (0.9990–1.0035)	0.178	0.9999 (0.9990–1.0009)	0.546	0.725

[a] estimated from core model + pollutant + season + pollutant × season
[b] estimated from <1> + co-pollutant + co-pollutant × season

for co-pollutants. As indicated by the statistics shown in the last column of Table 14.3, the between-season differences are statistically significant in estimating the effect of nitrogen dioxide on all non-accidental deaths and deaths caused by respiratory diseases if the impact of the co-pollutant is not adjusted. The between-season difference is also significant when the effect of PM_{10} on deaths due to cardiovascular diseases is estimated with no adjustment for the co-pollutant.

Discussion

The results reported in the previous section show that air pollution in Shanghai has a notable impact on population health and mortality. This is particularly the case for sulphur dioxide, which, after adjusting for auto-correlation and for co-pollutants, still significantly affects the relative risk for all non-accidental deaths and those caused by respiratory diseases. The health impact of air pollution is more observable in cold months, when the level of pollution concentration is markedly higher, than in warm months. As indicated by the results, in the cold season, the positive exposure–response relationships between sulphur dioxide, nitrogen dioxide and specified mortality outcomes are much stronger than in the warm season. These results are supported by evidence presented by other researchers, despite the variation in the reported magnitude of such effects between different studies (Kan et al. 2007; Chen et al. 2007). Epidemiological and medical research have also found that lower temperatures are associated with an increase in hospital admissions for congestive heart failure. 'Both lower temperatures and high air pollutant concentrations were related to increased blood viscosity. Changes in blood rheology may be caused by an inflammatory process in the lung induced by air pollution or by thermoregulatory adjustment to mild surface cooling in cold weather' (Wong et al. 2001:338; see also Keatinge et al. 1984; Peters et al. 1997; Tsai et al. 2003).

Shanghai's municipal government has made some progress in controlling air pollution. Our analysis of air quality in Shanghai between 1998 and 2007 shows that the concentration of PM_{10} reached its peak in 2003 and started to decline thereafter. The level of nitrogen dioxide also displayed a trend of slow decrease despite a rapid increase in the number of motor vehicles. The level of sulphur dioxide, however, seems to have increased slightly in recent years; this is particularly noteworthy given the adverse health impacts of sulphur dioxide. Further improving its air quality will not only make Shanghai a clean city, it will contribute greatly to improvements in the health of its population. Brajer and Mead (2004) recently estimated the number of deaths that could be averted through controlling air pollution in Chinese cities, including Shanghai.

While their estimation procedures are questionable, the issues addressed in their paper certainly deserve some attention.

While the conclusions drawn from this study are broadly similar to those reported by the study conducted in Hong Kong by Wong et al. (2001), there are a few differences in the results. For example, while the concentrations of sulphur dioxide and nitrogen dioxide (after controlling for auto-regression and co-pollutants) in both cities have a marked impact on major mortality outcomes, especially in the cold season, the best lag days of such impact are not the same. The impact of sulphur dioxide on major mortality outcomes is readily observable in both cities, but the level of its concentration is more closely related to deaths caused by respiratory diseases in Shanghai, while it shows a stronger association with deaths due to cardiovascular disease in Hong Kong. Furthermore, the health impact of the concentration of nitrogen dioxide is similar in the two cities in the cold season, but it has a stronger association with deaths caused by cardiovascular diseases in Hong Kong if the impact of the season is not taken into consideration in the model. Some of these differences can be explained by variations in the levels of air pollution between Shanghai and Hong Kong; others might be related to the different interaction between various climatic and environmental factors in the two cities—or the selection and specification of statistical models. These possibilities will be examined in our future investigations.

The results reported in this chapter were obtained from our preliminary data analysis of death records and weather and air-quality data gathered in Shanghai in 2000 and 2001. To make the results comparable with those reported by Wong et al. (2001) in their study conducted in Hong Kong, we have applied a method similar to the one they used. This approach has some advantages in facilitating comparative research, but it might have introduced or imposed some restrictions and limitations, especially in the selection and application of the most appropriate statistical methods that could be used effectively to examine the questions under investigation and to analyse the large amount of data collected from Shanghai. We are now exploring the possibility of using more effective statistical methods to analyse the data, and hope this will lead to a more comprehensive examination and contribute to a better understanding of the relationship between weather, air quality, population health and mortality.

References

Brajer, V. and Mead, R.W., 2004. 'Valuing air pollution mortality in China's cities', *Urban Studies*, 41(8):1567–85.

Chen, B., Kan, H. and Zhang, Y., 2007. 'Energy, air pollution and health: a case study in Shanghai', *Japanese Journal of Hygiene*, 62(2):125–9.

Erbas, B. and Hyndman, R.J., 2005. 'Sensitivity of the estimated air pollution–respiratory admission relationship to statistical model choice', *International Journal of Environmental Health Research*, 15(6):437–48.

Kan, H., Chen, B. and Jia, J., 2003. 'A case-crossover study of ambient air pollution and daily mortality in Shanghai', *Chinese Journal of Epidemiology* (in Chinese), 24(10):863–7.

Kan, H., London, S.J., Chen, G., Zhang, Y., Song, G., Zhao, N., Jiang, L. and Chen, B., 2007. 'Differentiating the effects of fine and coarse particles on daily mortality in Shanghai, China', *Environment International*, 33(3):376–84.

Keatinge, W.R., Coleshaw, S.R.K., Cotter, F., Mattock, M., Murphy, M. and Chelliah, R., 1984. 'Increases in platelet and red cell counts, blood viscosity, and arterial pressure during mild surface cooling: factors in mortality from coronary and cerebral thrombosis in winter', *British Medical Journal (Clinical Research Edition)*, 289:1405–8.

Kenney, P.L. and Ozkaynak, H., 1991. 'Associations of daily mortality and air pollution in Los Angeles County', *Environmental Research*, 54(2):99–120.

Lee, J., Shin, D. and Chung, Y., 1999. 'Air pollution and daily mortality in Seoul and Ulsan, Korea', *Environmental Health Perspectives*, 107(2):149–54.

Ostro, B., Sanchez, J.M., Aranda, C. and Eskeland, G.S., 1996. 'Air pollution and mortality: results from a study of Santiago, Chile', *Journal of Exposure Analysis and Environmental Epidemiology*, 6(1):97–114.

Peters, A., Doring, A., Wichmann, H.E. and Keonig, W., 1997. 'Increased plasma viscosity during an air pollution episode: a link to mortality?', *Lancet*, 349:1582–7.

Touloumi, G., Katsouyanni, K., Zmirou, D., Schwartz, J., Spix, C., Ponce de Leon, A., Tobias, A., Quennel, P., Rabcsenko, D. and Bacharova, L., 1997. 'Short-term effects of ambient oxidant exposure on mortality: a combined analysis within the APHEA project', *American Journal of Epidemiology*, 146(2):177–85.

Tsai, S., Goggins, W., Chiu, H. and Yang, C., 2003. 'Evidence for an association between air pollution and daily stroke admissions in Kaohsiung, Taiwan', *Stroke*, 34:2612–16.

Wong, C., Ma, S., Hedley, A.J. and Lam, T., 2001. 'Effect of air pollution on daily mortality in Hong Kong', *Environmental Health Perspectives*, 109(4):335–40.

World Health Organization (WHO), 2005. *World Health Organization air quality guidelines global update 2005*, Report on a Working Group Meeting, Bonn, Germany, 18–20 October.

Xu, X., Gao, J., Dockery, D.W. and Chen, Y., 1994. 'Air pollution and daily mortality in residential areas of Beijing, China', *Archives of Environmental Health*, 49(4):216–22.

Xu, Z., Yu, D., Jing, L. and Xu, X., 2000. 'Air pollution and daily mortality in Shenyang, China', *Archives of Environmental Health*, 55(2):115–20.

Acknowledgments

Researchers involved in preparing this chapter include Zhongwei Zhao, Xizhe Peng, Yuan Cheng, Xuehui Han, Guixiang Song, Feng Zhou, Yuhua Shi and Richard Smith. Correspondence should be directed to Professor Zhongwei Zhao, Australian Demographic and Social Research Institute, The Australian National University, ACT 0200, Australia. The Shanghai Mortality Research Team would like to thank the Wellcome Trust for its support for this project (Project Reference No.070318). We would also like to thank Pfizer Inc. and the Hong Kong Research Grants Council for their support for the research.

15

Energy and environment in China

Kejun Jiang and Xiulian Hu

Energy use in China

Due to rapid economic growth, China's total primary energy consumption increased from 400 mega-tonnes of oil equivalent (Mtoe) in 1978 to nearly 1,820 Mtoe in 2007, with an annual average rate of increase of 5.3 per cent (Figure 15.1) (NBS 2006a, 2006b, 2007a). Coal is the major energy source, providing 70.7 per cent of total primary energy use in 1978 and 71 per cent in 2006 (Figure 15.2). There has been a dramatic surge in recent years in the rate of increase of energy use in China, as well as widespread energy shortages.

China's energy development strategy gives high priority to conservation and improvements in energy efficiency, as well as to the efficient and clean use of coal and other fossil-fuel energy sources. The objective of developing clean-coal technology is to improve the efficiency of coal utilisation, reduce environmental pollution and promote economic development. High efficiency and clean technology will be crucial for China to achieve a low-emission development path.

Energy-efficiency improvements in the steel-making industry have been driven by the diffusion of advanced technology (Figure 15.3). Despite these improvements, steel making in China remains about 20 per cent less efficient than in Japan.

To realise China's sustainable development, the national energy development strategy includes a policy of energy conservation prioritisation, as well as vigorously developing renewable energy and new energies in China.

Figure 15.1 **Energy production and consumption in China, 1960–2007**

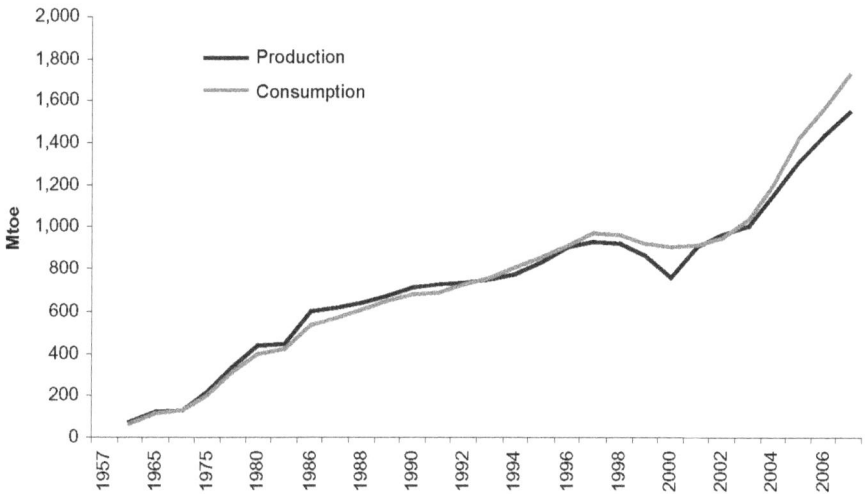

Source: National Bureau of Statistics (NBS), 2007b. *China Statistical Yearbook 2007*, China Statistics Press, Beijing.

Figure 15.2 **Primary energy use in China by energy type, 1957–2006**

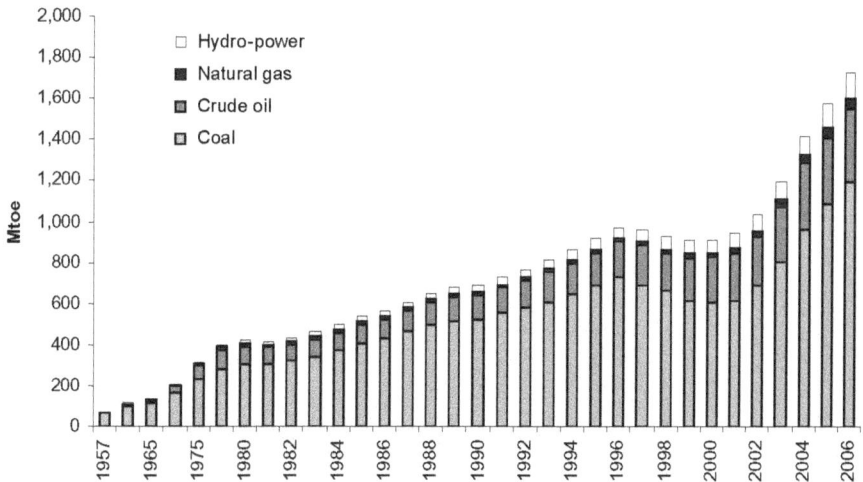

Source: National Bureau of Statistics (NBS), 2007b. *China Statistical Yearbook 2007*, China Statistics Press, Beijing.

Figure 15.3 **Introduction of energy-efficiency improvements in steel making in China, 1970–2000**

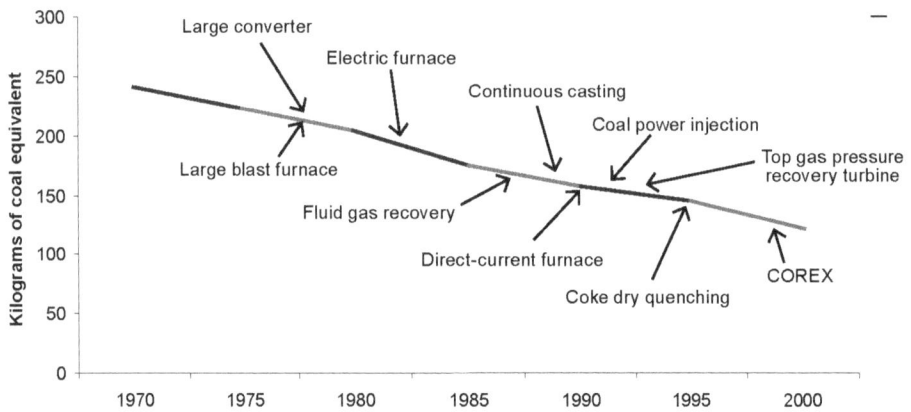

Source: Jiang, K., Morita, T., Masui, T. and Matsuoka, Y., 2006. 'Global long-term GHG mitigation emission scenarios based on AIM', *Environment Economics and Policy Studies*, 3.

Figure 15.4 **Energy-efficiency improvement in China, 1960–2006**

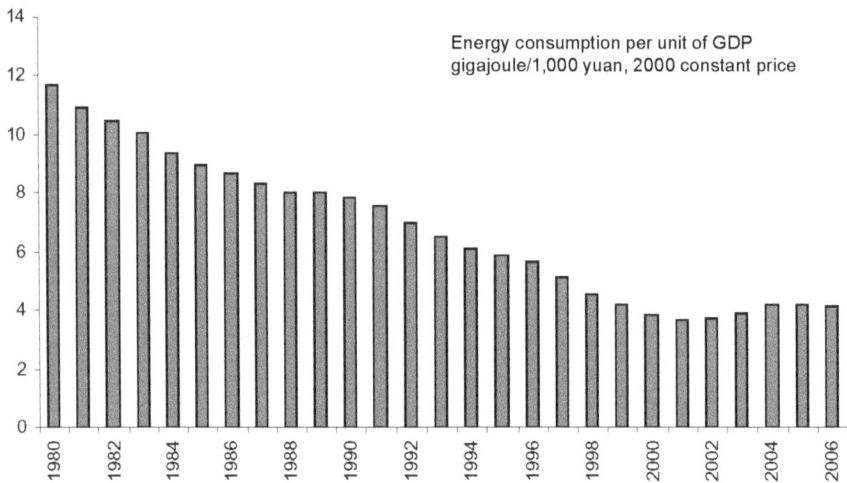

Source: Calculated by the authors based on data from National Bureau of Statistics (NBS), 2007b. *China Statistical Yearbook 2007*, China Statistics Press, Beijing.

China has implemented a series of economic and technological policies for energy conservation and has also established a three-tier system of energy saving within the central government, local governments and industry and enterprises since the 1980s. The 1998 Energy Conservation Law further established the energy-management system. China has implemented a series of energy-saving technological policies, including starting a national 'Energy Conservation Propaganda Week', energy-efficiency standards, a marking and authentication system, and is effectively advancing energy-saving practices and improving energy efficiency.

To promote the application of new and renewable energies in the long term, China has given financial subsidies and support to technical development. It also provides discount loans for the development and use of small hydro-power and wind-power plants, in addition to implementing tax preferences and protective price policies. With the implementation of these measures, the energy-consumption structure is being constantly optimised and the quality of energy supply has improved. The rapid development of clean energies and the growth of the proportion of high-quality energy in the total play an important role in enhancing energy efficiency and improving atmospheric quality.

As a result, energy efficiency improved significantly after 1980. From 1980 to 2000, the annual average energy improvement rate was 5.4 per cent (Figure 15.4). Due to the rapid development of industry in China after 2000, however, especially of energy-intensive products, improvements in energy efficiency with regard to gross domestic product (GDP) were negative.

Energy supply in China

Since the period of reform and 'opening-up', the energy industry has achieved rapid growth, contributing to the sustainable development of the national economy. China has set up an energy-supply framework that gives priority to coal and focuses on electricity generation and the development of renewable energy. It has built an integrated energy-supply system; there are many large coal-mines with an output capacity of more then 10 million tonnes and petroleum-producing bases have been set up in Daqing, Shengli, Liaohe and Talimu. At the same time, gas production has increased rapidly, the proportion of commercialised renewable resources in the primary energy structure has improved, and remarkable progress has been made in electricity generation.

In 2004, China replaced Russia as the second-biggest energy producing country in the world. In 2006, primary energy output was 2,201.56 mega-tonnes of coal equivalent (Mtce)—an increase of 251 per cent from 1978. Coal output was up to 2,373 Mt, maintaining China's first-place ranking in the world; gas output reached 58.55 billion cubic metres; and oil output was 185 Mt, while power generation was 2,865.7 terrawatt hours (TWh)—both of which ranked second in the world.

Due to the nature of China's energy-resource reserves, coal has long accounted for a large proportion of primary energy—up to 70 per cent—and, since 2001, the proportion has increased steadily, to 76.7 per cent in 2006. The elasticity coefficient for electric power became more than 1 after 2000.

Tables 15.1 and 15.2 show energy output by fuel from 1990 to 2006, the rank of Chinese energy output, and the gross and structure of Chinese energy production from 1978 to 2006.

After 2003, China became a net importer of oil, and, by 2007, a net importer of energy. At present, China is the third-largest oil importer in the world. In 2006, China imported 145.18 Mt of crude oil and 46.01 Mt of petroleum products, while it exported 6.34 Mt of crude oil and 15.51 Mt of petroleum products; therefore, net crude oil imports were 138.84 Mt and net imports of petroleum products were 30.5 Mt. In 2006, the volume of imports of crude oil and oil products was worth US$68 billion, accounting for 7.02 per cent of China's total volume of

Table 15.1   **Chinese energy output by fuel, 1990–2006**

Year	Raw coal (Mt)	Crude oil (Mt)	Gas (hundred million m³)	Electric power (TWh)	Hydro-power (TWh)
1990	1,080	138.3	153.0	621.2	126.7
1991	1,087	141.0	160.7	677.5	124.7
1992	1,116	142.1	157.9	753.9	130.7
1993	1,150	145.2	167.7	839.5	151.8
1994	1,240	146.1	175.6	928.1	167.4
1995	1,361	150.1	179.5	1,007.0	190.6
1996	1,397	157.3	201.1	1,081.3	188.0
1997	1,373	160.7	227.0	1,135.6	196.0
1998	1,250	161.0	232.8	1,167.0	198.9
1999	1,280	160.0	252.0	1,239.3	196.6
2000	1,299	163.0	272.0	1,355.6	222.4
2001	1,381	164.0	303.3	1,480.8	277.4
2002	1,455	167.0	326.6	1,654.0	288.0
2003	1,722	169.6	350.2	1,910.6	283.7
2004	1,992	175.87	414.6	2,203.3	353.5
2005	2,205	181.35	493.2	2,500.3	397.0
2006	2,373	184.77	585.5	2,865.7	435.8

Source: National Bureau of Statistics (NBS), 2007b. *China Statistical Yearbook 2007*, China Statistics Press, Beijing.

Table 15.2 **Gross and structure of Chinese energy production, 1978–2006**

Year	Gross output (10,000 tonnes of standard coal)	Proportion of energy production (%)			
		Raw coal	Crude oil	Gas	Hydro-power, nuclear power, wind power
1978	62,770	70.3	23.7	2.9	3.1
1980	63,735	69.4	23.8	3.0	3.8
1990	103,922	74.2	19.0	2.0	4.8
1995	129,034	75.3	16.6	1.9	6.2
2000	128,978	72.0	18.1	2.8	7.2
2001	137,445	71.8	17.0	2.9	8.2
2002	143,810	72.3	16.6	3.0	8.1
2003	163,842	75.1	14.8	2.8	7.3
2004	187,341	76.0	13.4	2.9	7.7
2005	206,068	76.5	12.6	3.2	7.7
2006	221,056	76.7	11.9	3.5	7.9

Source: National Bureau of Statistics (NBS), 2007b. *China Statistical Yearbook 2007*, China Statistics Press, Beijing.

imports, while the volume of exports of crude oil and oil products was worth US$17.48 billion, accounting for 2.21 per cent of China's total volume of exports. Since 2001, China has imported much more coal than ever before—up to 38.25 Mt in 2006, which was 20 times more than in 2000.

Energy and environmental development

Coal is the main energy source in China, and the coal-based energy structure will be difficult to change in the long term. The relatively backward modes of coal production and consumption have increased the pressure for environmental protection. Coal consumption is the main reason for China's air pollution and is the main source of the country's greenhouse gas emissions. With the constant increase of motor vehicles in China, air pollution in some cities has become a combination of soot and vehicle exhausts.

The main pollutant emissions

Pollution emissions from energy production in China. The impact of energy production on the environment is manifested in the dust pollution and gas emissions from coal-mining and processing, and atmospheric pollution from the spontaneous combustion of coal gangue. According to data from the *China*

Statistical Yearbook 2007 (NBS 2007b), China's coal-mining and washing industry in 2006 emitted 145,000 tonnes of sulphur dioxide, 122,000 tonnes of smoke and 176,000 tonnes of industrial dust, with removal rates of 38 per cent (sulphur dioxide), 88 per cent (smoke) and 40 per cent (industrial dust). In addition, in 2006, coal-mining and processing emitted 5 billion cubic metres of methane and 2.3 billion cubic metres of mine water, with mine-water utilisation of less than 40 per cent. In 2006, there were more than 1,600 waste dumps containing 4,200 Mt of coal gangue, covering 17,000 hectares. More than 700,000 hectares have been lost to land collapse caused by mining.

The other major atmospheric pollutants in the energy-production process are sulphur dioxide, nitrogen oxides and soot from thermal power plants. Because of the rapid increase in coal consumption in recent years, China's thermal power industry uses mostly unwashed steam coal, making the power industry the main source of air pollution. According to the *China Environmental Yearbook 2007* (SEPA 2007), in 2006, the thermal power industry emitted 12.041 Mt of sulphur dioxide (59 per cent of the total for China), 3.467 Mt of soot and 14,000 tonnes of industrial dust.

Pollutant emissions from energy consumption in China. The high growth of energy consumption in the industrial sector is still the main source of emissions of sulphur dioxide. In 2006, the industrial sector emitted 10.18 Mt of sulphur dioxide, 18 per cent of which came from the production of non-metallic mineral products, 15 per cent from ferrous and non-ferrous metal smelting and processing and 11 per cent from the chemical industry.

The main coal-consuming provinces are Shandong, Hebei, Shanxi and Jiangsu, and other high-sulphur coal-consuming provinces are Guizhou, Sichuan and Chongqing in the southwest, which are the leading provinces for sulphur dioxide emissions.

China's State Environmental Protection Administration (SEPA) monitored 522 cities in 2005: 77.4 per cent of the cities had an average annual concentration of sulphur dioxide at 'Standard II' (0.06 milligrams per cubic metre), while 6.5 per cent were above Standard III (0.10 milligrams per cubic metre). The cities with the highest levels of sulphur dioxide pollution were distributed mainly in the provinces, autonomous regions and municipalities of Shanxi, Hebei, Gansu, Inner Mongolia, Yunnan, Guangxi, Hubei, Shaanxi, Henan, Hunan, Sichuan, Liaoning and Chongqing. In addition, based on comparisons of coal consumption and urban air quality in key cities, the cities with the lowest sulphur dioxide concentrations and best urban environmental quality were Haikou, Sanya, Zhaoqing, Beihai, Zhanjiang and Zhuhai—all of which had low levels of coal consumption and a good-quality energy structure.

Emissions of nitrogen oxides and energy activities are closely related. The main sources for the emission of these gases are power-generation boilers, industrial boilers and kilns using natural gas, coal and heavy oil as their main fuel source, as well as the production of nitric acid, nitrogen fertiliser and explosives and vehicle exhausts. As there are a variety of factors related to nitrogen oxides emission and combustion processes, such as the technological level of the production process, it is difficult for SEPA to monitor these emissions, and there are no accurate data among the environmental statistics for the emissions of nitrogen oxides. According to a preliminary estimate, in 2006, China emitted a total of about 14 Mt of nitrogen oxides—an increase of 8.5 per cent compared with 2000. In the total amount of nitrogen oxides, thermal power accounted for 40 per cent and industrial boilers and kilns, chemical production processes and vehicle exhausts made up the remainder.

In 2006, the main sources of industrial dust emissions in China were as follows: thermal power, 45 per cent; non-metallic mineral production, 18 per cent; ferrous and non-ferrous metals smelting and pressing, 15 per cent; chemical raw materials and chemical production, 10 per cent.

In 2005, monitoring of acid rain was carried out in 696 Chinese cities. Acid rain was recorded in 357 cities, with the average annual rainfall pH value in the range of 3.87–8.35, accounting for 51.3 per cent of city statistics. The cities with an average annual rainfall pH value of less than 5.6 accounted for 38.4 per cent, with an increase of 1.8 per cent compared with 2004. The proportion of cities with an average annual rainfall pH value of 5.6 or less increased by 0.7 per cent, while the proportion of cities with a pH value of less than 4.5 increased by 1.9 per cent. The cities with frequency of acid rain over 80 per cent have increased by 2.8 per cent. The cities with low pH value of the average annual precipitation and a higher frequency of acid rain than that of 2004 shows that China's acid rain pollution became more serious in 2005. Table 15.3 shows the major pollution emissions in China from 1995 to 2006.

Greenhouse gas emissions

The rapid increase of energy use in China has caused large quantities of carbon dioxide emissions. Figure 15.5 presents recent carbon dioxide emissions in China. China is now the world's second-largest emitter of greenhouse gases, after the United States. It is believed that, if there is no change in the trend of increasing energy use, China's carbon dioxide emissions will overtake those of the United States in the near future—making China the largest emitter in the world.

Table 15.3   **Major pollution emissions in China, 1995–2006**

Year	Sulphur dioxide (Mt)	Dust (Mt)	Industrial dust (Mt)	Waste water (100 million m³)	Chemical oxygen demand (Mt)	Industrial solid waste (Mt)
1995	23.70	17.44	17.31	415.3	..	22.27
2000	19.95	11.65	10.92	415.2	14.45	31.86
2001	19.48	10.70	9.91	432.9	14.05	28.94
2002	19.27	10.13	9.41	439.5	13.67	26.35
2003	21.59	10.48	10.21	460.0	13.34	19.41
2004	22.55	10.95	9.05	482.4	13.39	17.62
2005	25.49	11.82	9.11	523.0	14.14	16.55
2006	25.89	10.89	8.08	536.8	14.28	13.02

Source: State Environmental Protection Administration (SEPA), 2007. *China Environmental Year Book 2007*, China Environmental Yearbook Press, Beijing.

Figure 15.5   **Carbon dioxide emissions in China, 1990–2006**

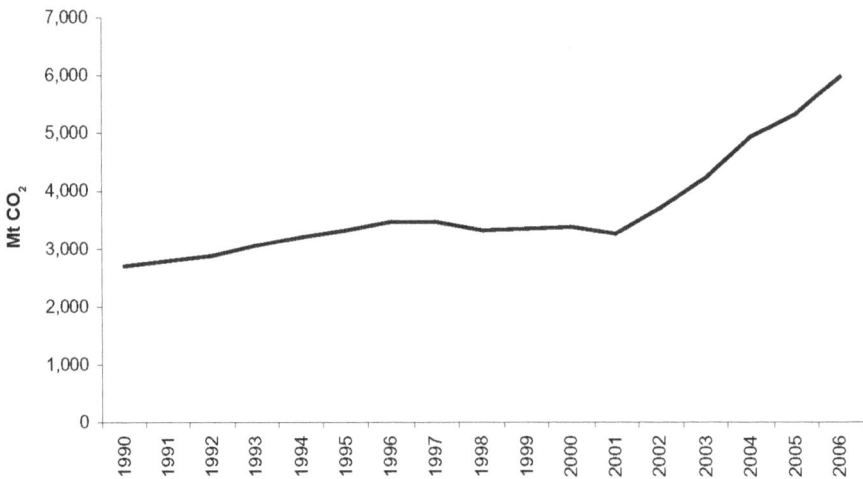

Source: State Environmental Protection Administration (SEPA), 2007. *China Environmental Year Book 2007*, China Environmental Yearbook Press, Beijing.

China's massive exploitation of fossil fuels has resulted in a rapid increase in emissions of methane, which is the second most important greenhouse gas after carbon dioxide. Natural processes and human activities can generate methane emissions. The main human-made sources of methane include rice cultivation, cattle and sheep breeding, decomposing garbage in landfill, and coal-mining, gas and oil extraction, processing and transportation. Global annual methane emissions caused by animal intestinal fermentation are 60–100 Mt, accounting for 22 per cent of the anthropogenic total; methane emissions from animal dung amount to 20–30 Mt, accounting for 5.5–8 per cent of the anthropogenic total. China is the leading country for animal breeding, and animal enteric fermentation and animal dung are China's largest sources of methane emissions. China's rice plantations account for about 21 per cent of the global total, and rice growing is one of the main sources of methane emissions.

China is the world's largest coal-mining country, most of which is underground mining, accounting for 95 per cent of the total coal production. Coal-mining is the key source of energy production-related methane emissions. According to the *China Climate Change Initial National Communication* submitted in 2004 by the Chinese government to the UN Framework Convention on Climate Change conference, in 1994, China's methane emissions were about 34.29 Mt—9.37 Mt (27 per cent) of which were caused by energy-generation activities. With the rapid increase in China's coal and natural gas production in recent years, methane emissions caused by energy activity have also been increasing greatly. It is estimated that gas produced in the coal-mining process (the main component of which is methane) in China increased from about 9.6 billion cubic metres in 2000 to about 12 billion cubic metres in 2004—an increase of 25 per cent. The large volume of coal-mine gas emissions is not only a waste of a valuable clean-energy resource, it has caused a rapid rise in methane gas emissions in China.

Policy measures for environmental and energy development

The Chinese Government attaches great importance to environmental protection, and strengthening environmental protection has become a basic national policy. Community awareness of environmental protection has generally improved. After the UN Conference on the Environment and Development in 1992, China formulated the *China Agenda 21*, and took comprehensive measures in the fields of law and economics and applied other means to strengthen environmental protection, achieving positive progress. The main components of China's energy policy are the reduction and effective management of environmental damage and pollution in the process of energy development and utilisation.

Adhering to the strategy of economical, clean and safe development

China's energy development adheres to a policy of economical, clean and safe development and solves problems through development and reform. The government implemented the scientific development concept, adheres to the people-centred policy or strategy, changes the concept of development to give priority to human development, innovates the mode of development, improves the quality of growth and development. The government insists that energy development follow a path of high technology, low resource consumption, less environmental pollution, good economic returns and public safety in order to achieve comprehensive, coordinated and sustainable development.

China's energy development adheres to the basic principles of relying on the domestic market and opening up to the global economy, while ensuring a steady supply of energy and promoting global energy development along with the steady growth of domestic energy. China's energy development will bring more development opportunities to other countries in the world, open space for development in the international market and make a positive contribution to the security and stability of the world's energy supplies.

The basic contents of China's energy strategy are: to give priority to conservation; to rely on the domestic market; to attach importance to diversified development and environmental protection; to rely on science and technology; to strengthen international cooperation; and to make efforts to build a stable, economical, clean and safe energy-supply system to support sustainable economic and social development.

Giving priority to conservation. Resource conservation is a basic national policy in China, combining energy development and conservation. Priority is given to conservation, changing the mould of economic development, adjusting the industrial structure and encouraging research and development of energy-saving technology. Priority is also given to the implementation of energy-saving products, improved energy management and energy-saving regulations and continuing improvements in energy efficiency.

Domestic supply. China will rely mainly on domestic energy supplies to meet its growing demand for energy by steadily improving the security and capacity of the domestic supply.

Diversified development. China will ensure a steady supply of energy through an orderly development of coal-generated electricity, accelerating the development of energy generation from oil and natural gas and encouraging the development of coal-bed methane production. It will also develop renewable energy sources such as hydro-power, and promote nuclear power development, the scientific development of alternative energy sources, the optimisation of energy structures and the implementation of multi-source energy.

Science and technology. China relies on the scientific and technological progress of energy development, while strengthening the capacity for independent innovation, improving the absorption of imported technology and re-innovation capabilities, achieving breakthroughs in the development of new energy technology, improving key technology and major equipment manufacturing, creating new methods of energy development and enhancing its development potential.

Environmental protection. China has a goal of building a resource-saving and environmentally friendly society. It actively promotes coordination between energy and environmental development, providing protection in the course of development and striving for sustainable development.

Mutually beneficial cooperation. China adheres to a principle of equality and mutual benefit. It has strengthened its energy cooperation with the International Energy Agency and the international community. It is working actively to improve the mechanisms for global cooperation to achieve and maintain international energy security and stability.

The seventeenth National People's Congress of the Communist Party of China, held in October 2007, identified a need to speed up the transformation of development, to quadruple per capita GDP from 2000 levels by 2020 by optimising the industrial structure, improving efficiency and lowering consumption while protecting the environment. Guidelines for the eleventh Five-Year Plan (2006–10) for National Economic and Social Development clearly propose that, by 2010, the unit of energy use of GDP be reduced by 20 per cent from 2005 levels, and the emission of major pollutants should be reduced by 10 per cent.

To achieve the goals of economic and social development, the energy development objectives in the eleventh Five-Year Plan are: by the end of the Five-Year Plan, the energy supply will meet the basic needs of the national economy and social development; progress will be made in energy conservation; energy efficiency will be increased; industrial structures will be further optimised; technological progress, economic efficiency and market competitiveness will be enhanced significantly; macroeconomic regulation and control of the energy sector, market supervision, laws and regulations, and the emergency warning system adapted for the socialist market economy will gradually be perfected; and coordination between energy, economic, social and environmental development will be achieved.

Promotion of energy conservation

China is a developing country with a large population and a relative shortage of resources. To achieve sustainable economic and social development, China must take the road of resource conservation. China planned and carried out energy-conservation efforts in the early 1980s, through the implementation of the policy of 'development and conservation simultaneously, conservation first', and, by the end of last century, it had achieved its goal of a doubling of economic and energy-consumption growth. To continue to promote energy conservation, the Chinese government has further proposed to adopt resource conservation as a basic national policy and it has issued a decision called 'work of the State Council on strengthening energy saving'. The Chinese government has always regarded energy conservation as the main component of macro-regulation and control, to change the development mould and optimise the industrial structure. In the course of promoting energy savings, the government emphasises relying on restructuring, on scientific and technological progress, strengthening management and the legal system, deepening reforms and the participation of all citizens. The government formulated and implemented a 'special long-term plan for energy saving', set goals to reduce energy consumption during the eleventh Five-Year Plan and to implement energy-saving tasks in the provinces, autonomous regions and municipalities, as well as in key enterprises. China is improving its system of GDP and energy-consumption indicators, and energy consumption will be included in comprehensive evaluations and annual assessments of economic and social development. China will implement a bulletin of units of GDP energy consumption indexes, implement energy-saving responsibilities and accountability, construct an energy-efficient industrial system and promote fundamental changes in economic development. China's measures for full implementation of energy conservation measures include the following.

- **Promote structural readjustment.** For a long time, the main reason for China's low-energy efficiency has been that the mode of economic growth is extensive and high-energy industries takes too large a part. China insists on changing the mode of development, adjusting the industrial structure and the internal structure of industry as the strategic focus for energy conservation, and creating an economic development mould with low input, low consumption, low emissions and high efficiency. China will accelerate the optimisation and upgrading of the industrial structure, actively developing hi-tech and service industries, strictly limiting the development of high-energy and high water use industries, and eliminating backward production capacity. It will promote fundamental changes in economic development to speed up construction of an energy-efficient industrial system.

- **Strengthen industrial energy conservation.** Industry is the key area of China's energy consumption. China has decided to take the road to high scientific and technological content, good economic returns, low resource consumption, less environmental pollution, less human resources, faster development of high-tech industries, and the use of new and advanced applicable technologies to transform traditional industries and improve the overall level of industrial development. The government will focus on strengthening energy savings in the production of iron and steel, non-ferrous metals, coal, electricity, oil, petrochemicals, chemicals and construction materials. Energy savings will be implemented in 1,000 enterprises, with a focus on improving the management of energy savings in industrial enterprises that consume more than 10,000 tonnes of standard coal. Their product structure will be adjusted, accelerating technological transformation and improving their management, to reduce energy consumption. The government supports a number of major energy-saving demonstration projects, promoting industrial energy-efficiency levels and eliminating backward, high-energy products, as well as improving access to an efficient energy market.

- **Implementation of energy-saving projects.** China is implementing 10 key energy-saving projects, such as oil conservation, cogeneration and waste-heat utilisation; demonstration projects focused on energy-saving construction; and encouraging the use of energy-saving products. The government will push for transformation of the existing energy-conservation measures in construction, and promote the extensive use of new building materials. It will implement measures for conservation and alternative petroleum engineering and scientific development of alternative fuels. It will speed up the phasing out of old automobiles and ships, while actively developing public transport, restricting high fuel use vehicles and developing energy-saving and environmentally friendly vehicles. China will speed up the transformation of coal-fired industrial boilers and furnaces and regional cogeneration and waste-heat capture programs to increase the efficiency of energy use. The government will promote 'green' lighting projects and the use of efficient electrical appliances, high-efficiency coal stoves and energy-saving housing technology. In rural areas, old high-energy farm machinery and fishing boats will be phased out, and agricultural and rural energy-saving measures will be promoted. The government and the community will play leading roles in energy conservation, monitoring and technical service systems.

- **Strengthening of energy management.** The Chinese government has established a system of compulsory purchases of energy-saving products, and actively promotes the prioritised procurement of energy-conservation products (included water-related products). Government procurement will play an active role in guiding these policies, leading the society towards the use of energy-saving products. It will study and formulate fiscal and taxation policies to encourage energy conservation, implement preferential taxation policies for comprehensive utilisation of resources and establish a multi-channel energy financing mechanism. China will deepen the reform of energy prices, create a situation favourable to the energy price formation mechanism, implement fixed-asset investment project assessment and verification of energy-saving systems, and restrict the source of energy-consumption growth. The government will establish new mechanisms for energy-saving enterprises, labelling of energy-efficient products, and promote energy management and energy-saving voluntary agreements. It will establish sound energy-saving laws and regulations, and step up law enforcement.

- **Creating an energy-saving society.** China publicises the importance of energy conservation in a variety of forms, continuously improving people's awareness of resource conservation. It promotes a culture of energy conservation, and works hard to create healthy, civilised and economical consumption patterns. Energy conservation is an integral part of basic education, vocational education, higher education and the technical training system. The government uses newspaper, radio, television and other media to promote knowledge of energy conservation; it holds an energy conservation publicity week, and mobilises the community to participate in establishing long-term energy-saving mechanisms throughout society.

Recent increases in energy demand have caused shortages and environmental problems. Recognising this, the Chinese government has made efforts to soften energy pressure, by introducing various policies and instructions. Energy has become one of the government's top concerns, especially since 2004. Recent energy policies include the Medium and Long-Term Energy Conservation Plan, the eleventh Five-Year Energy Plan, the Renewable Energy Law and the Fuel Efficiency Standard for Passenger Vehicles.

In 2005, the government set a target of reducing energy intensity by 20 per cent between 2005 and 2010. In order to reach the target, several programs were introduced, including 10 key energy-conservation projects (Table 15.4) and a monitoring program for 1,000 large energy users. The government regulates subsidies for renewable energy and a fuel tax and an energy tax are under discussion. The fuel tax will be implemented as part of the eleventh Five-Year Plan.

Promoting coordination of energy development with environmental protection

Climate change is a major global problem concerning the international community. It is not, however, just an environmental problem; ultimately, it is a development problem. The increasing development and utilisation of energy are the main causes of environmental pollution and climate change. All countries must find ways to correctly handle the relationship between the development and utilisation of energy, environmental protection and climate change.

China is a developing country in the early stages of industrialisation. China's carbon dioxide emissions from fossil fuels represent only 9.3 per cent of global

---

Table 15.4 **Energy-conservation projects approved by the Chinese government in 2005**

Program	Potential annual energy savings
Coal-fired industrial boiler conversion and increases in energy efficiency	70 Mtce (conversion) 35 Mtce (efficiency)
Heat–power cogeneration	5 Mtce
Residual heat and pressure usage	2.66 Mtce (steel industry) 3 Mtce (cement industry) 1.35 Mtce (coal-mining industry)
Oil conservation and substitution	35 Mt oil
Electrical machinery system energy conservation	20 billion kilowatt hours of electricity
Energy system optimisation	Strive to achieve international benchmarks for energy efficiency in steel, petrochemical and chemical industries
Energy conservation in construction	50 Mtce
Green lighting	29 billion kilowatt hours of electricity
Energy conservation by government organisations	Reduce energy consumption per capita and per area of office space by 20 per cent between 2002 and 2010
Energy conservation monitoring and technology services system construction	Implementation began in 2006

Source: National Development Research Commission.

emissions, while its per capita carbon dioxide emissions put it at number 92 in the world; its unit of GDP carbon dioxide emission elasticity coefficient is also very small. As a responsible developing country, China has paid great attention to environmental protection and global climate change.

Treating environmental protection as a basic national policy, the Chinese government has signed the UN Framework Convention on Climate Change, has set up a national body to coordinate climate change measures, submitted a 'national communication of information on initial climate change', established the 'clean development mechanism management approach', developed a national program to cope with the effects of climate change, and has implemented a series of policies and measures related to environmental protection and climate change.

Goals for the period of the eleventh Five-Year Plan include containing the basic trend of ecological deterioration, reducing the total discharge of major pollutants by 10 per cent and effectively controlling greenhouse gas emissions.

China is actively adjusting its economic and energy structure, comprehensively promoting energy conservation, focusing on the prevention and control of environmental pollution, and effectively controlling pollution emissions, to promote the coordinated development of energy and protection of the environment.

Climate change policies

On 4 June 2007, the *National Program for Climate Change* (*NPCC*) was released—the first such program in the developing world. This document specified China's objectives, basic principles, key activities and projects, as well as policies and measures for the country as a response to climate change up to the year 2010. China will commit to completing all the tasks set out in *NPCC*, while constructing a resource-conservative and environmentally friendly society, building national capacity to mitigate and adapt to the effects of climate change, and contributing further to the protection and understanding of the global climate system.

The *NPCC* includes the following
- China's current and future efforts to deal with climate change
- the impacts and challenges of climate change in China
- guidelines, principles and objectives for China to respond to climate change
- policies and measures to address climate change
- China's position on key climate change issues and the need for international cooperation.

In the short term, emission-mitigation policies will be implemented mainly through domestic energy-efficiency policies, renewable energy development, nuclear energy development, domestic sustainable development and energy security. In the long term, China's climate change policies will focus on further reductions in greenhouse gas emissions and policies such as a carbon tax, carbon pricing and so on. China will work together with other countries by joining an international emissions-reduction regime.

Energy and emission scenarios

According to the IPAC-Emission model, the primary energy demand in the baseline scenario will reach 4.5 billion tce in 2020 and 5.36 billion tce in 2030, with an annual growth rate of 3.6 per cent and an energy demand elasticity of 0.58. Coal still accounts for the major part of China's energy consumption (2.2 billion tce in 2030, or 58 per cent of the total primary energy demand) and the need for natural gas will have a rapid increase, from 4 per cent to 12.3 per cent of the total energy demand from 2000 to 2030 (Table 15.4).

By assuming the adoption of energy and environmental policy measures, the primary energy demand in policy scenario results is described in table 5. Compared to the baseline scenario, there is 385 million tce energy demand by 2020, 4280 mtce in by 2030. There are 668 million tce reduction in 2020 and 1082 million tce in 2030. There are pressures to apply these policy options in order to reach the lower energy demand scenario, and these should be introduced at the earliest opportunity to take advantage of the long lifespan of energy technologies.

Energy demands in the baseline and policy scenarios are given in Table 15.6 and Table 15.7. Coal use in final energy keep going up due to energy intensive products' increase, such as steelmaking, and demand for space heating in the service and residential sector. Natural gas and electricity increased quickly with a share 9.2 per cent and 23.9 per cent in 2030, increasing from 2.7 per cent and 17 per cent in 2005. Industry is major sector for energy consumption. The increase for energy demand in industry up to 2020 continues due to increasing of energy intensive sector, but the growth rate is smaller than between 2000–2005.

CO_2 emissions in 2020 and 2030 will reach 2.72 billion tons and 3.08 billion tons respectively. In the policy scenario, the CO_2 emission will be lowered by 28.3 per cent during the same period.

A package of policy options could be adopted now to reduce the growth rate of energy demand. Policies that would help China move to a low-energy demand scenario include promotion of the penetration of high energy-efficiency technologies; fiscal energy and environment policies, vehicle fuel taxes;

Table 15.5 **Primary energy demand in baseline scenario, mtce**

	Coal	Oil	N.Gas	Hydro	Nuclear	Wind	Biomass Power	Bio-Liquid	Bio-Diesel	Total
2000	923.2	278.6	30.2	82.0	6.1	0.4	1.0	0.0	0.0	1321.4
2005	1555.9	451.4	60.2	131.5	17.6	0.8	1.9	1.8	0.6	2221.8
2010	2339.3	645.3	108.1	208.3	42.9	8.7	14.70	9.7	0.6	3377.6
2020	2688.1	1092.2	238.2	323.5	106.8	22.6	26.5	21.5	3.1	4522.6
2030	2631.8	1582.4	448.3	397.2	178.7	43.7	38.9	33.4	7.9	5362.1

Source: Authors' projections.

Table 15.6 **Primary energy demand in policy scenario, mtce**

	Coal	Oil	N.Gas	Hydro	Nuclear	Wind	Biomass Power	Ethonal	Bio-Diesel	Total
2000	909.2	327.9	32.8	84.6	6.3	0.4	1.0	0.0	0.0	1362
2005	1555.9	465.0	60.2	131.5	17.6	0.9	1.7	1.5	0.6	2234
2010	2033.7	537.8	114.6	178.8	24.4	10.4	3.5	7.8	1.0	2912
2020	2167.1	818.5	337.6	333.9	121.4	27.4	16.1	25.9	5.8	3853
2030	1937.5	985.9	513.0	414.2	263.6	75.3	43.3	35.1	12.0	4280

Source: Authors' projections.

subsidies for renewable energy; emission taxes; resource taxes; promotion of public involvement (Table 15.8).

A package of policy options could be adopted now to reduce the growth rate of energy demand. Policies that would help China move to a low-energy demand scenario include promotion of the penetration of high energy-efficiency technologies; fiscal energy and environment policies, vehicle fuel taxes; subsidies for renewable energy; emission taxes; resource taxes; promotion of public involvement (Table 15.8).

Conclusion

China's energy demand is increasing rapidly as a result of rapid economic development. This trend is expected to continue for several decades.

The Chinese government has announced a package of energy-efficiency policies, renewable energy policies and environmental emission-control policies to reach its environmental targets. There is, however, still significant pressure to provide energy security and to control environmental pollution. The increase in energy demand means China will have to work hard to maintain its energy supply and energy imports. The increase in demand also increases the emissions of pollutants.

Table 15.7   **Final energy demand in baseline scenario**

	Coal	Coke	Coal gas	Oil	N.Gas	Heat	Electricity	Total
2000	421.6	81.2	12.8	265.2	22.2	59.5	148.7	1011.2
2005	670.5	196.8	26.7	405.5	44.9	90.5	269.8	1704.7
2010	934.1	224.8	31.2	614.3	77.8	158.8	407.3	2448.3
2020	1087.6	211.3	33.0	1039.2	145.9	273.1	566.4	3356.6
2030	1136.6	164.7	23.8	1504.5	246.0	400.3	720.4	4196.3

Source: Authors' projections.

Table 15.8   **Final energy demand in policy scenario**

	Coal	Coke	Coal gas	Oil	N.Gas	Heat	Electricity	Total
2000	394.8	82.9	12.8	271.3	22.2	54.0	153.3	991.4
2005	616.7	192.1	26.7	405.1	45.0	83.4	280.2	1649.2
2010	684.8	212.8	26.3	501.6	67.0	110.7	411.2	2014.4
2020	801.9	187.8	29.7	823.7	178.0	174.4	588.1	2783.6
2030	757.1	125.0	18.9	951.4	283.1	202.9	735.4	3073.8

Source: Authors' projections.

Table 15.9 Policy options used in the modelling study

Policy option	Explanation
Technology-promotion policy	Efficiency of end-use technologies increases as a result of new technologies
Energy-efficiency standards for buildings	New buildings reach 75 per cent of energy-efficiency standard by 2030
Renewable energy development	Policy includes subsidies for wind power and biomass power generation, as well as government support for village biogas supply systems
Energy tax	Vehicle tax introduced by 2005; energy tax introduced by 2015
Public transport	Urban public transport's share of traffic volume will be 10–15 per cent higher in 2030 than in 2000
Increases in transport efficiency	High fuel-efficiency vehicles, including hybrid vehicles, compact cars and advanced diesel cars, used widely
Increases in power generation efficiency	Efficiency of coal-fired power plants increases to 40 per cent by 2030
Natural gas incentives	Natural gas supply enhanced, technology localised to reduce cost
Nuclear power development	Target setting in national promotion program, enhanced government investment, technology development

Source: Jiang, K., Morita, T., Masui, T., and Matsuoka, Y., 2006. 'Global long-term GHG mitigation emission scenarios based on AIM', *Environment Economics and Policy Studies*, 3.

China is paying great attention to climate change issues. Domestic action on energy efficiency and renewable energies has contributed to China's mitigation of greenhouse gas emissions.

Analysis of various scenarios for the future shows that such a high level of energy demand and imports will put heavy pressure on China's energy-supply system. A well-designed strategy for the energy system and energy-industry development should therefore be prepared. That strategy should consider the following options.

- Because technological progress is key to reducing energy demand and ensuring a clean future, much more emphasis should be placed on new-generation technologies. In the simulations, technological progress contributes much of the energy savings while having no negative effects on public welfare.

- Export taxes are imposed on energy-intensive products. Energy, resource and similar taxes have significant effects on energy saving and optimisation of the economic structure. They should be given much more attention.

- Like other developing countries with high levels of energy imports, China should establish an energy security system. The size of strategic storage should be determined based on a global oil market situation.

- If there is no strong control of emissions, pollution will increase quickly in the short term. In the longer term, some pollutant emissions could be controlled by implementing various measures adopted by the government.

- Clean-coal technology should be emphasised to mitigate emissions from coal combustion. Only a few countries in the world are using coal on a large scale; development of clean-coal technologies therefore depends on them. China is the largest user of coal in the world, and its use of coal will increase in the future (by 2020, China could account for more than 40 per cent of global coal consumption). Therefore, clean-coal technology is crucial. China should have a clear development plan to promote clean-coal technology, working closely with other countries to develop a new generation of clean-coal technologies.

Because of its low production costs, China is likely to become a major manufacturing centre, producing energy and resource-intensive products. China must ensure it does not become excessively reliant on raw materials, the extraction of which causes damage to the environment. External costs should be included in production costs. Planning for energy and resource-intensive products should include measures to avoid environmental and economic damage.

References

Asian Integrated Model(AIM) Project Team, 1996. *A guide to the AIM/end-use model*, AIM Interim Paper, IP-95-05, Tsukuba.

China Environment Year Book Editing Committee, 2007. *China Environment Year Book*, Beijing.

China Statistics Publishing House, 2007. *China Energy Statistic Yearbook, 2007*, China Statistics Publishing House, Beijing.

Edmonds, J. and Reilly, J., 1983. 'A long-term global energy-economic model of carbon dioxide release from fossil fuel use', *Energy Economics*, 5:75–88.

Hu, X. and Jiang, K., 2001. *Greenhouse Gas Mitigation Technology Assessment* [in Chinese], China Environment Science Publishing House, Beijing.

Hu, X., Jiang, K. and Liu, J., 1996. *Application of AIM/emission model in China and preliminary analysis on simulated results*, AIM Interim Paper, IP-96-02, Tsukuba.

Intergovernmental Panel on Climate Change (IPCC), 2007a. *Climate Change 2007: Mitigation, Working Group III*, Cambridge University Press, Cambridge.

——. 2001b. *IPCC Special Report on Emission Scenario*, Cambridge University Press, Cambridge.

——. 2007. *Synthesis Report of TAR*, Cambridge University Press, Cambridge.

Jiang, K. and Xiulian, H., 2006. 'Energy demand and emission in 2030 in China: scenarios and policy options', *Environment Economics and Policy Studies*, 7(3):233–50.

Jiang, K., Hu, X., Matsuoka, Y. and Morita, T., 1998. 'Energy technology changes and CO_2 emission scenarios in China', *Environment Economics and Policy Studies*, 1:141–60.

Jiang, K., Masui, T., Morita, T. and Matsuoka, Y., 2000. 'Long-term GHG emission scenarios of Asia Pacific and the world', *Technological Forecasting and Social Change*, 61:2–3.

Jiang, K., Morita, T., Masui, T., and Matsuoka, Y., 1999. 'Long-term emission scenarios for China', *Environment Economics and Policy Studies*, 2:267–87.

——, 2006. 'Global long-term GHG mitigation emission scenarios based on AIM', *Environment Economics and Policy Studies*, 3.

Li, J., 2007. *Renewable Energy Development in 2006*, Energy Research Institute, Beijing.

Liu, J., Ma, F. and Fang, L., 2002. *China Sustainable Development Strategy* [in Chinese], China Agriculture Publishing House, Beijing.

Lu, Z., Zhao, Y. and Shen, Z., 2003. *Can China Become a Global Factory*? in Chinese], Economic Management Publishing House, Beijing.

Power Industry Information, 2005. *China Power*, 38:3.

Qu, K., 2003. *Energy, Environment Sustainable Development Study,* China Environment Science Publishing House, Beijing.

State Statistical Bureau, 2007. *China Energy Year Book 2007*, China Statistical Publishing House, Beijing.

——. 2005. *China Year Book 2005*, China Statistical Publishing House, Beijing.

Zheng Y., Zhang, X., and Xu, S., 2004. *China Environment and Development Review*, Social Science Documentation Publishing House, Beijing.

16

Chinese urban household energy requirements and CO_2 emissions

Jane Golley, Dominic Meagher and Xin Meng

Much of the literature on Chinese energy focuses on the insatiable demand of China's rapidly growing industrial sector. Given that in 2005 industry accounted for close to 70 per cent of China's energy demand that was consumed directly by the various sectors of the economy, this focus is not unwarranted. Likewise, given that residential consumption accounted for only 11 per cent of this direct energy demand in 2005, perhaps it is not surprising that there has been so little research on this aspect of China's total energy demand.[1] What tends to be overlooked, however, is that whatever is produced within China is ultimately used for household or government consumption, investment or exports. This gives rise to a number of interesting questions regarding the end users of energy. This paper focuses on one of those end users, urban households, and considers the consequences of different household consumption patterns for energy requirements and carbon dioxide emissions (henceforth carbon emissions).

Drawing on national-level energy and output data for all production sectors and on the Urban Household Income and Expenditure Survey (National Bureau of Statistics 2005), we address three related issues. First, using national-level data, we extend the notion of household energy consumption to include 'indirect' energy requirements, defined in terms of the energy inputs used in the production of goods consumed ultimately by households. Adding this to direct energy consumption yields considerably higher total energy requirements and, therefore, higher contributions to aggregate emissions, than accounting for direct energy requirements alone, supporting the findings of many others

(Wei et al. 2007; Pachauri and Spreng 2002; Cohen et al. 2005). This suggests that understanding household consumption patterns could be more important for understanding China's future energy demand than has been recognised previously.

Second, using the urban household survey data, we examine the extent of variation in total energy requirements and emissions across households with different income levels. It seems obvious that richer households should generate more emissions, simply because they have more money to spend and virtually no expenditure is emission free. More crucially though, we are concerned with household variations in energy requirements and emissions per yuan—that is, energy and emission intensities. Given that households with different levels of per capita income are likely to have different consumption bundles, and that different goods clearly require different quantities of energy in order to produce one yuan of output, variations are highly likely. A simple examination of how expenditure and emission shares of each sector vary across income brackets enables us to identify that while richer households do indeed emit more per capita, poorer households tend to be more emissions intensive—that is, generating higher emissions per yuan spent. Their heavy reliance on coal as a source of direct energy also makes poorer households significantly less 'energy efficient', as reflected in higher emissions/energy ratios.

Third, econometric analysis provides estimates of the extent of variations in energy requirements and emissions that can be attributed to household variations in per capita income and a variety of other household-specific characteristics. Beyond per capita income, other variables of interest include household size and the level of education. Smaller households might be expected to emit more per capita based on economies of scale in the direct use of energy: electricity, fuel for cooking and so on. It is possible—although perhaps overly optimistic—that higher levels of education could result in successful efforts to reduce per capita emissions. Other factors that might cause variations in household energy requirements include dwelling size, the age of household members and the province in which households are located. The combined findings that emerge give rise to some critical policy implications.

Background

First, it is necessary to clarify some key terms. 'Direct' energy requirements of households are defined as the consumption of energy carriers (coal, petroleum, natural gas and electricity) purchased by households in order to cater for energy services, such as space heating, heating tap water, lighting, appliances, cooking and motor fuel. 'Indirect' energy requirements refer to the consumption of energy that occurs during the production process of a good or service before its use (Weber and Perrels 2000; Wei et al. 2007). Direct energy requirements are relatively easy to calculate. As long as physical quantities consumed of, say, petroleum, can be observed (which they can in the survey data used below), it is possible to convert those quantities into tonnes or grams of 'coal equivalent' using the appropriate conversion factors. The carbon emissions associated with each form of energy can be derived readily using 'carbon coefficients', which are constants that vary across different energy forms (and across countries and over time as well).[2] Multiplying the petroleum quantity—expressed in tonnes of coal equivalent (tce)—by the appropriate carbon coefficient provides a measure of the total amount of carbon emissions associated with that quantity of petroleum consumption.

Calculating indirect energy requirements and their associated carbon emissions is more complex. Kok et al. (2006) survey the relevant literature and note the obvious but important point that different methods, all based on input–output (IO) analysis but using different data sources and different levels of aggregation, produce different results. They identify three basic methods: first, basic IO energy analysis draws on monetary data based entirely on national accounts. The indirect energy requirements of households are calculated by multiplying 'sectoral cumulative energy intensities' with monetary data on the final demand of households, which are available from the IO table. Energy intensities do not include direct energy deliveries to households, which are calculated separately using physical energy data from energy statistics at the national level. Second, 'IO plus household expenditure' differs from the basic method in that it relies on expenditure surveys rather than IO tables to determine household expenditure. Expenditure on each consumption item is multiplied with the corresponding value of energy intensity for that item. The direct energy requirements of households are also calculated separately, but here they are based on energy-use data collected at the household level. This is essentially the methodology adopted below, and has been used by many others (Weber and Perrels 2000; Pachauri 2004; Cohen et al. 2005; Bin and Dowlatabadi 2005). The third method is called 'hybrid energy analysis', which involves a combination of process and IO analysis. In process analysis, the life cycle of a product is described in physical terms, delving in great detail into the various stages of the production process, such as distribution, storage, transport, waste and recycling, and so on. It is the most accurate and time-consuming method and, as a consequence, appears to have been used only for European analysis where such detailed data are available (see, for example, Reinders et al. 2003). This method is simply out of the question at this stage for a China-focused analysis.

This variety of methods (often referred to by different names) has been applied to address a range of issues in single-country and cross-country analyses.[3] Weber and Perrels (2000) assess the energy and emissions consequences of consumer activities in West Germany, the Netherlands and France, considering how alternative lifestyles—such as slower economic growth or the adoption of cleaner technologies—impact on projected household consumption patterns. Pachauri and Spreng (2002) use a 115-sector classification of IO tables for India in three different years, enabling them to conduct cross-sectional and time-series analyses. They find that total energy consumption by Indian households is divided evenly between direct and indirect energy requirements and accounts for 75 per cent of total energy consumption in India. Their time-series analysis enables them to identify the main drivers of increasing energy requirements in total and per capita terms—namely, growing expenditure per capita and increasing energy intensity in the food and agricultural sectors. Pachauri (2004) builds on this to consider cross-sectional variations in household energy requirements in 1993–94, finding that the size of the household dwelling, the number of members in the household and the literacy level of the head of the household all effect household energy requirements—the first two positively, the last one negatively. Cohen et al. (2005) investigate the picture of equity that emerges within a developing country, using expenditure data on different income levels across 11 cities in Brazil. They show that 61 per cent of household energy is consumed indirectly and find that there are fairly constant energy intensities across income classes. Reinders et al. (2003) conduct a detailed examination of household energy intensities for a range of consumer goods in 11 European Union countries, showing that the direct energy share varies across countries, ranging from 34 to 64 per cent.

The analysis of China by Wei et al. (2007) is closest in spirit to this paper. They first quantify the direct and indirect energy requirements and related carbon emissions of urban and rural residents between 1999 and 2002. They find that 26 per cent of China's total energy consumption and 30 per cent of its emissions are the 'consequence of residents' lifestyles and economic activities to support these demands' (Wei et al. 2007:247). Using the IO plus expenditure approach, they find that for urban residents, the indirect contribution of energy consumption is 2.44 times greater than the direct contribution (accounting for 71 per cent of the total), while for rural residents it is smaller (accounting for 35 per cent of the total). The sectors covered are food, clothing, residence, household facilities and services, medicine and medical services, transport and communication services, education, cultural and recreation services and miscellaneous commodities and services. While these sectors differ significantly from the ones used below, and the methodology is not identical, this chapter provides a useful benchmark for comparison.

Before turning to a detailed description of the methodology adopted in this chapter, a few caveats should be noted. One regards the problem of using expenditure as the key variable for calculating energy and emission intensities. As noted by Weber and Perrels (2000), calculating an average energy requirement per dollar spent on 'food' does not allow a distinction between high and low-quality product choices within that category. They argue that high quality usually comes at a high price, while the embodied energy increase is not commensurate, resulting in a lower energy intensity per monetary unit spent on high-quality products. Regardless, to improve on this would require a much more detailed treatment that is very difficult to combine with a dynamic description of consumption and production in a society (Weber and Perrels 2000). Aside from the quality issue, high levels of aggregation are also problematic. If we had data on the energy inputs for, say, rice, pork and tofu, it would be (relatively) reasonable to assume that each of these is a homogeneous group and that price is also homogeneous within each group. Given that instead all we can do is aggregate these into 'food', because that is the extent of energy data available, it is true that we could overestimate or underestimate the energy and emission intensities associated with each sub-group: rice, pork and tofu. Given data constraints, there is little to be done about this problem other than to emphasise that the discussion is necessarily about *average* intensities, and to concede the shortcomings of this approach.[4]

Other problems raised in the literature include the high level of uncertainty for countries that rely heavily on imports, where imports are assumed to be produced using the same technology and structure as domestic industries, and also the incompleteness of sectoral environmental statistics (for example, small and medium enterprises might be registered only in part and fail to be recorded accurately) (Suh et al. 2004). Hondo et al. (2002) identify numerous sources of uncertainty in carbon intensity calculations using IO tables, including data errors, multiple goods in one sector, multiple prices of one good, multiple technologies of one good and multiple producers of one good, presumably using different technologies. The household survey data used below do not indicate the share of imports in household expenditure and, likewise, it is beyond the scope of this paper to assess just how important these various sources of uncertainty might be. Instead, we simply acknowledge these issues and concede that the analysis is necessarily approximate rather than precise.

National-level energy requirements and carbon emissions

China's direct energy consumption was 2.14 billion tonnes of coal equivalent (tce) in 2005 (Table 16.1). Industry (including mining and quarrying, manufacturing and the production and supply of electric power, heat power and water) accounted for the lion's share of this consumption, at 69.8 per cent. Residential consumption accounted for the next highest share, at 10.9 per cent, with urban and rural residents accounting for 6.6 and 4.3 per cent respectively. This places aggregate residential direct energy consumption above transport, storage, postal and telecommunication services, agriculture, construction and wholesale, retail trade and catering services. The dominance of industry in terms of direct energy consumption explains why much of the literature has focused on the relationship between China's rapid industrial development and energy demand.[5] This, however, overlooks a critical point, which is that China's households ultimately consume a large share of industrial output, and indeed the output of other sectors as well. This is where the notion of 'indirect' energy consumption comes into play.

Input–output tables provide the value of each productive sector used (either as a final product or in the production of another product) by the various sectors of the economy, which are: each of those productive sectors (agriculture, mining and quarrying, foodstuff, and so on); rural, urban and government consumption expenditure; gross fixed-capital formation and changes in inventories; and net exports. The sum of these values is equal to the total output value for each productive sector. The shares of each sector's total output consumed by rural and urban households are listed in Table 16.2. Our notion of indirect energy consumption follows from observing these shares, which are as high as 49.5 per cent of the total output in the case of foodstuffs and average 16.8 per cent across all sectors. Given that the foodstuff sector, for example, used 43.3 million tce as direct energy inputs in the production process, households indirectly consumed 21.4 million tce through their consumption of foodstuffs. In terms of total energy requirements, then, in addition to the 10.9 per cent of total energy consumed directly by urban and rural households (as shown in Table 16.1), another 16.8 per cent of China's total energy—or 360 million tce—was consumed through their expenditure on other products (Table 16.3). This implies that 27.7 per cent of China's total energy can be traced to household consumption (very close to Wei et al.'s [2007] figure of 26 per cent). For urban households, indirect energy requirements of close to 248 million tce are 1.75 times higher than their direct energy requirements (accounting for 64 per cent of total urban household energy consumption), while for rural households indirect requirements are 1.2 times higher (accounting for 55 per

Table 16.1   **China's direct energy consumption in 2005**

	'000 tonnes of coal equivalent (tce)	Share (per cent)
Total	2,144,794	100.0
Agriculture	79,783	3.7
Industry	1,496,389	69.8
Mining and quarrying	118,283	5.5
Foodstuff	43,277	2.0
Textile, sewing, leather and fur products	58,416	2.7
Other manufacturing	58,500	2.7
Production and supply of electric power, heat power and water	109,278	5.1
Coking, gas and petroleum refining	104,715	4.9
Chemical industry	244,744	11.4
Building materials and non-metal mineral products	215,151	10.0
Metal products	462,770	21.6
Machinery and equipment	81,255	3.8
Construction	34,111	1.6
Transport, postal and telecommunication services	162,793	7.6
Wholesale and retail trades, hotels and catering services	50,311	2.3
Other services	86,912	4.1
Residential consumption	234,495	10.9
Urban	142,205	6.6
Rural	92,290	4.3

Note: 'Other services' combines real estate, leasing and business services, banking and insurance and other services.
Source: National Bureau of Statistics (NBS), 2006. *China Energy Statistical Yearbook*, China Statistics Press, Beijing.

cent of the total). As urbanisation progresses in China, this is indicative of the fact that indirect energy demand will become an increasingly dominant source of China's consumer-driven emissions trends.

In order to look more closely at the energy requirements and carbon emissions of households in China in the next section, it is first necessary to calculate the 'energy intensity', defined as the energy requirements per yuan of output in each sector, and the 'carbon intensity', defined as the carbon emissions per yuan of output in each sector. Dividing the total energy consumed in each sector by the gross value of output in that sector gives sector-level energy intensities, which are provided in Table 16.4 in terms of grams of coal equivalent (gce) per yuan for each of the sectors for which data are available for 2005.[6] One point that is immediately clear is that energy intensities vary substantially across sectors, ranging from 1,178 gce per yuan in the 'mining of other ores', to just 5.47

Table 16.2 **Shares of household consumption expenditure in total output** (per cent)

	Rural	Urban	Subtotal
Total	5.2	11.6	16.8
Agriculture	17.3	19.4	36.6
Mining and quarrying	0.8	2.0	2.8
Foodstuff	17.2	32.3	49.5
Textile, sewing, leather and fur products	4.2	16.5	20.7
Other manufacturing	1.8	7.8	9.6
Production and supply of electric power, heat power and water	2.7	12.3	15.0
Coking, gas and petroleum refining	0.6	3.9	4.5
Chemical industry	2.3	4.4	6.7
Building materials and non-metal mineral products	1.7	7.9	9.6
Metal products	0.4	1.5	1.8
Machinery and equipment	1.4	5.3	6.6
Construction	0.0	0.0	0.0
Transport, postal and telecommunication services	3.2	8.9	12.1
Wholesale and retail trades, hotels and catering services	5.8	20.0	25.8
Real estate, leasing and business services	12.8	18.9	31.7
Banking and insurance	7.3	12.9	20.2
Other services	5.4	21.1	26.5

Source: National Bureau of Statistics (NBS), 2007. *China Statistical Yearbook*, China Statistics Press, Beijing.

Table 16.3 **Indirect energy requirements for rural and urban households** ('000 tonnes of coal equivalent)

	Rural	Urban	All
Total	111,347	248,397	359,744
Agriculture	13,771	15,441	29,212
Mining and quarrying	893	2,377	3,269
Foodstuff	7,440	13,998	21,439
Textile, sewing, leather and fur products	2,428	9,643	12,071
Other manufacturing	1,037	4,551	5,588
Production and supply of electric power, heat and water	2,952	13,471	16,423
Coking, gas and petroleum refining	674	4,066	4,741
Chemical industry	5,653	10,774	16,427
Building materials and non-metal mineral products	3,689	17,061	20,750
Metal products	1,640	6,875	8,514
Machinery and equipment	1,128	4,272	5,399
Transport, postal and telecommunication services	5,220	14,516	19,736
Wholesale and retail, hotels and catering services	2,931	10,054	12,985
Other services	6,952	16,773	23,726

Note: 'Other services' combines real estate, leasing, business, banking, insurance and other services.
Source: Tables 16.1 and 16.2 and authors' calculations.

gce per yuan in the 'manufacturing of communication, computers and other electronic equipment'. In aggregate, the construction sector has the lowest energy requirements per yuan of output at 9.87 gce, compared with 20.22 gce in agriculture, 58.25 gce in manufacturing and 79.36 gce in mining.[7]

Calculating the carbon intensities is a little more involved because not all forms of energy are associated with the same carbon emissions. To resolve this, we first need to know how much of each type of energy is used by each production sector. These data are available from the *China Energy Statistical Yearbook* (National Bureau of Statistics 2006) and provide the energy requirements (that is, direct energy inputs) for production in 48 sectors. Eighteen different fuels are specified, which we then aggregate into six categories: coal, coking products, petroleum products, natural gas, heat and electricity.[8]

Since the data are measured in consistent units, it is possible to calculate the share of energy from each category used in production in each of the 48 sectors.[9] For example, agriculture consumed 79.8 million tce in 2005. Of that, 30.1 per cent was coal, 1.5 per cent was coking products, 50.4 per cent was petroleum products, 0 per cent was natural gas, 0.05 per cent was heat and 17.9 per cent was electricity. Based on data from the Energy Information Administration (EIA 2008) of the US Department of Energy, Table 16.5 provides the carbon emission coefficients for China in 2005, measured in tonnes of carbon equivalent (tcae).[10] It is then straightforward to calculate the carbon intensity for each sector by multiplying each emission coefficient by the proportion of each fuel used in production and summing over all fuels, giving the emissions intensities in grams of carbon per yuan reported in Table 16.4. Given that different sectors use different types of energy in different proportions, there is not a perfect correlation between energy intensity and emissions intensity. Although the correlation at this sectoral level is extremely high, at 0.999, a closer look at the household data below reveals important differences in across-income deciles.

Urban household energy requirements and carbon emissions

The data used in this section are drawn from the Urban Household Income and Expenditure Survey (National Bureau of Statistics 2005). Although the survey covered all 31 provinces in China, we have access only to the data for 16 provinces. In the survey, household expenditure is divided into seven categories: current-period consumption; expenditure on housing building or purchasing; transfer expenditure (including tax, donations, buying lottery tickets, paying for expenditure of non-coresiding family members [parents or children], expenditure on non-saving insurance, and other transfer expenditure); interest-only mortgage repayments; social welfare savings; saving and

Table 16.4   Energy and carbon intensities by sector in 2005

	Energy intensity (grams/yuan)	Carbon intensity (grams/yuan)
Agriculture	20.22	11.51
Industry	59.47	37.68
Mining and quarrying	79.36	48.13
Mining and washing of coal	98.95	66.32
Extraction of petroleum and natural gas	57.09	28.57
Mining and processing of ferrous metal ores	95.75	57.54
Mining and processing of non-ferrous metal ores	58.27	34.63
Mining and processing of non-metal ores	114.13	72.04
Mining of other ores	1,178.23	665.47
Manufacturing	58.25	37.09
Processing of food from agricultural products	19.21	12.36
Manufacture of foods	30.97	20.24
Manufacture of beverages	28.50	18.90
Manufacture of tobacco	8.38	5.34
Manufacture of textiles	39.33	25.22
Manufacture of textile apparel, footwear and caps	11.00	6.74
Manufacture of leather, fur, feather and related products	8.97	5.34
Processing of timber, manufacture of wood products	37.83	24.43
Manufacture of furniture	9.04	5.29
Manufacture of paper and paper products	78.80	51.97
Printing, reproduction of recording media	19.04	11.00
Manufacture of articles for culture, education and sport activity	13.23	7.47
Processing of petroleum, coking, processing of nuclear fuel	87.26	47.17
Manufacture of raw chemical materials and chemical products	134.52	80.93
Manufacture of medicines	26.43	17.17

Manufacture of chemical fibres	51.50	32.12
Manufacture of rubber	49.20	30.40
Manufacture of plastics	28.59	16.71
Manufacture of non-metallic mineral products	206.47	136.47
Smelting and pressing of ferrous metals	171.72	117.28
Smelting and pressing of non-ferrous metals	90.49	55.30
Manufacture of metal products	33.91	19.90
Manufacture of general-purpose machinery	19.49	12.13
Manufacture of special-purpose machinery	20.43	12.80
Manufacture of transport equipment	12.41	7.53
Manufacture of electrical machinery and equipment	8.58	4.92
Manufacture of communication and other electronic equipment	5.47	3.01
Manufacture of measuring instruments and office machinery	6.99	4.02
Manufacture of artwork and other manufacturing	61.17	37.31
Recycling and disposal of waste	11.67	6.99
Electric power, gas and water production and supply	57.88	35.85
Production and distribution of electric power and heat power	54.94	34.46
Production and distribution of gas	90.05	49.53
Production and distribution of water	119.52	76.78
Construction	9.87	5.25

Source: Authors' own calculations and National Bureau of Statistics (NBS), 2006. *China Energy Statistical Yearbook*, China Statistics Press, Beijing; National Bureau of Statistics (NBS), 2007. *China Statistical Yearbook*, China Statistics Press, Beijing.

Table 16.5   **Carbon coefficients by fuel**

Coal	0.7018
Oil	0.4876
Gas	0.3999
Electricity	0.5631

Source: Authors' calculations and Energy Information Administration (EIA), 2008. *International Energy Annual*. Available from www.eia.doe.gov.iea (accessed 15 January 2008).

mortgage repayments (principles only); and end-of-period cash in hand. The part of household expenditure that we are able to convert into direct and indirect energy consumption and carbon emissions is the first category: current-period consumption, which, on average, accounts for 75 per cent of total household expenditure (see column 5 in Appendix Table A16.1). To the extent that any of the other categories require energy inputs and therefore generate emissions, the calculations below will underestimate households' indirect contributions to China's energy demand. The survey provides details of household current-period consumption on a range of highly disaggregated goods. We aggregate these according to the sectors in Table 16.4, and are therefore able to calculate the urban household energy requirements and carbon emissions that occur indirectly via each household's consumption bundle.[11] This is the notion of indirect energy introduced above, but now applied to household-level expenditure data. In addition to these indirect energy requirements, each household also consumes energy directly, in the form of coal, gas, petroleum and electricity. The survey data provide the quantities of each of these consumed by each household, which are readily converted into tonnes of coal equivalent (tce), from which the carbon emissions can be obtained. We also have expenditure data for each of these forms of direct energy consumption, from which energy and emission intensities can be obtained.

Three important assumptions are made in calculating household energy consumption and carbon emissions. First, a sizeable amount of household food consumption comes from eating out. We assume that one-third of the cost of eating out is due to food consumption, while the remainder is service cost.[12] We further assume that half of the cost of food eaten out stems from food processed from agricultural products, while the other half stems from manufactured food. Second, we assume that 25 per cent of total expenditure on public transportation fees, such as expenditure on taxi, bus, air and train tickets, is related to petrol consumption, while the rest is attributed to services.

Third, all service fees are assumed to be energy free, due to data limitations. Excluding service costs, the proportion of current-period consumption that is counted for as direct and indirect energy consumption is about 73 per cent of current-period consumption and 55 per cent of per capita total household expenditure (see Table A16.1). As with the above excluded categories in total household expenditure, omitting services from the analysis will mean that we underestimate households' indirect contributions to China's energy demand.

Table 16.6 presents summary statistics by income deciles for urban households' total energy consumption, carbon emissions, direct and indirect emissions, per capita energy consumption and carbon emissions, and direct emissions by different energy sources. The table shows that the urban Chinese households surveyed consume an average of 1.71 tce per annum, which in turn generates an average of 0.9 tonnes of carbon emissions (tcae). Average per capita energy consumption across all households is 0.61 tce per annum, which generates 0.32 tcae. Energy consumption and carbon emissions are much higher for richer households than for poorer ones. Households in the top-10 percentile income group (the tenth decile) consume an average of 2.48 tce, some 86 per cent more than the amount consumed by households in the poorest group (the first decile), at 1.34 tce. Similarly, the richest decile emits, on average, 66 per cent more carbon than the poorest.

By separating total energy consumption into direct and indirect sources, it is revealed that only 32 per cent of household energy consumption is from direct sources.[13] Nevertheless, with regard to carbon emissions, indirect sources account for a larger proportion, especially for the richest decile, in which close to 40 per cent of carbon emissions are from indirect sources.[14] This compares with just 20 per cent for the poorest decile (see Table A16.3). Why does indirect energy demand account for a higher proportion of total emissions generated by richer households than by poorer ones? When comparing carbon emissions across income groups, it is clear that rich households emit more per capita than poor households; however, if comparing direct emissions, the poor generate only slightly less carbon than the rich. Further investigation shows that the high direct emissions for poor households are due mainly to their relatively high levels of coal consumption—the most emission-intensive form of energy. The poorest income decile's carbon emissions from coal consumption are about seven times as high as the richest decile's. This, in turn, is due to the fact that the price of coal is one-third of the price of gas, one-ninth of the price of electricity and one-tenth of the price of petrol, as shown in Table 16.7.

A closer look at the data reveals some interesting points regarding variations in the level of carbon emissions per unit of energy consumed across households.

Table 16.6 Average energy consumption and carbon emissions by income decile

	Household emissions (tcae)	Household energy (tce)	Per capita emissions (tcae)	Per capita energy (tce)	Direct energy (tce)	Indirect energy (tce)
First decile	0.77	1.34	0.24	0.42	1.09	0.24
Second decile	0.81	1.46	0.26	0.47	1.15	0.31
Third decile	0.84	1.54	0.28	0.51	1.18	0.35
Fourth decile	0.80	1.51	0.28	0.52	1.13	0.38
Fifth decile	0.85	1.61	0.30	0.57	1.19	0.41
Sixth decile	0.86	1.63	0.31	0.59	1.19	0.44
Seventh decile	0.90	1.74	0.33	0.64	1.26	0.48
Eighth decile	0.93	1.80	0.35	0.68	1.28	0.53
Ninth decile	1.00	1.95	0.39	0.76	1.36	0.59
Tenth decile	1.28	2.48	0.50	0.97	1.68	0.80
Average	0.90	1.71	0.32	0.61	1.25	0.45

	Direct emissions (tcae)	Indirect emissions (tcae)	Emissions (coal) (tcae)	Emissions (oil) (tcae)	Emissions (electricity) (tcae)	Emissions (gas) (tcae)
First decile	0.61	0.16	0.35	0.02	0.07	0.17
Second decile	0.61	0.20	0.28	0.03	0.08	0.22
Third decile	0.61	0.23	0.24	0.04	0.09	0.24
Fourth decile	0.56	0.24	0.17	0.04	0.10	0.25
Fifth decile	0.59	0.26	0.17	0.05	0.10	0.27
Sixth decile	0.57	0.28	0.13	0.06	0.11	0.28
Seventh decile	0.59	0.30	0.11	0.06	0.12	0.31
Eighth decile	0.60	0.34	0.09	0.06	0.12	0.32
Ninth decile	0.63	0.37	0.07	0.09	0.13	0.34
Tenth decile	0.77	0.50	0.05	0.18	0.16	0.38
Average	0.61	0.29	0.17	0.06	0.11	0.28

Source: Authors' calculations and National Bureau of Statistics (NBS), 2005. *Urban Household Income and Expenditure Survey*, China Statistics Press, Beijing.

Table 16.7 **Price per tonne of coal equivalent**

	Yuan/tonne	Proportion of coal price
Coal	374	1
Gas	1,070	2.9
Electricity	3,246	8.7
Petrol	4,033	10.8

Source: Derived from National Bureau of Statistics (NBS), 2005. *Urban Household Income and Expenditure Survey*, China Statistics Press, Beijing.

This emissions/energy ratio provides a measure of the extent of the carbon intensity of energy, or energy efficiency, with lower levels implying that energy use is relatively energy efficient. To sum up the discussion of Table 16.6 and further examine this energy efficiency, we present a group of graphs that provide visual relationships between energy and emissions on the one hand, and income on the other. Figure 16.1 presents per capita total energy consumption, carbon emissions and the ratio of carbon emissions over energy consumption, by income percentile. The figure shows that while rich households consume more energy and produce more carbon emissions per capita, their emissions/energy ratio is lower than poor households'. That is, richer households are relatively energy efficient.

We then plot separately direct and indirect energy, emissions and their ratio in Figures 16.2 and 16.3a. The two figures indicate that the major source of lower energy efficiency for the poor is direct energy consumption, which generates substantially more emissions for the poor than for the rich. For indirect energy consumption and emissions, there is almost no difference across income groups.

The data presented in Table 16.6 indicate that the main reason for high levels of direct energy consumption and carbon emissions for people in poor households is their high level of coal consumption. We therefore exclude coal from direct energy consumption and consequent emissions and plot the remaining direct energy and emissions by income group again (Figure 16.3b). The new figure shows almost the same level of energy efficiency for poor and rich households. In fact, there is a slightly higher emissions/energy ratio for the top income decile compared with the other income groups.

The next question of interest is whether there is any variation in rich and poor households' per yuan carbon emissions. Figure 16.4 plots average emissions

Figure 16.1 **Total energy, carbon emissions and energy efficiency, by income**

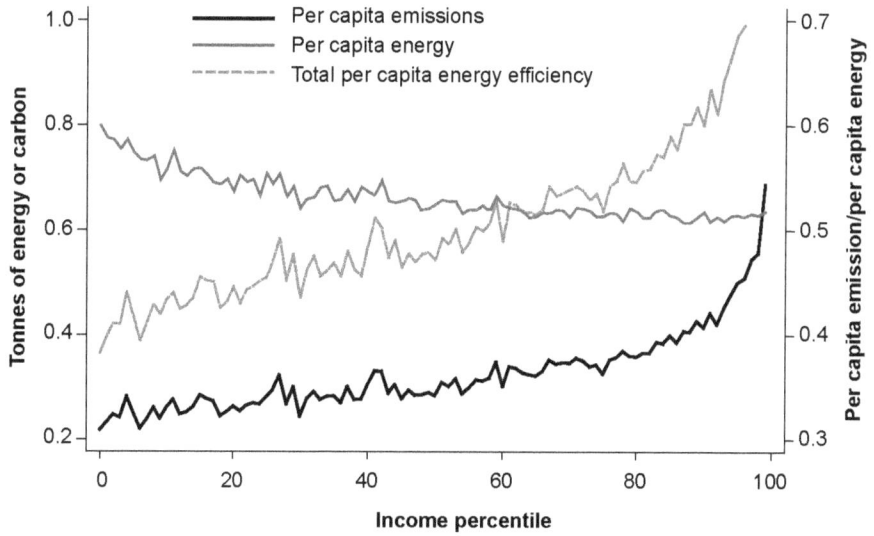

Figure 16.2 **Indirect energy, carbon emission and energy efficiency, by income**

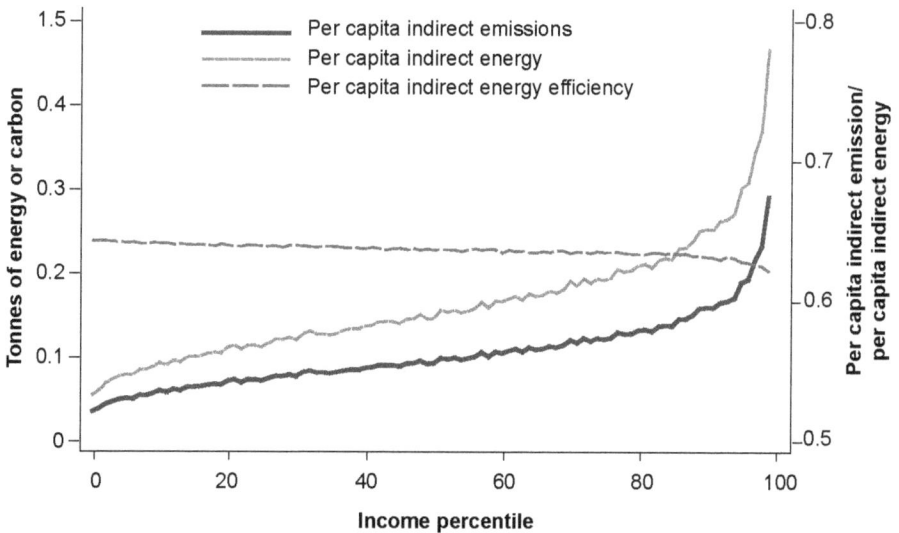

Figure 16.3a **Direct energy, carbon emission and energy efficiency, by income**

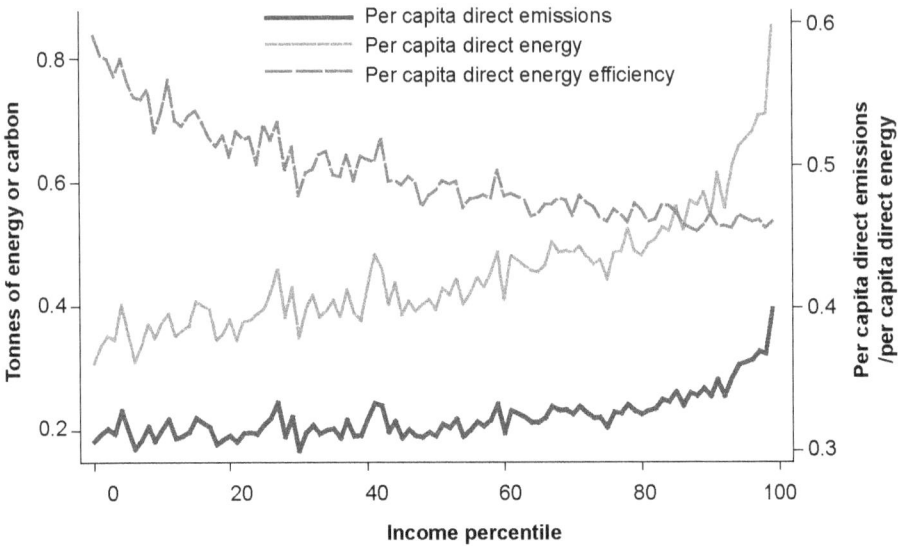

Figure 16.3b **Direct energy, carbon emission and energy efficiency excluding coal, by income**

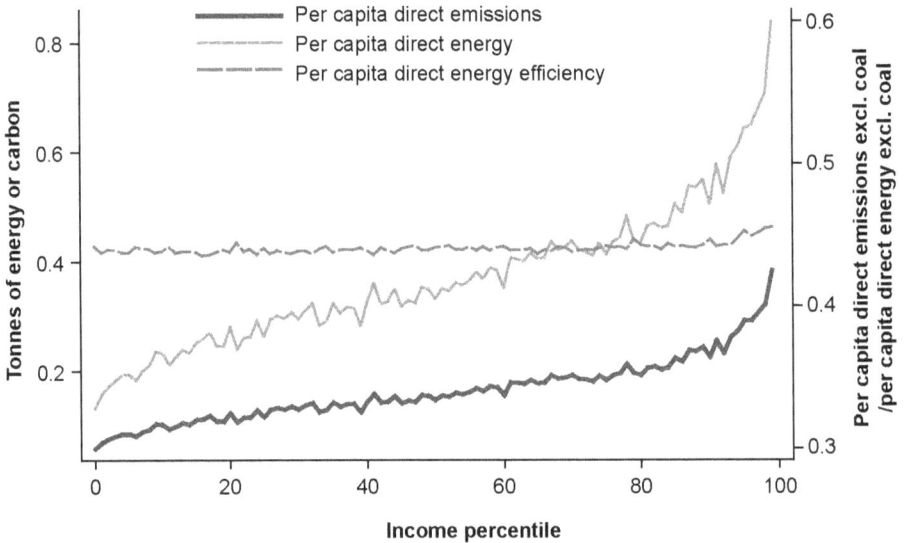

divided by total expenditure per household. The top solid line in Figure 16.4 is total emissions per yuan. The figure indicates that every yuan spent by the poorest household generates 120 grams of carbon equivalent (gcae), while one yuan spent by the richest household generates only 20 gcae. Almost the entire difference in these emissions intensities is due to direct energy consumption, as can be seen by the fact that the curve for direct emissions per yuan follows the total emissions curve closely while the curve for indirect emissions per yuan is almost flat. This, in turn, is due to the high level of coal consumption by the poor (as seen by the curve for direct emissions excluding coal, which is also very flat). In other words, the most important discrepancy between rich and poor households lies in the different shares of expenditure allocated to direct coal consumption and the emissions generated as a result.

This is not to say that the sources of indirect emissions do not vary across income brackets. For example, food (processed from agriculture, manufactured and eating out combined) accounts for 75 per cent of the poorest decile's emissions, compared with just 48 per cent for the richest decile, while 10 per cent of the richest decile's indirect emissions stem from the manufacturing of raw chemicals and chemical materials, compared with 5 per cent for the poorest decile. These differences do not, however, amount to significant differences in emissions per yuan spent for each of these income groups. Instead, it is all

Figure 16.4 Yuan carbon emissions by income percentile (kg)

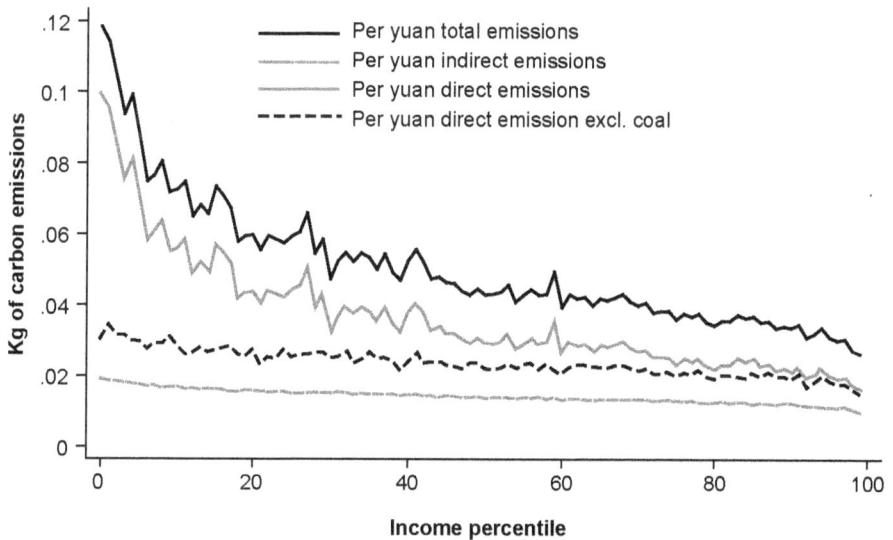

about direct emission shares, and those generated from coal consumption in particular: 58 per cent of the poorest households' direct emissions stem from coal consumption compared with just 6.5 per cent for the richest households.[15] While rich households therefore consume more energy and produce more carbon emissions than poor households, poor households produce more emissions per unit of energy consumption and more emissions per yuan spent than rich households.

An econometric analysis

The above analysis is based entirely on simple averages by income group. Energy consumption and emissions can also be affected by other factors, such as household composition and dwelling size. Further, households' residential location can also have a significant effect due to weather conditions and regional energy price variations. In this section, we estimate simple energy and emissions determination equations, which allows us to tease out other factors and to identify the pure income effect of urban household energy requirements and emissions.

In our econometric model, in addition to per capita income and its squared term, we control for household size, the gender, age and years of schooling of the household head, a group of household composition variables relating to gender and age, dwelling size, and dummy variables for each of the 15 provinces other than Beijing, which is the benchmark province and therefore the omitted dummy variable.

Tables 16.8 and 16.9 present the results for the determinants of per capita energy consumption and per capita emissions respectively. We observe almost the same pattern of determinants in both and therefore focus the remaining discussion on the energy consumption results in Table 16.8. On average, larger households consume less energy per capita, verifying an economy-of-scale effect. Cooking for five people should consume less energy than cooking for one person five times. Similarly, five people watching TV requires the same amount of energy as one person watching TV. Our results also show that males and older household heads are associated with higher energy consumption, while households with more educated heads consume less energy. This last effect is, however, applicable only to total energy and direct energy consumption.[16] This lends some support to the optimistic claim that people with higher levels of education are more aware of the impact of their energy consumption on total emissions and account for this in their consumption decisions. The household composition variables indicate that relative to the omitted category (male and female adults aged 20–65), children in most age groups consume less energy, while older people (male and female above 65) consume more energy. This could

Table 16.8 **Determinants of household per capita energy consumption**

	Per capita total energy	Per capita direct energy	Per capita direct energy excluding coal	Per capita indirect energy
Per capita income/104	0.203 (0.006)***	0.114 (0.006)***	0.175 (0.005)***	0.089 (0.001)***
(Per capita income/104)2	-0.006 (0.001)***	-0.003 (0.001)***	-0.009 (0.001)***	-0.003 (0.000)***
Number of household members	-0.084 (0.003)***	-0.068 (0.003)***	-0.046 (0.003)***	-0.016 (0.001)***
Number of male household heads	0.009 (0.005)*	0.007 (0.005)	0.016 (0.004)***	0.002 (0.001)**
Household head age/103	4.724 (0.294)***	4.491 (0.280)***	3.731 (0.227)***	0.233 (0.057)***
Household head years of schooling	-0.002 (0.001)*	-0.002 (0.001)**	0.005 (0.001)***	0.000 (0.000)
Percentage of household members aged 0–5	-0.043 (0.032)	-0.038 (0.030)	-0.042 (0.024)*	-0.005 (0.006)
Percentage of household members aged 6–10	-0.083 (0.025)***	-0.079 (0.024)***	-0.075 (0.019)***	-0.005 (-0.005)
Percentage of household members aged 11–15	-0.099 (0.022)***	-0.108 (0.021)***	-0.077 (0.017)***	0.009 (0.004)**
Percentage of household members aged 16–19, female	-0.148 (0.027)***	-0.134 (0.026)***	-0.069 (0.021)***	-0.014 (0.005)***
Percentage of household members aged 16–19, male	-0.109 (0.026)***	-0.115 (0.025)***	-0.049 (0.020)**	0.006 (0.005)
Percentage of household members aged above 65, female	0.090 (0.021)***	0.090 (0.020)***	0.067 (0.016)***	0.000 (0.004)
Percentage of household members aged above 65, male	0.133 (0.022)***	0.132 (0.021)***	0.052 (0.017)***	0.002 (0.004)
Housing size (square metres/103)	0.443 (0.073)***	0.355 (0.069)***	-0.108 (0.056)*	0.088 (0.014)***
Shanxi	0.356 (0.017)***	0.412 (0.016)***	0.319 (0.013)***	-0.056 (0.003)***
Liaoning	0.06 (0.015)***	0.074 (0.014)***	0.151 (0.011)***	-0.014 (0.003)***
Heilongjiang	0.126 (0.015)***	0.17 (0.015)***	0.035 (0.012)***	-0.044 (0.003)***
Shanghai	0.324 (0.016)***	0.308 (0.015)***	0.359 (0.012)***	0.016 (0.003)***
Jiangsu	-0.082 (0.015)***	-0.063 (0.014)***	-0.019 (0.012)	-0.019 (0.003)***
Anhui	0.040 (0.015)***	0.068 (0.014)***	-0.006 (0.011)	-0.029 (0.003)***
Jiangxi	-0.023 (0.017)	0.009 (0.016)	0.050 (0.013)***	-0.031 (0.003)***
Shandong	-0.003 (0.014)	0.036 (0.013)***	-0.006 (0.011)	-0.040 (0.003)***
Henan	0.062 (0.015)***	0.110 (0.015)***	0.049 (0.012)***	-0.049 (0.003)***
Hubei	-0.010 (0.015)	0.020 (0.014)	0.036 (0.011)***	-0.030 (0.003)***
Guangdong	-0.026 (0.014)*	-0.046 (0.014)***	0.014 (-0.011)	0.020 (0.003)***

Chongqing	0.190 (0.017)***	0.201 (0.016)***	0.279 (0.013)***	-0.011 (0.003)***
Sichuan	0.165 (0.014)***	0.184 (0.014)***	0.249 (0.011)***	-0.019 (0.003)***
Yunnan	0.000 (0.017)	0.035 (0.016)**	0.094 (0.013)***	-0.035 (0.003)***
Gansu	-0.067 (0.020)***	-0.031 (0.019)	-0.001 (0.015)	-0.036 (0.004)***
Constant	0.355 (0.027)***	0.236 (0.025)***	0.003 (0.021)	0.119 (0.005)***
Observations	33,358	33,358	33,358	33,358
R-squared	0.24	0.18	0.26	0.46

* significant at 10 per cent
** significant at 5 per cent
*** significant at 1 per cent
Note: Standard errors in parentheses.

Table 16.9 Determinants of household per capita carbon emissions

	Per capita total emissions	Per capita direct emissions	Per capita direct emissions excluding coal	Per capita indirect emissions
Per capita income/104	0.092 (0.003)***	0.036 (0.003)***	0.079 (0.002)***	0.055 (0.001)***
(Per capita income/104)2	-0.001 (0.001)***	0.000 (0.000)	-0.004 (0.000)***	-0.002 (0.000)***
Number of household members	-0.046 (0.002)***	-0.036 (0.002)***	-0.020 (0.001)***	-0.010 (0.000)***
Number of male household heads	0.001 (0.003)	0.000 (0.003)	0.007 (0.002)***	0.001 (0.001)**
Household head age/103	2.214 (0.166)***	2.041 (0.157)***	1.508 (0.094)***	0.173 (0.035)***
Household head years of schooling	-0.003 (0.000)***	-0.003 (0.000)***	0.002 (0.000)***	0.000 (0.000)
Percentage of household members aged 0–5	-0.018 (0.018)	-0.015 (0.017)	-0.017 (0.010)*	-0.003 (0.004)
Percentage of household members aged 6–10	-0.037 (0.014)***	-0.034 (0.013)**	-0.031 (0.008)***	-0.003 (0.003)
Percentage of household members aged 11–15	-0.048 (0.012)***	-0.053 (0.012)***	-0.032 (0.007)***	0.005 (0.003)*
Percentage of household members aged 16–19, female	-0.084 (0.015)***	-0.075 (0.015)***	-0.029 (0.009)***	-0.009 (0.003)***
Percentage of household members aged 16–19, male	-0.063 (0.015)***	-0.066 (0.014)***	-0.019 (0.008)**	0.003 (0.003)
Percentage of household members aged above 65, female	0.044 (0.012)***	0.044 (0.011)***	0.028 (0.007)***	0.000 (0.002)
Percentage of household members aged above 65, male	0.079 (0.012)***	0.077 (0.012)***	0.021 (0.007)***	0.002 (0.003)
Housing size (square metres/103)	0.367 (0.041)***	0.313 (0.039)***	-0.012 (-0.023)	0.054 (0.009)***
Shanxi	0.156 (0.009)***	0.191 (0.009)***	0.126 (0.005)***	-0.036 (0.002)***
Liaoning	0.002 (0.008)	0.011 (0.008)	0.065 (0.005)***	-0.008 (0.002)***
Jilin	0.084 (0.009)***	0.112 (0.008)***	0.017 (0.005)***	-0.028 (0.002)***
Shanghai	0.118 (0.009)***	0.107 (0.008)***	0.143 (0.005)***	0.011 (0.002)***
Jiangsu	-0.052 (0.008)***	-0.040 (0.008)***	-0.009 (0.005)*	-0.012 (0.002)***
Anhui	0.031 (0.008)***	0.049 (0.008)***	-0.003 (0.005)	-0.018 (0.002)***
Jiangxi	-0.031 (0.009)***	-0.011 (0.009)	0.018 (0.005)***	-0.020 (0.002)***
Shandong	0.002 (0.008)	0.027 (0.007)***	-0.003 (-0.004)	-0.025 (0.002)***
Henan	0.031 (0.009)***	0.062 (0.008)***	0.019 (0.005)***	-0.031 (0.002)***
Hubei	-0.014 (0.008)*	0.004 (-0.008)	0.016 (0.005)***	-0.019 (0.002)***
Guangdong	-0.017 (0.008)**	-0.031 (0.008)***	0.011 (0.005)**	0.013 (0.002)***

Chongqing	0.052 (0.010)***	0.059 (0.009)***	0.114 (0.005)***	-0.007 (0.002)***
Sichuan	0.042 (0.008)***	0.054 (0.008)***	0.100 (0.005)***	-0.012 (0.002)***
Yunnan	-0.025 (0.009)***	-0.003 (-0.009)	0.038 (0.005)***	-0.022 (0.002)***
Gansu	-0.044 (0.011)***	-0.021 (0.011)**	0.000 (0.006)	-0.023 (0.002)***
Constant	0.248 (0.015)***	0.171 (0.014)***	0.008 (-0.009)	0.077 (0.003)***
Observations	33,358	33,358	33,358	33,358
R-squared	0.19	0.13	0.27	0.47

* significant at 10 per cent
** significant at 5 per cent
*** significant at 1 per cent
Note: Standard errors in parentheses.

be related to the fact that old people spend more time at home. Further, the larger the size of the house, the more energy consumed. On average, every 100 square metres requires 44.3 kg of coal equivalent energy. This is due mainly to direct energy consumption and, in particular, coal consumption; however, larger housing is associated with higher indirect energy consumption as well.

The final part of Table 16.8 presents results for the provincial dummy variables. Relative to Beijing, the following provinces (cities) consume more energy (in descending order): Shanxi, Shanghai, Chongqing, Sichuan, Heilongjiang, Henan, Liaoning and Anhui. This order is almost the same for direct energy consumption. For non-coal direct energy consumption, there is a slight difference in consumption orders. For example, Shanghai ranks first, while Shanxi ranks second. In terms of indirect energy consumption, only Guangdong and Shanghai exceed Beijing, while the remaining provinces all consume less direct energy than Beijing. These energy consumption differences across provinces could be related to the differences in energy price levels and resource reserves across regions. For example, it is understandable that households in Shanxi would consume more energy, especially coal, given its abundance in Shanxi. Richer regions appear to be the largest consumers of indirect energy, as evidenced by the positive coefficients on Shanghai and Guangdong in the final column. This compounds the earlier claim that as Chinese households and regions become richer, indirect energy patterns become increasingly important.

To capture potential diminishing marginal energy requirements and carbon emissions as income increases, we include a squared term for the latter in the regressions. The results indicate that the relationship between energy consumption/carbon emissions and income for urban Chinese households is essentially linear across the income range of our sample. For example, with regard to total per capita energy consumption, every additional yuan of income increases energy consumption by 20 gde, but the rate of increase falls by just 0.00006 gce. For indirect energy consumption, the diminishing point is not reached until 158,000 yuan per capita, which is more than five times the current average per capita income level. For total energy consumption, the points of diminishing marginal energy requirements and carbon emissions are reached at income levels of 174,000 and 363,000 yuan respectively. To visualise the estimated relationship between energy consumption/carbon emissions and income, Figures 16.5 and 16.6 plot the predicted relationship for 99 per cent of the range of income levels in our sample (100–42,100 yuan). The figures show that within our sample income range, the relationship between income and all types of energy/emissions is almost linear. One of the interesting results is that income has a stronger effect on non-coal direct energy consumption and carbon emissions than on total direct energy consumption and carbon emissions, confirming yet again that coal is mainly a poor peoples' energy.

Conclusions

This chapter focused on three interrelated issues. First, we drew on national-level data to establish the notion that the total energy requirements of households are substantially higher when their indirect energy requirements are added to their direct energy consumption. Indeed, the calculations showed that this allocated an additional 16.8 per cent of China's total energy demand to households, 11.2 per cent of which was attributed to urban households. If the eleventh Five-Year Plan succeeds in its objective of shifting the Chinese economy towards consumption-led growth, indirect energy shares are likely to rise in the future, thereby becoming increasingly important for understanding future energy demand and consumer-driven emissions trends.

Second, using urban household survey data, we examined the extent of variation in total energy requirements and emissions across households with different income levels. While there were clear differences in the shares of indirect emissions attributable to the different consumption bundles of rich and poor households, these were overshadowed by overwhelming evidence that the share of emissions attributed to direct energy consumption matters most: poorer households are more emissions intensive because of their heavy dependence on coal. The most critical implication is, therefore, that poor households are given the opportunity to reduce this coal dependence. This could occur as a consequence of income growth, given our observation that richer households consume less coal; therefore, policies targeting poor household income growth remain crucial. It is equally important, however, to ensure that adequate investment is directed towards cleaner energy alternatives in the near future so that the switch away from coal is a feasible option.

The survey-based analysis produced a share of indirect energy for urban households of 32 per cent, just half the share that we found using national-level aggregate data. This discrepancy stemmed from a number of issues, including the need to use per capita consumption rather than expenditure and the exclusion of services from the survey-based analysis. Moreover, throughout the chapter, we overlooked the processes that deliver final goods to household doorsteps, including the storage and distribution of goods—all of which require considerable energy inputs. These inadequacies only strengthen the claim that the indirect energy requirements of Chinese households matter more to the future of China's contributions to global environmental pressures than

Figure 16.5   Predicted relationship between household energy consumption per capita and income

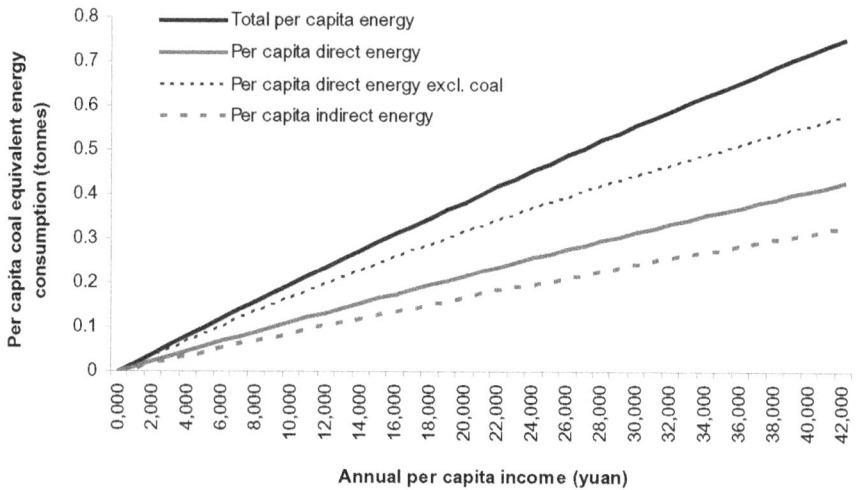

Figure 16.6   Predicted relationship between household carbon emissions per capita and income

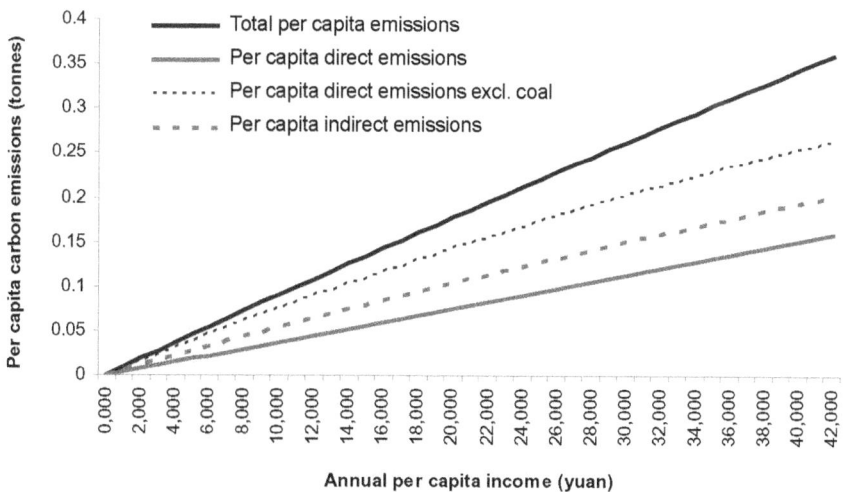

many people realise, and that more detailed research in this area is therefore warranted.

Third, the econometric analysis focused on the determinants of variation in per capita energy requirements and emissions. The virtually linear relationship between per capita income and energy demand is indicative of the dilemma China is facing as per capita incomes continue to grow, since there is no indication of diminishing marginal energy requirements happening over the wide range of current income levels. This finding stresses the need for policymakers to promote ways to reduce the emissions intensities of all of the goods consumed by Chinese households and, for that matter, used by the other end users as well. The higher energy requirements of richer provinces even after controlling for per capita income further compound this need, as poorer provinces continue to grow and strive towards living standards comparable with their richer neighbours. While there is some evidence that households with higher levels of education consume less energy per capita, this result does not hold up in the regression using indirect energy only. Whether or not education has a role to play in promoting cleaner consumption patterns in the future therefore remains an open question, although it is hard to think that such efforts would prove detrimental. This point extends to thoughts about whether 'green' consumption policies stand any chance of reducing China's emissions path below current trajectories.

To end on a bright note, assuming that services are relatively energy efficient (which is certainly the case compared with direct coal consumption) and noting the above evidence that richer Chinese households spend a higher share of their income on services, the future for China could well look cleaner if incomes can be raised across the board to the point at which services dominate consumption patterns. That future cannot arrive quickly enough.

Notes

1   Wei et al. (2007) and Liang et al. (2007) are the only English-language papers we have found, after an extensive literature search, that focus explicitly on Chinese household energy consumption.

2   By focusing on carbon intensities only, we overlook other greenhouse gases, such as methane, nitrous oxide and chlorofluorocarbons. While these other greenhouse gases could be equally, or even more, important for certain environmental problems, data restrictions prevent a more holistic analysis at this stage.

3   For example, Kok et al.'s IO plus household expenditure approach is referred to by Bin and Dowlatabadi (2005) as 'environmental IO life-cycle analysis', while Weber and Perrels (2000) refer to the same method as the 'mixed monetary energetic approach'.

4   See Suh et al. (2004) for further discussion of this point.

5   See, for example, Liao et al. (2007) and Fisher-Vanden et al. (2006).

6 Unfortunately, this precludes 'transport, postal and telecommunication services', 'wholesale and retail trades, hotels and catering services' and 'other services', as the gross value of output for these sectors is not available. Expenditure on transport will be accounted for in the direct energy calculations, discussed further below. Services are, however, omitted from the analysis as there is no alternative given data availability at this stage.

7 As mentioned earlier, the sectors are not comparable directly with those used by Wei et al (2007). Note, however, that they calculate energy and carbon intensities in their 'food' sector of 28 grams and 23 grams per yuan respectively. This sector incorporates the food processing, manufacturing and beverages sectors used above and, taking simple averages across these sectors, reassuringly puts our calculations in the same ballpark as theirs.

8 Coal includes raw coal, clean coal and other washed coal. Coking products include coke, coke oven gas, other gas and other coke products. Petroleum products include crude oil, gasoline, kerosene, diesel oil, fuel oil, petroleum liquid gas, refinery gas and other petroleum products. Natural gas, heat and electricity stand alone.

9 To complicate matters slightly, the relevant table in the *China Energy Statistical Yearbook* (NBS 2006) reports total final energy consumption in each sector in two consistent forms: coal equivalent calculation and calorific value calculation. It provides, however, only the breakdown of energy sources in terms of calorific value calculations, so we use these to calculate the shares of each source. In terms of shares, the results are equivalent whichever measure is used.

10 The electricity coefficient is a weighted sum of the coal, oil and gas coefficients, based on the shares of coal, oil and gas used in China's electricity generation being 78 per cent, 2.4 per cent and 1 per cent respectively (IEA 2007). The remainder of China's electricity is fuelled by hydro-power (16 per cent), nuclear (2.1 per cent) and wind, biomass and other renewables (0.5 per cent). These are all assumed to have zero carbon coefficients. Coking products and heat are assumed to have the same carbon coefficient as coal.

11 See Table A16.2 in the Appendix for details. Due to the data limitations discussed in previous footnotes, not all sectors in Table 16.4 are included in the analysis in this section.

12 Although the nutrition literature normally assumes a 50 per cent premium to reflect processing margins, studies also find that this assumption is not sensitive to the estimated income elasticity of nutrition intake (see, for example, Subramanian and Deaton 1996 and Gibson and Rozelle 2002).

13 This is quite a departure from the aggregate figures in the preceding section, which show that direct energy demand accounts for only 36 per cent of urban households' total energy demand. Two obvious reasons why we understate indirect energy demand are that we do not incorporate all sectors or all expenditure in our analysis. The figure also differs significantly from Wei et al. (2007), who find that direct energy accounts for just 29 per cent of urban household energy requirements. Their figure is at the lowest end of the literature surveyed in the previous section, while ours here is towards the high end. One reason for the discrepancy could be the different consumption categories that we assess, particularly as we divide consumption into more than 20 categories compared with Wei et al.'s eight. Another clear difference is that in Wei et al. (2007), the 'residence' category dominates urban indirect energy demand, accounting for more than 60 per cent of it. Included in this category are the production and supply of electric power, steam, hot water and gas, all of which are assumed to be direct energy demand in our analysis.

14 In addition, given that richer households spend a higher proportion of their income on services than poorer households, and these are largely excluded from our story here, the share of energy and emissions that can be attributed to indirect sources is likely to be understated more for rich than for poor. Our sample households spend, on average, 2,243 yuan per capita on services, which accounts for 27 per cent of per capita current-period consumption. For the poorest group, service expenditure is 727 yuan per capita, accounting for 23 per cent of the total current-period consumption, while for the richest group this figure is 5,592 yuan, accounting for 30 per cent of the current-period consumption (see columns 3 and 6 in Appendix Table A16.1).

15 See Appendix Table A16.3 for details.

16 When, however, only non-coal direct energies are included, the effect of the household head's education switches sign and becomes positive and statistically significant. These contradictory results are puzzling, but are perhaps related to high correlations between per capita income and education levels in combination with the high coal content of poorer households' direct energy consumption.

References

Bin, S. and Dowlatabadi, H., 2005. 'Consumer lifestyle approach to US energy use and the related CO_2 emissions', *Energy Policy*, 33:197–208.

Cohen, C., Lenzen, M. and Schaeffer, R., 2005. 'Energy requirements of households in Brazil', *Energy Policy*, 33:555–62.

Energy Information Administration (EIA), 2008. *International Energy Annual*. Available from www.eia.doe.gov/iea (accessed 15 January 2008).

Fisher-Vanden, K., Jefferson, G.H., Jingkui, M. and Jianyi, X., 2006. 'Technology development and energy productivity in China', *Energy Economics*, 28:690–705.

Gibson, J. and Rozelle, S., 2002. 'How elastic is calorie demand: parametric, non-parametric and semiparametric results for urban Papua New Guinea', *Journal of Development Studies*, 38(6):23–46.

Hondo, H., Sakai, S. and Tanno, S., 2002. 'Sensitivity analysis of total CO_2 emission intensities using an input–output table', *Applied Energy*, 72:689–704.

International Energy Agency (IEA), 2007. *World Energy Outlook 2007: China and India insights*, International Energy Agency, Paris.

Kok, R.D., Benders, R.M.J. and Moll, H.C., 2006. 'Measuring the environmental load of household consumption using some methods based on input–output energy analysis: a comparison of methods and a discussion of results', *Energy Policy*, 34:2744–61.

Liang, Q.-M., Fan, Y. and Yi-Ming, W., 2007. 'Multi-regional input–output model for regional energy requirements and CO_2 emissions in China', *Energy Policy*, 35:1685–700.

Liao, H., Fan, Y. and Yi-Ming, W., 2007. 'What induced China's energy intensity to fluctuate: 1997–2006?', *Energy Policy*, 35:4640–9.

National Bureau of Statistics (NBS), 2005. *Urban Household Income and Expenditure Survey*, China Statistics Press, Beijing.

——, 2006. *China Energy Statistical Yearbook*, China Statistics Press, Beijing.

——, 2007. *China Statistical Yearbook*, China Statistics Press, Beijing.

Pachauri, S., 2004. 'An analysis of cross-sectional variations in total household energy requirements for India using micro survey data', *Energy Policy*, 32:1723–35.

Pachauri, S. and Spreng, D., 2002. 'Direct and indirect requirements of households in India', *Energy Policy*, 30:511–23.

Reinders, A.H.M.E., Vringer, K. and Blok, K., 2003. 'The direct and indirect energy requirement of households in the European Union', *Energy Policy*, 31:139–53.

Subramanian, S. and Deaton, A., 1996. 'The demand for food and calories', *Journal of Political Economy*, 104 (1):133–62.

Suh, S., Lenzen, M., Treloar, G.J., Hondo, H., Horvath, A., Huppes, G., Jolliet, O., Klann, U., Krewitt, W., Moriguchi, Y., Munksgaard, J. and Norris, G., 2004. 'System boundary selection in life-cycle inventories using hybrid analysis', *Environmental Science and Technology*, 38(3):657–64.

Weber, C. and Perrels, A., 2000. 'Modelling lifestyle effects on energy demand and related emissions', *Energy Policy*, 28:549–66.

Wei, Y.-M., Liu, L.-C., Fan, Y. and Wu, G., 2007. 'The impact of lifestyle on energy use and CO$_2$ emission: an empirical analysis of China's residents', *Energy Policy*, 35:247–57.

Table A16.1 Breakdown of per capita expenditure

	Total per capita expenditure	Current period per capita consumption	Per capita service cost	Per capita consumption excluding services	Per capita current period consumption as % of per capita expenditure	Per capita service cost as % of per capita current period consumption	Per capita current period consumption excluding services as % of per capita total expenditure
	(1)	(2)	(3)	(4)	(5) = (2)/(1)	(6) = (3)/(2)	(7) = (4)/(1)
Decile 1	3,655	3,095	727	2,369	0.85	0.23	0.65
Decile 2	5,200	4,323	1,039	3,284	0.83	0.24	0.63
Decile 3	6,438	5,211	1,285	3,926	0.81	0.25	0.61
Decile 4	7,402	5,934	1,513	4,421	0.80	0.25	0.60
Decile 5	8,694	6,872	1,790	5,082	0.79	0.26	0.58
Decile 6	9,973	7,687	2,021	5,666	0.77	0.26	0.57
Decile 7	11,639	8,843	2,421	6,422	0.76	0.27	0.55
Decile 8	13,562	10,015	2,723	7,292	0.74	0.27	0.54
Decile 9	16,390	11,827	3,316	8,511	0.72	0.28	0.52
Decile 10	27,303	18,845	5,592	13,252	0.69	0.30	0.49
Average	11,026	8,265	2,243	6,022	0.75	0.27	0.55

Table A16.2 **Aggregation of survey sectors into Table 16.4 sectors**

Table 16.4 sector	Survey sectors included
Processing of food from agricultural products	Grain, starch, meat, poultry, eggs, seafood, vegetables, fruits
Manufacture of foods	Manufacture of grain, beans, meat, poultry, eggs, seafood, sugar, vegetables, fruits, milk, cake, canned food and semi-cooked food plus one-third of eating-out expenses
Manufacture of beverages	Alcohol and drinks
Manufacture of tobacco	Tobacco
Manufacture of textiles	Clothing materials
Manufacture of textile apparel, footwear and caps	Clothes, shoes, bedclothes, textile decoration
Processing of timber and wood products	Furniture materials
Manufacture of furniture	Furniture
Manufacture of paper and paper products	Paper and stationery
Printing, reproduction of recording media	Newspapers, magazines, books, textbooks, education software
Manufacture of articles for culture, education and sports	Sports equipment, electronic dictionaries, audio and video products, culture decorations
Manufacture of raw chemical materials and chemical products	Detergent and cosmetic goods for hairdressing and bathing
Manufacture of medicines	Medicine, nutriments
Manufacture of metal products	Lighting, cutlery, tea sets, small tools
Manufacture of transport equipment	Cars, bikes, motors and their components
Manufacture of electrical machinery and equipment	Electrical, medical and health equipment
Manufacture of measuring instruments and office machinery	Pianos, other musical instruments, body-building equipment, watches
Manufacture of artwork and other manufacturing	Jewellery and other miscellaneous items
Production and distribution of water	Water
Construction	Housing decoration and building materials
Services	Services: sewing, housekeeping, medical treatment, communication fees, recreation fees, education fees, rent, imputed rent, housing management fees; transportation fees * 0.75; eating-out expenditure * 0.67

Table A16.3 **Emission shares of consumption goods by income decile**

	Share of total emissions		Share of indirect emissions	
	First	Tenth	First	Tenth
Indirect emissions				
Processing of food from agricultural products	0.045	0.046	0.223	0.117
Manufacture of foods	0.104	0.123	0.513	0.312
One-third of eating out	0.003	0.019	0.015	0.048
Manufacture of beverages	0.004	0.011	0.021	0.027
Manufacture of tobacco	0.002	0.003	0.008	0.007
Manufacture of textiles	0.000	0.001	0.002	0.002
Manufacture of textile apparel, footwear, caps	0.007	0.026	0.036	0.065
Processing of wood products	0.000	0.000	0.000	0.001
Manufacture of furniture	0.000	0.004	0.001	0.009
Manufacture of paper and paper products	0.002	0.002	0.009	0.006
Printing and recording media	0.002	0.003	0.008	0.009
Manufacture of articles for culture, education and sports	0.000	0.003	0.002	0.009
Manufacture of raw chemical materials and products	0.011	0.041	0.052	0.103
Manufacture of medicines	0.001	0.014	0.005	0.036
Manufacture of metal products	0.001	0.003	0.003	0.008
Manufacture of transport equipment	0.001	0.023	0.002	0.059
Manufacture of electrical machinery and equipment	0.001	0.005	0.002	0.012
Manufacture of communication, computers and electronics	0.000	0.004	0.001	0.011
Manufacture of instruments, cultural and office machinery	0.000	0.001	0.000	0.002
Manufacture of artwork and other manufacturing	0.009	0.038	0.045	0.097
Production and distribution of water	0.010	0.015	0.050	0.039
Construction	0.001	0.009	0.003	0.024
Direct emissions			Share of direct emissions	
Coal	0.459	0.040	0.576	0.065
Petrol	0.029	0.144	0.036	0.237
Gas	0.222	0.295	0.279	0.487
Electricity	0.087	0.128	0.109	0.211
Share of direct emissions in total	0.797	0.606		

17

Can China's coal industry be reconciled with the environment?

Xunpeng Shi

Coal provides 70 per cent of China's primary energy, therefore it is no surprise that China's air pollution is caused mainly by coal use. It has been reported that 85 per cent of the sulphur dioxide, 70 per cent of the smoke and 60 per cent of the nitrogen oxides emitted into the atmosphere in China come from the burning of coal (Wang and Feng 2003). The correlation between increased sulphur dioxide levels and coal consumption is above 95 per cent (He et al. 2002). Emerging concerns about climate change add further pressure to the use of fossil-fuel energy in general and coal in particular. As the world's second-largest consumer of energy, China's demand for energy will increase due to its growing population and the rapid increase in living standards. Given that China is the world's largest producer and consumer of coal, this issue is of great concern and importance to the global community.

To mitigate the environmental impacts of rapid economic growth, proposals to reduce waste-gas emissions (WGEs) usually include changing the economic structure, reducing energy intensity and enforcing waste-gas treatment (Liang and Zhou 2008). These measures often lead to reductions in use of energy, particularly coal. Globally and in China—a country with serious environmental pollution, a growing demand for a clean environment, serious concerns about climate change and coal-dominated air pollution—many people think that the coal industry has no future; in China, the coal industry is sometimes called *xiyang gongye* ('the sunset industry') (Coal Enterprise Management 2001).

Many others are, however, arguing that there is a promising future for the coal industry in general and the Chinese industry in particular (Coal Enterprise Management 2001; Li 2003; Shi 2003, 2006; Wang 1999; Huang 2001). A key argument for this is that there is a decreasing trend of pollution emissions per unit of coal, or emission intensity, and therefore the coal industry can harmonise with the environment (Shi 2003, 2006). The argument suggests that environmental pressure will induce innovations in clean-coal technologies. The ultimate level of cleanliness will be decided by technical progress and socio-economic conditions. Many coal experts are also arguing that coal can be a truly clean energy;[1] however, no empirical evidence has been provided with which to examine this issue.

This study will test the evolving pattern of coal emissions intensity using industry WGE data for the period 1996–2006. The focus on industrial WGEs only is appropriate because industrial pollution plays a dominant role in total emissions (Table 17.1): for example, 86 per cent of total sulphur dioxide emissions and 79 per cent of total smoke emissions came from industrial sources in 2005 (SEPA various years). The study will focus on three air pollutants: sulphur dioxide emissions (SO_2), industrial smoke emissions (*Smoke*) and dust emissions (*Dust*). Carbon dioxide emissions, although a popular topic and probably significant, are not included because there are no current data for them. Furthermore, although greenhouse gas emissions are a major concern, the most immediate environmental phenomenon is local ambient air pollution, which is predicted to cause health damage worth 13 per cent of China's gross domestic product (GDP) by 2020 (OECD 2007, cited in IEA 2007). The literature (Ang and Pandiyan 1997; Ang et al. 1998; Wang et al. 2005; Wu et al. 2006) often infers carbon dioxide emissions by assuming a constant emission intensity for each fuel, including coal—something that will be challenged in this study.

The rest of this chapter is organised as follows. The second section provides a general background of China's environmental regulations for air pollution. The third section introduces the two study methods, followed by a section describing the data. The fifth section presents the empirical results and the sixth section discusses some technical issues. The last section concludes the chapter.

Environmental regulations and the coal industry

The Chinese government attaches great importance to environmental protection. The first national conference on environmental protection in 1973 set up the Environmental Protection Office under the State Council and stipulated the 'three synchronisations' system.[2] The 1978 Chinese Constitution stated the central government's intention to protect and improve the environment, and to prevent and control pollution. Environmental protection was declared one of two Chinese 'national fundamental policies' in 1983.[3] The Environmental Protection Law (Trial Implementation) was promulgated in 1979, which proposed the Environmental Impact Assessment (EIA) System and the polluter-pays principle. It was revised and enacted officially in 1989. Between 1949 and 2005, China legislated nine environmental laws and 15 resource-protection laws, it formulated and promulgated more than 50 administrative regulations and 660 ministerial and local environmental rules and regulations related to environmental protection (State News Office 2006).

With the development of environmental protection, the control of WGEs is evolving. The National Ambient Air Quality Standards were published in 1982, specifying standards for air pollutants such as sulphur dioxide, total suspended particulate matter (TSP), nitrogen oxides and carbon (SEPA 1982). In 1998, the Chinese government (State Council 1998) approved the delimiting of two 'control areas' (the 'sulphur dioxide control area' and the 'acid rain control area'), which led to a significant improvement in environmental quality (State News Office 2006). In June 2007, China's State Council approved a national plan to address the challenges of climate change, symbolising that climate change was becoming increasingly important on China's national policy agenda (NDRC 2007a).

Some financial incentive mechanisms for reducing air pollution are already in place. Tax reductions or exemptions are given to enterprises engaged in environmental protection (State News Office 2006). A pollution levy system based on the polluter-pays principle was implemented nationally in 1982. Between 1995 and 2005, the collection of sulphur dioxide discharge fees was expanded to include all related enterprises and the rate was raised from 0.2 to 0.63 yuan per kilogram (State News Office 2006). The rate has been increased further in some provinces.[4]

Massive pollution control measures have also been implemented: 84,000 small enterprises that caused serious waste and pollution were closed during the ninth Five-Year Plan (1996–2000); eight resource pollution-intensive industries, including iron and steel, cement, electrolytic aluminium and coking, were restructured; and the construction of more than 1,900 projects was either stopped or postponed (State News Office 2006).

With these efforts in place, the control of air pollution has been considerable. The volume of removed industrial sulphur dioxide reached 10.9 million tonnes in 2005, up from 2.34 million tonnes in 1998, while the volumes for smoke and dust reached 205.87 million tonnes and 64.54 million tonnes respectively in 2005, compared with 86.70 million tonnes and 30.98 million tonnes in 1998. The amount of industrial sulphur dioxide, industrial smoke and dust discharged in the generation of one unit of GDP in China in 2004 dropped by 42 per cent, 55 per cent and 39 per cent, respectively, from the levels in 1995 (State News Office 2006).

With increasing implementation of environmental regulations in China, the future of coal is under suspicion as it is the most polluting of the fossil fuels and the key source of pollution in China. After 1998, Chinese coal production experienced a three-year depression[5] and state-owned coal mines (SOCMs) suffered heavy deficits, which led to increasingly widespread pessimism about the industry's future (Shi 2003, 2006; Coal Enterprise Management 2001). Amid the concerns and dire predictions about the future of the coal industry, coal production and consumption increased dramatically after 2000.

Although coal consumption has been soaring in recent years, dust and smoke emissions are declining. There has been a slight increase in sulphur dioxide emissions, but the speed of the increase is far slower than that of coal consumption. Coal consumption increased nearly 92 per cent between 1997 and 2006, yet emissions of sulphur dioxide increased only slightly, by 10.35 per cent, during the same period.

Figure 17.1 **Coal industry development and air pollution emissions, 1997–2006**

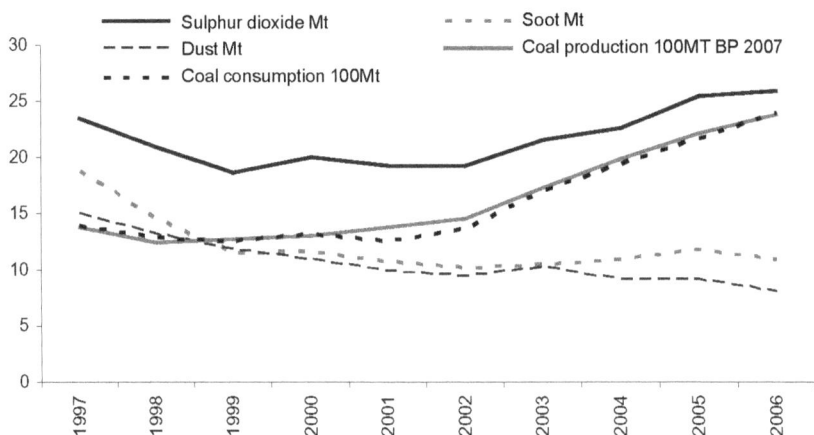

Sources: State Environmental Protection Administration (SEPA), 1996–2006. *State Public Release of Environmental Protection*, State Environmental Protection Administration; National Bureau of Statistics (NBS), various years. *China Statistical Yearbook*, China Statistics Press, Beijing. To avoid the underreporting of production due to the mine closure policy, the production data were extracted from British Petroleum (BP), 2007. *BP Statistical Review of World Energy*.

Emission intensity trends

Most studies of emission intensity focus on carbon dioxide emissions (see Ang and Zhang 2000 for a review). Only a few studies address sulphur dioxide emissions (Lin and Chang 1996; Shrestha and Timilsina 1997; Viguier 1999); among these, two (Shrestha and Timilsina 1997; Viguier 1999) address the emission intensity of individual energy sources, including coal, as a constant over time and focus only on changes in the energy mix—that is, the inter-energy composition of oil, gas, coal and non-fossil fuel energies. Lin and Chang (1996), however, find a continuous decrease in pollution intensity and attribute it to substitution for imported low-sulphur coal and 'desulphurisaton' of fuel oil, in the case of Taiwan. No study has, however, been done from the perspective of a clean future for fossil-fuel energy, and coal in particular. Emission intensities for smoke and dust have also not been addressed.

In this section, two alternative methodologies are presented to test the hypothesis that coal emission intensity is declining. The econometric technique is used to test the emission intensity of coal with a focus on a long-term dynamic aspect of this intensity. The index decomposition (ID) approach is used to identify various factors that determinate the final emissions and their individual contributions.

The fixed-effects panel data model

The first method to be used is a fixed-effects panel data model, which, together with econometric data, will test the general trend of emission intensity—that is, at time t, in province i, the jth WGE, is

$$WGE_{itj} = \beta_0 + (\beta_1 + \beta_2 T)FuelC_{it} + (\beta_3 + \beta_4 T)MatC_{it} + \beta_5 X_{it} + \alpha_{ij} + u_{itj} \qquad (1)$$

in which WGE denotes pollutant emissions; $FuelC$ and $MatC$ denote consumption of fuel coal and material coal, respectively; T is a general time trend; α_{ij} is the province-specific effect in the case of the jth pollutant; u_{itj} is a normally distributed error term; $j = 1,2,3$ is sulphur dioxide (SO_2), smoke and dust respectively. Of particular interest are the signs of β_1 and β_4. If β_2 or β_4 were significantly negative, we would find evidence of decreasing emission intensity. Due to different characteristics between fuel and material coal in different cases of air pollution, β_2 and β_1 are expected to be different.

X is a vector of exogenous variables such as population (POP), average GDP and environmental regulation and implementation variables, to investigate the impact of various exogenous variables on the emissions function. The more stringent the environmental regulation and enforcement, the less are the WGEs because polluters are more likely to be punished or charged. Several variables have been used to test the impacts of legislation and enforcement. Bao and Peng (2006) used the number of cumulative environmental standards issued by provincial governments to measure the effect of environmental policy on emissions, but they failed to find a positive role for such policies on air pollution. The compliance cost was used as a proxy for enforcement based on the assumption that the two had a positive correlation (Gray 1987). Gray (1987) uses the number of penalties (fines) to study safety regulations. In this study, the cumulative number of environmental standards (*Standard*) is used to approximate the effect of legislation; the operating cost of waste-gas treatment equipment (*Cost*) is used to approximate the stringency of enforcement. Environmentally related research and development expenditure (*R&D*) is included to measure the technology progress effect, as in Bao and Peng (2006). The GDP deflator is used on all monetary value terms to the 1996 constant price of 10,000 yuan.

Decomposed factors determining emissions

Since most WGEs in China come from the consumption of fuel, they are affected by factors such as economic structure, energy intensity, economic development and population growth. It would be helpful if the WGE changes were broken down into various factors. Index decomposition has been a popular tool in the past 40 years to assess quantitatively various factors affecting WGEs and energy demand. Ang and Zhang (2000) found that it had been applied in at least 124 studies by 2000. This approach has advantages over econometric estimations in that it can apply to small samples and does not need any assumptions about distribution. This method is applied initially to studies of industrial energy decomposition and energy demand. Torvanger (1991) was the first to apply the index-decomposition methodology to study energy-related gas emissions, and was followed by many other studies (Ang and Zhang 2000). The types of gas emissions studied included carbon dioxide, sulphur dioxide and nitrogen oxides. Compared with energy-demand decomposition, the emission decomposition includes more factors, including sectoral fuel share and fuel emissions (Ang and Zhang 2000).

The Laspeyres method and the Divisia index (LMDI) are the most frequently used and preferred decomposition methods in energy-induced gas emission studies (Ang and Zhang 2000). In this study, the LMDI approach is applied because it has the time-reversal property of an ideal index and can perform a perfect decomposition and accommodate zero values in the data set, which is preferable to the refined Laspeyres method (Ang and Zhang 2000).

Two kinds of indicators are often used in the index-decomposition studies of energy and environmental issues. The first is a quantity indicator such as total energy consumption or total gas emissions; the other is a ratio or index indicator, including aggregate energy intensity and aggregate gas emissions intensity (Ang and Zhang 2000). In this study, we chose the first, with total WGE as the indicator. Total consumption of coal (TC), fossil-fuel energy (FE) and energy consumption (TE) are evaluated at 10,000 tonnes of standard coal equivalent (SCE), which will avoid heterogeneity of different coal qualities. Y is GDP and P is population. The explicit introduction of coal into the emissions function is an extension of the previous literature.

Similar to Wang et al. (2005), the WGE is expressed as an extended Kaya Identity (IPCC 2001; Kaya 1990)—that is

$$WGE_i = \frac{WGE}{TC}\frac{TC}{FE}\frac{FE}{TE}\frac{TE}{Y}\frac{Y}{P}P = ECFIGP \tag{2}$$

in which i is the type of emission, including sulphur dioxide, smoke and dust; E is the mean WGE emission intensity of coal (which is the core interest of this study); C is the share of coal in total fossil-fuel energy, or the fossil-fuel composition factor; F is the share of fossil-fuel energy in total energy consumption, or the energy component factor; I is energy intensity; and G is GDP per capita.

As shown by Wang et al. (2005), using the LMDI (Ang et al. 1998), the difference in WGEs between two periods, t and T, can be expressed as

$$\Delta WGE_{tT} = WGE_{iT} - WGE_{it} = E_{iT}C_{iT}F_{iT}I_{iT}G_{iT}P_{iT} - E_{it}C_{it}F_{it}I_{it}G_{it}P_{it}$$
$$= \sum_K \Delta WGE_{K-effect}, \quad K = E, C, F, I, G, P \tag{3}$$

In which, the K-effect on emission reduction is

$$\Delta WGE_{K-effect} = L(WGE_{it}, WGE_{iT})\ln(K_T / K_t) \tag{4}$$

in which

$$L(x,y) = (x - y) / \ln(x / y) \tag{5}$$

We also define the case of no change in emission intensity (*Non–CEI*) as the basis for comparison with the current study. Following Equation 3, the *Non–CEI* of WGEs can be derived by dropping the emission intensity effect as

$$Non \text{ - } CEI = \sum_K \Delta WGE_{K-effect}, \ K = C, F, I, G, P \qquad (6)$$

Data description

We study the time trend and determinants of WGEs from coal consumption using China's provincial panel data from 1996–2006. We chose this period because data for coal consumption broken down to combusting and material inputs were available only from 1996. Data for three kinds of air pollutants, consumption of two kinds of coal and various environmental variables were drawn from the various issues of the *China Environmental Yearbook* (SEPA various years). Table 17.1 provides a summary.

The data for national total coal production (in physical quantity and in standard coal equivalent), energy consumption and its mix, GDP and its deflators and population are drawn from the various issues of the *China Statistical Yearbook* (NBS various years). The national GDP data are deflated to 1996 constant prices. When there is more than one set of data, we prefer the most up-to-date one because China's State Statistical Bureau significantly adjusted energy-use data in 2006. A history of the evolution of these aggregate data is shown in Figure 17.2.

Empirical estimation and analysis of results

Econometric results: evidence from the industrial sector

As in the literature (Shadbegian and Gray 2006), the 'seemingly unrelated regressions' (SUR) model is employed to allow for correlations in the residuals across equations for the three air pollutants. SUR is used because factors such as environmental legislation, environmental policy and changes in enforcement will affect the outcomes for all air pollutants simultaneously. To accompany both the fixed-effects, or unobserved regional heterogeneity and SUR, we use a dummy variable version of the SUR model—that is, we create 31 dummy variables for 31 provinces and include 30 of them into the regression functions to accompany the fixed effect, or unobserved regional heterogeneity. Estimation results of the SUR model are shown in Table 17.2.

The results demonstrate that there is a significant (at the 1 per cent level) decline in emission intensity in the case of material coal. For fuel coal, there is

Table 17.1 **Summary statistics of the panel data**

Variable	Obs	Mean	Std dev.	Min	Max
SO2 emissions (tonnes)	340	542,871.8	392,335.2	734	1,760,057
Smoke emissions (tonnes)	340	286,477.8	228,046.8	936	1,432,735
Industrial dust emissions (tonnes)	339	287,156.3	233,446.3	1,132	1,005,809
Fuel coal (104 tonnes)	340	3,235.429	2,679.42	2	14,956
Material coal (104 tonnes)	333	1,319.634	1,750.11	1	16,297
Operational cost (104 yuan)	339	27,400.54	32,351.93	8.071749	365,706
Research and development (104 yuan)	319	372.2674	499.9261	1.356945	2,739.663
GDP per capita (yuan)	340	5,960.203	3,754.129	1,918.126	20,463.64
Population (10⁴ persons)	340	4,079.382	2,619.882	244	11,430
Cumulated standards	341	1.95	4	0	22

Note: All monetary terms are converted to 1995 constant prices by the GDP deflator.
Source: Author's own calculations.

Figure 17.2 **Historical changes in the five factors in the index-decomposition study, 1990 – 2006**

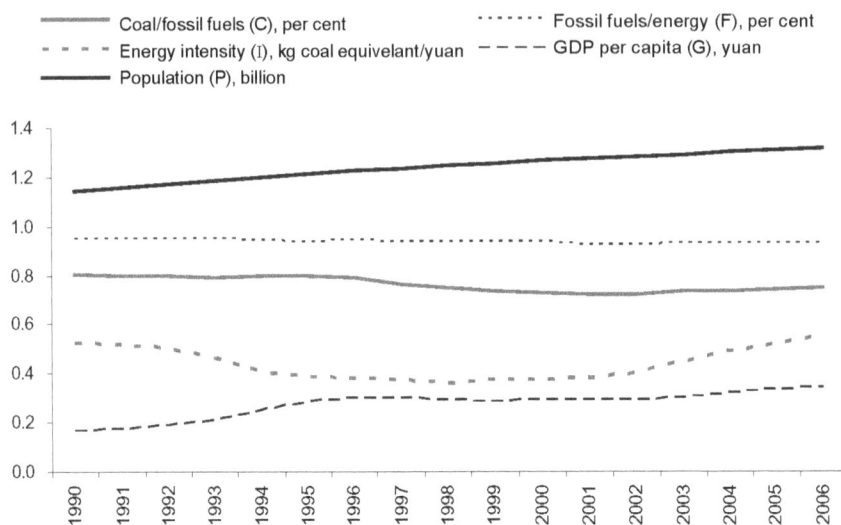

— Coal/fossil fuels (C), per cent
- - - - Energy intensity (I), kg coal equivelant/yuan
— Population (P), billion
······ Fossil fuels/energy (F), per cent
- - - - GDP per capita (G), yuan

Note: Units are rescaled to fit into the same frame.
Source: Author's own calculations.

also a significant (at the 1 per cent level) decline in emission intensity, except in the case of dust. We can calculate the emission intensity of these two kinds of coal by taking the partial derivative of WGE with respect to coal consumption. To test the robustness of this conclusion, in the second specification, we remove regulatory and economic variables and find the conclusion unchanged (Table 17.2).

We can also calculate the time when overall emission intensity will decline to zero, which is called the 'zero emissions point' (Table 17.3).

The decline in emission intensity is unlikely to be linear in time, as assumed here, because that would lead to negative emission intensity. The actual decline is likely in a non-linear pace with diminishing rate. This hypothetical zero emissions point, however, provides a relative comparison among different cases in terms of the two kinds of coal and the three types of air pollution emissions. It can be seen that in the case of sulphur dioxide, the zero emissions point for material coal will be reached more quickly than that for fuel coal. The reasons for this could be: 1) economic, as material-coal users can reduce WGEs with a

Table 17.2   Estimated results of the fixed-effects SUR model

| | SO_2 | | | | Smoke | | | | Dust | | | |
| | Specification 1 | | Specification 2 | | Specification 1 | | Specification 2 | | Specification 1 | | Specification 2 | |
	Coefficient	P>z	Coefficient	P>z	Coefficient	P>z	Coefficient	P>z	Coefficient	P>z	Coefficient	P>z
Fuel coal	112.61	0.00	123.63	0.00	59.63	0.00	43.36	0.01	-14.03	0.50	-30.44	0.14
T*Fuel coal	-3.49	0.01	-4.74	0.00	-3.09	0.00	-2.63	0.02	0.36	0.81	0.68	0.64
Material coal	56.14	0.00	56.79	0.00	128.38	0.00	148.56	0.00	141.36	0.00	166.82	0.00
T*Material coal	-4.18	0.01	-3.19	0.04	-8.28	0.00	-9.89	0.00	-8.32	0.00	-10.38	0.00
Time trend	13,695.29	0.00	14,188.33	0.00	5,905.04	0.00	3,694.51	0.09	9,645.54	0.00	7,702.12	0.01
Standard	-6,730.50	0.00	n.a.	n.a.	-540.82	0.71	n.a.	n.a.	274.65	0.91	n.a.	n.a.
Cost	-0.39	0.08	n.a.	n.a.	-0.44	0.01	n.a.	n.a.	-0.53	0.04	n.a.	n.a.
Research and development	-54.27	0.00	n.a.	n.a.	-28.15	0.02	n.a.	n.a.	-37.86	0.05	n.a.	n.a.
Average GDP (aveGDP)	124.89	0.00	n.a.	n.a.	-50.42	0.03	n.a.	n.a.	-74.10	0.04	n.a.	n.a.
aveGDP2	-0.0047	0.00	n.a.	n.a.	0.0016	0.05	n.a.	n.a.	0.0023	0.09	n.a.	n.a.
Population	66.27	0.01	n.a.	n.a.	-46.12	0.01	n.a.	n.a.	-34.75	0.23	n.a.	n.a.
constant	-809,755.90	0.00	-90,396.33	0.02	404,876.10	0.03	-73,420.19	0.02	647,503	0.03	-31,147.8	0.44

Source: Author's own estimations.

Table 17.3   Emission intensity and the appearance of the zero emissions point

		SO_2	Smoke	Dust
For fuel coal	Emission intensity	112.61-3.49*T	56.63-3.09*T	0
	Zero emissions point (year)	T=32.3	T=18.3	n.a.
For material coal	Emission intensity	56.14-4.18*T	128.38-8.28*T	144.36-8.32*T
	Zero emissions point (year)	T=13.4	T=15.5	T=17.4

Note: T is time trend and equal to one in 1996.
Source: Author's own calculations.

lower marginal cost than fuel-coal users; or 2) technical, as material-coal users, usually chemical producers, can produce by-products, such as sulphur, from WGEs. The second reason could also explain why sulphur dioxide has the shortest zero emissions point for material coal among the six cases.

Legislation, approximated by cumulative environmental standards, has a significant impact on reducing sulphur dioxide emissions, while increased use of WGE-treatment machines can lead to a reduction of all three air pollutants. The variable of cumulative environmental standards is significant only at the 1 per cent level in the case of sulphur dioxide. Implementation, approximated by the operating costs of WGE-treatment machines, is significant for smoke and dust at the 5 per cent level and for sulphur dioxide at the 10 per cent level. The significance of the effect of regulations on the removal of sulphur dioxide could be the reason why environmental regulations have led to the increased application of desulphurisation equipment. It could also be due to the fact that environmental regulations have spurred a demand for low-sulphur coal, as noted by Darmstadter (1999) in the US case. Whatever the reason, we can conclude that implementation is more important than legislation alone. The insignificant effect of legislation could also be due to the fact that much environmental legislation is not related to air pollution and this variable is therefore not a good proxy for this study.

The technological effect, approximated by environmentally related research inputs, is, however, negative and significant at the 5 per cent level in all three emissions cases, which is consistent with the literature (Bao and Peng 2006). This demonstrates that technical change has played a role in reducing emissions; therefore, investment in environmentally related research in general, and in clean-coal technology in particular, is one way to reduce air pollution.

Our findings about the pollution–income relationship are not consistent with the literature. Bao and Peng (2006) found that all three air pollutants—sulphur dioxide, smoke and dust—had an inverted-U shaped relationship with economic growth. Grossman and Krueger (1991) found that two pollutants—sulphur dioxide and smoke—exhibited an inverted-U shaped relationship. We found an inverted-U shaped relationship only between economic growth and sulphur dioxide emissions—a finding made also by Kaufmanna et al. (1998) and Markandya et al. (2006). The reason for this could be that we have controlled the consumption of coal, which captures a majority effect of economic growth on air pollution emissions.

The variable POPULATION is estimated to have opposite signs between the case of sulphur dioxide and smoke. The reason for this could be that population is not itself a decisive factor in air pollution emissions when we control the use of coal.

Table 17.4   **Estimated results of the provincial dummy of the SUR model***

	SO$_2$		Smoke		Indust	
	Coefficient	P-value	Coefficient	P-value	Coefficient	P-value
Tianjin	-74,712.42	0.23	-20,084.44	0.65	-71,001.17	0.32
Hebei	247,410.00	0.12	218,990.70	0.05	400,382.70	0.03
Shanxi	278,565.60	0.08	-75,897.98	0.50	-318,093.00	0.08
Neimeng	441,203.70	0.00	-28,164.28	0.79	-156,646.80	0.35
Liaoning	-10,285.15	0.93	239,425.30	0.00	257,120.80	0.05
Jilin	-15,106.05	0.91	49,665.25	0.59	-158,403.50	0.28
Heilongjiang	-218,583.70	0.07	189,376.10	0.03	-90,686.61	0.51
Shanghai	160,503.90	0.03	-76,023.68	0.15	-57,286.46	0.50
Jiangsu	-14,575.92	0.93	219,807.00	0.05	340,339.90	0.05
Zhejiang	-36,566.59	0.72	139,597.00	0.05	357,680.80	0.00
Anhui	50,817.54	0.76	50,837.26	0.66	52,659.95	0.78
Fujian	-41,882.53	0.67	4,805.48	0.95	48,122.38	0.67
Jiangxi	238,842.00	0.12	-463.16	1.00	-57,798.59	0.74
Shandong	227,807.40	0.23	280,067.40	0.04	473,812.70	0.03
Henan	-49,629.20	0.81	441,930.90	0.00	489,336.40	0.04
Hubei	141,033.80	0.33	114,727.10	0.27	106,951.40	0.52
Hunan	285,509.20	0.07	250,142.70	0.03	338,133.90	0.06
Guangdong	-12,313.66	0.94	293,509.10	0.01	580,014.50	0.00
Guanxi	607,458.40	0.00	249,493.20	0.02	117,233.20	0.51
Hainan	174,040.40	0.22	-175,703.70	0.08	-346,651.30	0.03
Chongqing	563,091.50	0.00	-80,074.48	0.44	-138,723.30	0.40
Sichuan	417,942.00	0.03	530,490.10	0.00	213,966.10	0.34
Guizhou	725,534.20	0.00	-61,527.27	0.63	-244,188.10	0.24
Yunnan	232,539.60	0.13	-40,925.44	0.71	-220,664.40	0.21
Xizang	359,338.70	0.06	-285,079.70	0.03	-481,698.40	0.03
Shaanxi	528,531.30	0.00	43,023.70	0.69	-91,783.74	0.59
Guansu	448,379.10	0.01	-134,572.20	0.26	-272,718.80	0.16
Qinghai	293,903.00	0.08	-220,338.80	0.06	-383,854.40	0.04
Ningxia	383,930.10	0.02	-203,961.60	0.08	-339,471.70	0.07
Xinjiang	181,035.40	0.15	-71,052.97	0.42	-162,888.40	0.25

* based on estimation results from specification 1.
Source: Author's own estimations.

In term of regional diversity, several provinces achieve the same results as Beijing in all three emissions cases: Tianjin, Jilin, Anhui, Fujian, Jiangxi, Hubei, Yunnan and Xinjiang. No province differs from Beijing for all three emissions.

In the case of sulphur dioxide emissions, the level in Inner Mongolia, Shanghai, Guangxi, Chongqing, Sichuan, Guizhou, Shaanxi, Gansu and Ningxia are estimated significantly higher than Beijing, while only Helongjiang is estimated lower than Beijing (at 10 per cent significant level). It can be seen that western China performs worse than Beijing for sulphur dioxide emissions. This could be because Beijing started to promote the use of low-sulphur coal in 1998 (Cao 1998). In the case of smoke, Xizang (significant at the 5 per cent level), Qinghai, Ningxia and Henan (significant at the 10 per cent level) were all less polluted than Beijing. The majority of the remaining provinces were more polluted than Beijing. A similar pattern exists in the case of dust.

Index decomposition results

These results show that emission intensity has the biggest impact on emissions changes among all six factors during the period 1996–2006. As theory predicts, economic development and population growth are key factors driving the increase in emissions.

The decrease in smoke and dust emissions from 1996 to 2006 was due primarily to the decrease in emission intensities. In the case of sulphur dioxide emissions, even though overall emissions are increasing, emission intensity is decreasing and this has the most significant impact. The changing structure of coal among fossil-fuel energy and the energy mix also contributes to the decrease in emissions.

The fossil-fuel composition effect and the energy composition effect were mostly negative and were small for the entire period, except for a few years. The positive sign means that coal has recaptured some of its share of fossil-fuel energy, which is due to the high demand for energy in recent years. The negative sign reflects the improvement in the atmospheric environment due to the switch from coal to other fossil-fuel energies and from fossil-fuel energy to non-fossil fuel energy; this impact is usually desirable. In contrast, however, the small value reveals a relatively weak impact on WGE changes from the change of fuel composition and the use of renewable energy, mainly because the composition of energy consumption is stable.

It is noteworthy that China's energy intensity increased from 0.37 kg of coal equivalent per renminbi (RMB) at the 1990 price (kgce/RMB) in 1997 to 0.55 kgce/RMB in 2006—an increase of 48.65 per cent for the period. As a result, the contribution of energy intensity to emissions increases is higher than that of traditional contributors such as economic development and population growth. The energy intensity effect results in increases in emissions of 947.65 Mt of sulphur dioxide, 512.16 Mt of smoke and 438 Mt of dust. The major reason for high and increasing energy intensity could be rapid industrialisation, which increases not only total energy consumption, but energy intensity. The biggest decrease in emission intensities occurred between 1998 and 1999, which might have been due to improvements in average coal quality when there was an oversupply of coal and consumers would not accept coal with high sulphur and ash content.

Compared with industrialised countries, where decreases in the aggregate energy intensity and aggregate carbon dioxide intensity are explained mainly by declines in energy intensity (Ang and Zhang 2000; Torvanger 1991), China's changes in industrial WGEs are to a large extent the result of the decline in coal emissions intensity.

To simplify the discussion, the results for the three WGEs are normalised to the year 1997. The cumulative change of emissions between 1997 and 2006 is decomposed to change each factor.

In all three cases, it is clear that the non-CEI increase of emissions is much higher than the real emissions, which demonstrates that the decreasing emission intensity is a significant contributor to emissions reduction. Energy intensity is next to emission intensity in terms of scale, but in the opposite direction. Differing from the findings in the literature (Shalizi 2007; Lin and Chang 1996), this study shows that energy intensity replaces economic growth as the major driver of increased emissions.

This result demonstrates that the real contribution of decreased emission intensity has been covered by the popular idea that emissions will be reduced by the decrease in energy intensity (Lin and Chang 1996; Shalizi 2007). This finding is significant because it shows another way of reducing emissions. If

Table 17.5 **Summary of factors that affect the changes in sulphur dioxide, smoke and dust emissions, 1997–2006**

	WGE	$WGE_{E\text{-effect}}$	$WGE_{C\text{-effect}}$	$WGE_{F\text{-effect}}$	$WGE_{I\text{-effect}}$	$WGE_{G\text{-effect}}$	$WGE_{P\text{-effect}}$
SO2	322.80	-1,005.54	-53.83	-25.98	947.65	311.02	148.70
Smoke	-495.00	-1,212.90	-29.09	-14.04	512.15	168.09	80.37
Dust	-697.50	-1,311.46	-24.88	-12.01	438.00	143.75	68.73

Source: Author's own summary.

Figure 17.3 **Decomposition of sulphur dioxide emissions changes in China, 1997–2006**

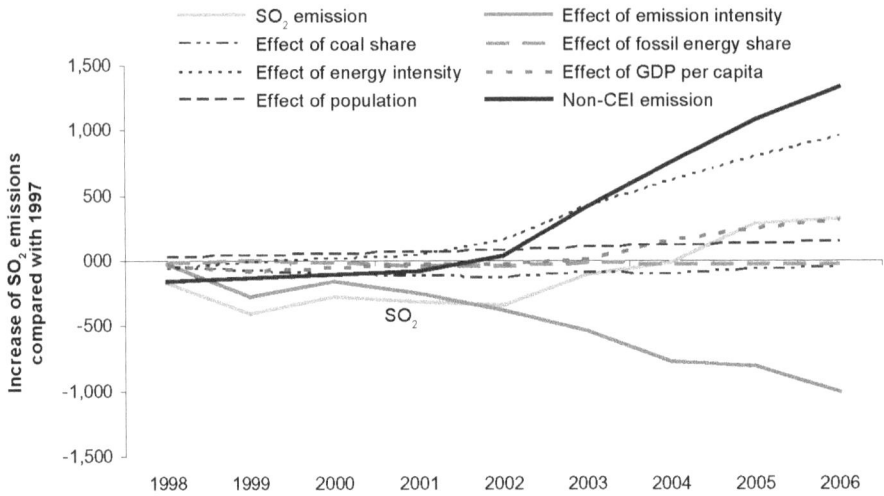

Source: Author's own calculations.

Figure 17.4 **Decomposition of smoke emissions changes in China, 1997–2006**

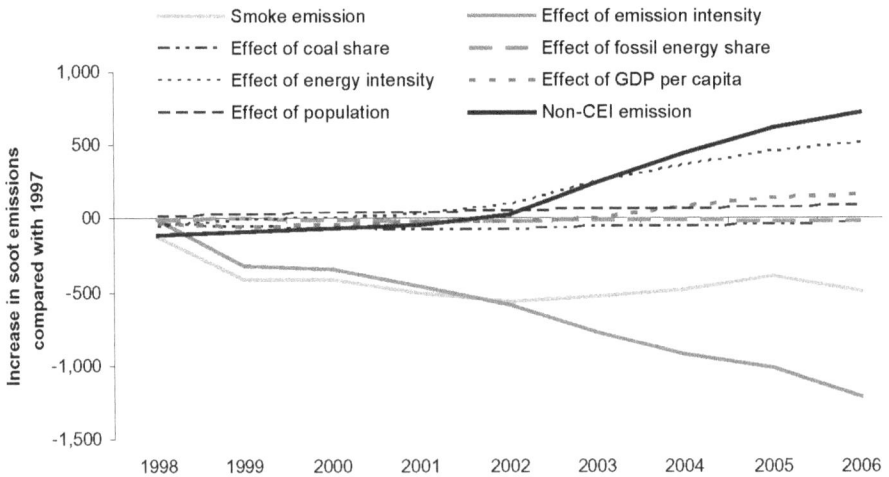

Source: Author's own calculations.

Figure 17.5 **Decomposition of dust emissions changes in China, 1997–2006**

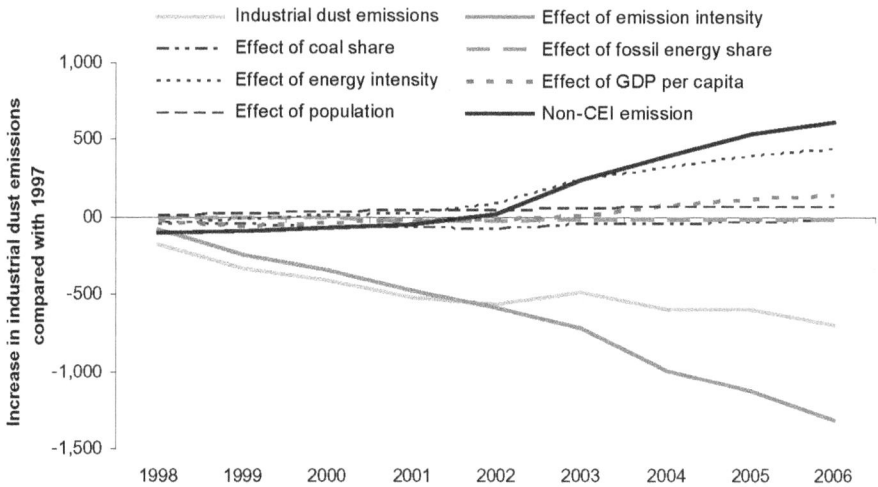

Source: Author's own calculations.

energy intensity could be reduced, as has happened in industrialised countries or as it was in China before 1998 (Figure 17.1), the decrease in China's WGEs could occur at double speed with the co-movement of these two drivers.

Reduced emission intensity is, however, found to be insignificant in China in the case of carbon dioxide in the decade after 1980, but has increasing significance in India in the 1990s, as reported by Shalizi (2007).

Discussion: the role of technology

It should be noted that the decrease in emission intensity of coal would not occur automatically. In some cases, this could be due to the switch from high-sulphur to low-sulphur coal; however, the major contributor is technical change, which is stimulated more or less by environmental regulations.

The decrease in coal pollution has been achieved significantly through three basic methods: coal cleaning or washing, a pre-combustion method that can reduce sulphur content by as much as 30 per cent; post-combustion treatments, such as flue-gas desulphurisation (FGD) systems; and the use of electrostatic precipitators to remove airborne ash (National Energy Foundation 2007). The flue-gas clean-up systems that are available commercially and that have long been used in power plants can remove 99.95 per cent of particulates, 95 per cent of sulphur dioxide and 90 per cent of nitrogen oxides (Energy Committee of the ASME Council of Engineering 2005). Some of these methods, although straightforward now, seemed unrealistic in the 1970s when the US government started to introduce air-quality standards and regulations (EPA 1971).

Declining carbon dioxide emission intensity of coal is also technically possible, even though it has not yet been commercialised. Zero or near-zero emissions from coal-fired power plants are technologically feasible (Energy Committee of the ASME Council of Engineering 2005; Keay 2003; Shimkus 2005). Many international cooperation programs and some experimental projects have been initiated (IEA 2007). Carbon capture is thought most promising for integrated gasification combined-cycle (IGCC) coal-fired plants, even though the cost and reliability of IGCC have not been proved (Sachs, 2008). The 'FutureGen' project, which is dedicated to building near-zero emissions coal-fired power plants, is ready to demonstrate carbon dioxide capture and storage (CCS) technology on a commercial scale; integrated gasification combined-cycle (IGCC) coal power plants, which will be the cleanest coal-fired plants in the world, are expected to be operational in 2015 (DOE 2008). Rio Tinto and BP are working together on decarbonised energy projects, which can generate almost carbon-free electricity from coal (Macalister 2007). Further progress might mean that such techniques could be applied commercially. Carbon dioxide can also be separated and stored or put into other commercial use in the process of coal liquefaction. The Shenhua Group is cooperating with some oil companies to separate carbon dioxide and inject it into oil fields to increase the oil-recovery rate, which will be a double win: carbon dioxide will not be released into the atmosphere and more oil will be extracted from fields.

The Chinese government treats CCS technology seriously; it is documented in China's eleventh Five-Year Plan under the National High Technologies Program and in the *National Medium and Long-Term Science and Technology Plan Towards 2020*. The first 'green' power project, a 250-megawatt (MW) IGCC power station, located in Tianjin, is due to start operating by the end of 2009 (GreenGen 2008).

Since technology improvements provide ways for the coal industry to harmonise with the environment, it is important to popularise and utilise clean-coal technology to make coal use cleaner; it is also important to study and test CCS technology.

One particular problem for China is not a shortage of feasible technology, but a lack of application—the result of insufficient incentives and external pressures. For example, most of China's electricity is produced from coal and most coal-fired plants are far dirtier than those found in Organisation for Economic Cooperation and Development (OECD) countries (IEA 2007). Although China has regulations for the installation of FGD requirements in power plants, in 2005, only 45 of 389 gigawatts (GW) of installed thermal capacity had an FGD unit installed, which helps explain why the target of the nation's tenth Five-Year Plan for a 10 per cent reduction in sulphur dioxide in 2005 compared with 2000 levels has not been achieved (IEA 2007).

As a cheaper alternative to FGD, coal washing has not been used well. Washed coal accounted for only 32 per cent of total coal consumption in 2005 (NDRC 2007b). Furthermore, coal-washing efficiency in China has achieved a removal rate of only 45 per cent (Watson et al. 2000). Since the use of washed coal depends on demand, it is necessary to create an institutional environment that stimulates and forces coal users—particularly large-scale users such as steel makers and power generators—to use washed coal.

Among conversion technologies, coal gasification has been used for the production of town gas and its increasing use is helping to reduce local air pollution in cities. This process is further extended to coal liquefaction, which provides a clean way to use coal. The technology is available and economical energy use has found increasing favour in recent years due to surging oil prices. Many Chinese companies are working to produce coal from oil, during the process of which polluting elements are separated and utilised and emissions are minimised. The Shenhua Group, the largest coal producer in China, with the best safety record, is building the first direct coal-to-oil plant. The plant aims to discharge zero pollution, other than carbon dioxide, and is dedicated to researching the technology of carbon dioxide capture and storage. The Shenhua Group and Southern African Sasol Energy Company Construction are planning to construct several indirect coal liquefaction plants for Shaanxi and Ningxia. The Shenhua Group plans to produce 100 million tonnes of coal to oil per annum by 2020. Other liquefying technologies are also being tested. Furthermore, coal gasification will reduce a large amount of pollution from coal combustion. The Shenhua Group has launched the world's first coal-to-olefin plant in Baotou City, in the Inner Mongolia Autonomous Administration Region.

Improvements in combustion efficiency for large numbers of industrial boilers in China—where emissions reduction efficiency is just 65 per cent compared with 80 per cent in Europe (Watson et al. 2000)—could produce even more significant environmental benefits.

Table 17.6 Coal-based power generation technology in China

Technology	Technological availability	Cost ($/per kW)	Efficiency (%)	Market share in China
Sub-critical		500–600	30–6	Main base of China's current Generating fleet
Supercritical	Now	600–900	41	About half of current new orders
Ultra-supercritical	Now but needs further research and development to increase efficiency	600–900	43	Two 1,000-MW plants in operation
IGCC	Now but faces high costs and needs more research and development	1,100–400	45–55	Twelve units awaiting NDRC's approval

Source: International Energy Agency (IEA), 2007. *World Energy Outlook 2007: China and India insights*, International Energy Agency, Paris.

Conclusion

Relying on current technology, the continuing use of coal will create increased waste-gas and carbon dioxide emissions. The key factor in resolving the contradiction between environmental protection and coal industry development is to use coal in a clean way. This could take a long time; however, if it is possible, we must see a decline in WGE intensity from coal during the transitional process.

This research clarifies the misunderstanding that the same unit of coal will always cause the same amount of environmental pollution. The empirical analysis using the data for Chinese coal consumption and WGEs shows that there is a trend of declining WGE intensity, which is demonstrated by using two alternative methods. The index-decomposition method further demonstrated that declining emission intensity was the most important factor affecting overall WGEs. This study reveals that energy intensity in China has been increased and thus, besides of economic and population growth, has led to an increase in final emissions. With a fall in emission intensity, the coal industry can be developed while improvements are made to the environment—providing that emission intensity continues to fall without limit. Declining emission intensity, which is the key factor that alleviates the tension between the use of coal and the environment, has often been omitted from previous studies.

This finding sheds light on the future of the coal industry. In China's case, environmental regulations are needed and are welcome, even though they seem to impose some constraints on the coal industry's development initially. A harmonised future for the coal industry relies on the application of clean-coal technologies. The current coal-dominant pollution in China is due in part to insufficient use of available clean-coal technologies, such as coal washing and dust precipitation. With proper environmental regulations, China's environment can be improved because more clean-coal technologies will be applied. The confidence about the future of coal industry also comes from the fact that non-fossil alternatives have not been available in a necessary scale in the foreseeable future (Shi 2006).

The coal industry is not necessarily incompatible even with the worst-case scenario for the future: a carbon-constrained world. The biggest challenge for the future development of the coal industry is how to deal with carbon dioxide emissions and climate change. The technical possibilities have been explored. Energy and carbon prices can further accelerate the pace of technical application. If energy prices rise due to the scarcity of fossil-fuel resources, or a price is put on carbon emissions, these carbon dioxide reduction technologies will become economically viable and thus can lead to an accelerated occurrence of zero emission.

Notes

1 This opinion is held by coal experts such as Yin Wu, director-general of the Energy Bureau at China's National Development and Reform Commission, Michael Han at Rio Tinto's Beijing representative office, and Weier Pan at the State Administration of Work Safety.

2 This system entailed: 1) designing anti-pollution measures; 2) constructing anti-pollution equipment simultaneously with the construction of industrial plants; and 3) operating anti-pollution equipment simultaneously with the operation of industrial plants.

3 The other national fundamental policy is aimed at controlling population growth.

4 In 2007, Jiangsu Province doubled its pollution discharge fees to 1.26 yuan/kg for sulphur dioxide and to 1.2 yuan/kg for other WGEs (Jiangsu EPA 2007).

5 Total coal production dropped dramatically, which was magnified by the deliberate underreporting of production as a countermeasure against the national policy that caps output for each province. The initial reported output for 2000 was 998 Mt, which was corrected in 2006 to 1,299 Mt.

References

Ang, B.W. and Pandiyan, G., 1997. 'Decomposition of energy-induced CO_2 emissions in manufacturing', *Energy Economics*, 19:363–74.

Ang, B.W. and Zhang, F.Q., 2000. 'A survey of index decomposition analysis in energy and environmental studies', *Energy*, 25:1149–76.

Ang, B.W., Zhang, F.Q. and Choi, K.-H., 1998. 'Factorizing changes in energy and environmental indicators through decomposition', *Energy*, 23:489–95.

Bao, Q. and Peng, S., 2006. 'Economic growth and environmental pollution: a panel data analysis', in R. Garnaut and L. Song (eds), *The Turning Point in China's Economic Development*, Asia Pacific Press and ANU E Press, The Australian National University, Canberra.

British Petroleum (BP), 2007. *BP Statistical Review of World Energy*.

Cao, Q., 1998. 'Beijing enforces the use of low sulfur coal', *Living Time*, Beijing.

Coal Enterprise Management, 2001. 'Coal industry is not a "sunset industry"', *Coal Enterprise Management*, 2001:5–9.

Darmstadter, J., 1999. 'Innovation and productivity in US coal mining', in R.D. Simpson (ed.), *Productivity in Natural Resource Industries: improvement through innovation*, Resources for the Future, Washington, DC.

Department of Energy (DOE), 2008. *FutureGen Clean Coal Projects*, US Department of Energy, Washington, DC.

Energy Committee of the ASME Council on Engineering, 2005. *The Need for Additional US Coal-Fired Power Plants*, ASME International.

Environmental Protection Agency (EPA), 1971. *EPA Sets National Air Quality Standards*, US Environmental Protection Agency, Washington, DC.

Gray, W.B., 1987. 'The cost of regulation: OSHA, EPA and the productivity slowdown', *American Economic Review*:998–1006.

GreenGen, 2008. *Development Plan of the GreenGen Co.*, The GreenGen Company, Beijing.

Grossman, G.M. and Krueger, A.B., 1991. *Environmental impacts of a North American Free Trade Agreement*, Working Paper 3914, National Bureau of Economic Research, Cambridge, Mass.

He, K., Huo, H. and Zhang, Q., 2002. 'Urban air pollution in China: current status, characteristics and progress', *Annual Review of Energy and the Environment*:397–431.

Huang, Y., 2001. 'Coal will still be China's key energy in the 21st century', *China Energy*, 3.

Intergovernmental Panel on Climate Change (IPCC), 2001. *Special Report on Emissions Scenarios*, Cambridge University Press, Cambridge.

International Energy Agency (IEA), 2007. *World Energy Outlook 2007: China and India insights*, International Energy Agency, Paris.

Jiangsu Provincial Environmental Protection Agency (Jiangsu EPA), 2007. *Jiangsu Will Adjust Standards of Pollution Discharge Fees*, Jiangsu Provincial Environmental Protection Agency, Nanjing.

Kaufmanna, R.K., Davidsdottira, B., Garnhama, S. and Paulyb, P., 1998. 'The determinants of atmospheric SO_2 concentrations: reconsidering the environmental Kuznets curve', *Ecological Economics*, 25:209–20.

Kaya, Y., 1990. Impact of carbon dioxide emission control on GNP growth: interpretation of proposed scenarios, Presented to the Intergovernmental Panel on Climate Change Energy and Industry Subgroup, Response Strategies Working Group, Paris.

Keay, M., 2003. 'The view from Europe—and elsewhere', *Oxford Energy Forum*, 52.

Li, J., 2003. 'Coal industry is not a "sunset industry"', *Journal of Coal Economics Research*, 2003:10–12.

Liang, D. and Zhou, Y., 2008. 'Waste gas emission control and constraints of energy and economy in China', *Energy Policy*, 36:268–79.

Lin, S.J. and Chang, T.C., 1996. 'Decomposition of SO_2, NO_x and CO_2 emissions from energy use of major economic sectors in Taiwan', *The Energy Journal*, 17:1–17.

Macalister, T., 2007. 'Rio Tinto and BP plan project to clean up', *The Guardian*.

Markandya, A., Golub, A. and Pedroso-Galinato, S., 2006. 'Empirical analysis of national income and SO_2 emissions in selected European countries', *Environmental and Resource Economics*, 35:221–57.

National Bureau of Statistics (NBS), various years. *China Statistical Yearbook*, China Statistics Press, Beijing.

National Development and Reform Commission (NDRC), 2007a. *China's National Climate Change Program*, National Development and Reform Commission, Beijing.

——, 2007b. Opinions on work towards energy conservation and emission reduction in the coal industry, National Development and Reform Commission, Beijing.

National Energy Foundation, 2007. *Coal and the Environment,* National Energy Foundation, Beijing.

Sachs, J.D., 2008. 'Keys to climate protection (extended version)', *Scientific American Magazine*, 18 March.

Shadbegian, R. and Gray, W., 2006. 'Assessing multi-dimensional performance: environmental and economic outcomes', *Journal of Productivity Analysis*, 26:213–34.

Shalizi, Z., 2007. *Energy and Emissions: local and global effects of the rise of China and India*, World Bank, Washington, DC.

Shi, X., 2003. 'The future of the coal industry under the context of strict environmental regulations', *Journal of Coal Economics Research*, 2003:6–11.

——, 2006. 'Harmonising the coal industry with the environment', in R. Garnaut and L. Song (eds), *The Turning Point in China's Economic Development*, Asia Pacific Press and ANU E Press, The Australian National University, Canberra.

Shimkus, J., 2005. Press release from the office of Congressman John Shimkus, 14 September.

Shrestha, R.M. and Timilsina, G.R., 1997. 'SO_2 emission intensities of the power sector in Asia: effects of generation-mix and fuel-intensity changes', *Energy Economics*, 19:355–62.

State Council, 1998. 'Official reply of the State Council concerning acid rain control areas and sulfur dioxide pollution control areas', *Guohan*, 5.

State Environmental Protection Administration (SEPA), 1982. *National Ambient Air Quality Standards*, State Environmental Protection Administration.

——, 1996–2006. *State Public Release of Environmental Protection*, State Environmental Protection Administration.

——, various years. *China Environmental Yearbook*, China Environmental Yearbook Press, Beijing.

State News Office, 2006. *Environmental Protection in China (1996–2005)*, Beijing.

Torvanger, A., 1991. 'Manufacturing sector carbon dioxide emission in nine OECD countries, 1973–87', *Energy Economics*, 13:168–86.

Viguier, L., 1999. 'Emissions of SO_2, NO_x and CO_2 in transition economies: emission inventories and Divisia index analysis', *Energy Journal*, 20:59–87.

Wang, C., Chen, J. and Zou, J., 2005. 'Decomposition of energy-related CO_2 emissions in China: 1957–2000', *Energy*, 30:73–83.

Wang, Q., 1999. 'Coal industry is not a "sunset industry"', *China Coal*, 25:7–9.

Wang, X. and Feng, Z., 2003. 'Energy consumption with sustainable development in developing countries: a case in Jiangsu, China', *Energy Policy*, 31:1679–84.

Watson, J., Xue, L., Oldham, G., Mackerron, G. and Thomas, S., 2000. *International Perspectives on Clean Coal Technology Transfer to China*, Final Report to the Working Group on Trade and Environment, China Council for International Cooperation on Environment and Development, Beijing.

Wu, L., Kaneko, S. and Matsuoka, S., 2006. 'Dynamics of energy-related CO_2 emissions in China during 1980 to 2002: the relative importance of energy supply-side and demand-side effects', *Energy Policy*, 34:3549–72.

Acknowledgment

The author acknowledges the financial support for my PhD research provided through the Rio Tinto-ANU China Partnership.

18

Emissions and economic development
Must China choose?

Peter Sheehan and Fiona Sun

China's energy use has surged since the turn of the century, almost doubling between 2000 and 2007. The rate of growth in energy use during this period (9.2 per cent per annum) has been more than twice that of the previous two decades (4.5 per cent), in spite of similar rates of economic growth. Many factors were undoubtedly responsible for this sharp change in trend. One was the post-2001 development pattern, with a focus on heavy industry, exports and fixed-asset investment. Another was the fundamental change in the structure, ownership and operation of the energy industry that took place during the 1990s, which rendered the controls of the planned economy no longer applicable. These and other factors on the demand side were supported by a massive expansion of energy supply, with electricity-generation capacity, for example, more than doubling in the period.

The scale of this energy use, and of the industrial and infrastructure investment associated with it, is now such as to have major ramifications within China and globally. Internally, the environmental and social costs of heavy industry, construction and fossil-fuel use have become onerous, and a major concern of the Chinese people. As a result, a central priority in the eleventh Five-Year Plan (2006–10) is to change the pattern of development and to reduce energy use per unit of gross domestic product (GDP) by 20 per cent. Greenhouse gas emissions from China are also now of major significance for the global climate: in spite of its low per capita emissions, China has now passed the United States as the largest emitter of carbon dioxide, and these emissions continue to rise rapidly. If present trends continue, China's annual greenhouse gas emissions by 2030 could be comparable with total global emissions in 2000.

There can be no dispute that the current atmospheric concentration of greenhouse gases is due mainly to the past activities of the industrialised countries, and that they have the prime responsibility for addressing this problem. The scale of China's fossil-fuel use, however, and of its emissions, is now such that no solution to global warming can be found without the active involvement of China. More specifically, it is now clear that, if rapid global warming imposing severe damage on many communities is to be avoided, global emissions must peak within the next one to two decades at least, and then decline. The scale of China's activities now means that, even given an urgent response by the industrialised world, China's emissions must also peak and then begin to fall within the foreseeable future.

Here the issue that China faces, along with other developing countries, is often put in terms of a choice between economic development and emissions. In spite of rapid growth, the bulk of China's population remains poor, with average GDP per capita in China still at only 15 per cent of Organisation for Economic Cooperation and Development (OECD) levels in 2005, according to the new International Comparison Program estimates at purchasing power parity prices (World Bank 2007). With increasing energy use seen as essential to growth, and with coal forming the majority of China's energy resources, it is argued that China can stabilise emissions within, say, 25–30 years only at the cost of sharply slowing growth, and hence giving up on its historical goal of lifting its people out of poverty.

This chapter addresses the question: does China have to choose between rapid economic growth and stabilising emissions? This question is posed with regard to the priority being given in the eleventh Five-Year Plan to the reduction of energy use per unit of output, and more generally to the Chinese government's strong enunciation of the need to move to a new development model—one that is socially and environmentally sustainable and that contributes to maintaining a harmonious society. Could such a new model, with the many policy initiatives that it contains, be consistent with continued rapid growth and a stabilisation of greenhouse gas emissions?

To build a basis for providing a tentative answer to this question, we undertake three tasks in the next three sections: we review China's energy use and emissions in recent decades, with special reference to the break in the trend about 2001; we discuss the new development model and some of its implications; and we outline the scoping model that will be used to project and analyse China's future emissions. We then use this simple model to estimate the potential impact on energy use and carbon dioxide emissions to 2030 from interventions in six policy areas consistent with the new development model, with conclusions presented in the final section.

China's energy use and carbon dioxide emissions, 1979–2007

As noted above, the explosive growth in energy use after 2001 was in sharp contrast with earlier trends. From the 'opening to the market' in 1979 to 2001, energy use grew at a much lower rate than GDP, with average rates of growth of 4.1 per cent and 9.7 per cent for energy use and GDP, respectively, implying that the energy intensity of China's GDP fell continuously through to 2001 and the elasticity of energy use with respect to GDP was less than 0.5 on average for the period (Figure 18.1). This decline in reported energy intensity was especially marked in the second half of the 1990s, so that the shift to rates of growth in energy use in excess of GDP growth after 2001 had profound and unexpected implications in energy markets, and led to severe shortages in 2003 and subsequent years. This apparent structural shift in the relationship between GDP growth and energy use in China raises serious questions in several areas. For example, is this a temporary aberration or a fundamental structural shift and, if so, what caused it? What basis should we use to interpret China's likely future energy demand on existing policies? What policies are likely to be most effective in reducing the rate of growth of China's demand for energy, in the light of this shift?

There is an extensive literature on the reasons for the low income elasticity of energy use in China during 1979–2001, and hence the rapidly declining overall energy intensity. This literature is reviewed in Sheehan and Sun (2007). Much of the debate has been about the relative roles of structural change in the pattern of output and energy demand and of changes in energy intensity within sectors in explaining China's energy use. Although some earlier studies emphasised structural change, all of the empirical studies from Huang (1993) onwards found a major role for reduced energy intensities at the sectoral level. In terms of structural effects, the results are much more mixed, with some studies finding negative effects of structural change on overall energy intensity but others (such as Lin and Polenske 1995; Garbaccio et al. 1999) finding positive effects. Overall, it seems clear that widespread declines in energy intensity rather than a shift in the pattern of growth to less-intensive sectors was the main factor driving the fall in overall energy intensity during 1979–2001.

The reasons for the increased energy efficiency within sectors in China in this period are another matter. Three main reasons have been given in the literature for these sectoral effects: the impact of rationing and energy conservation programs in a planned economy with an initially high level of energy use and limited growth in energy supplies; the impact of technology, broadly defined, on energy use; and the impact of higher energy prices on the demand for energy (Sinton and Levine 1994; Sinton et al. 1998; Andrews-Speed 2004; Lin 2005). Our interpretation of the literature therefore is that, during the 1980s, the fall in sectoral intensities is to be ascribed to a combination of energy conservation programs and technological change being driven by a planned economy with energy rationing (Sheehan and Sun 2007). In the 1990s, those factors continued to be of importance, while rising relative energy prices also began to play a significant role as the economy was freed up. The literature does not give an unequivocal answer to the question about the role played by structural change in reinforcing or partially offsetting these declining sectoral intensities.

Figure 18.1 **Energy intensity and the energy elasticity of GDP in China, 1979–2006**

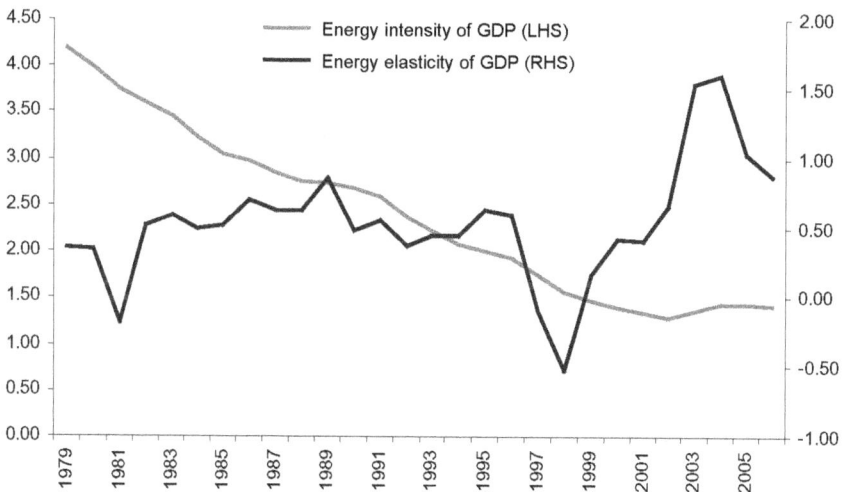

Note: Energy intensity is measured in terms of units of energy use (in hundred thousand tonnes of standard coal equivalent) per unit of GDP (in billion yuan in 2000 values).
Source: National Bureau of Statistics (NBS), 2007. *China Statistical Yearbook 2007*, China Statistics Press, Beijing.

It is likely therefore that several interrelated factors are responsible for the widespread fall in sectoral energy intensities in the first two decades of rapid economic growth in China: a combination of energy rationing and strong energy conservation programs in a planned economy with limited energy supplies; continuing technological upgrading, in part spurred by these circumstances; and, in the 1990s, the impact of rising relative prices for energy. By the end of the 1990s, however, fundamental changes in the structure, ownership and operation of Chinese industry, including in the energy sector, had taken place, so that the mechanisms of the planned economy were no longer relevant. Energy use per unit of GDP reduced by two-thirds between 1980 and 2000, and much technological upgrading took place. Without the control mechanisms of the planned economy, however, further major reductions in energy intensity could be achieved only by market forces and by strong policy initiatives. There are many reasons why, in developing countries, energy is normally a superior good,[1] as the development process shifts the pattern of production and of lifestyles towards more energy-intensive activities and products. Beyond the command economy, strong market and policy effects would be needed in China to offset this development effect.

In the aftermath of China's entry to the World Trade Organization (WTO) in 2001, demand for China's products was very strong, the supply of energy increased rapidly and the policy focus on energy efficiency was limited. With structural effects contributing to rising energy use and the command-economy mechanisms generating falling sectoral intensities no longer operational, it is not surprising that an aggregate energy intensity of 1 or more re-emerged. This analysis implies that, going ahead, there is no reason to expect a return to an aggregate energy elasticity of 0.5–0.6 to occur 'naturally'; achieving an aggregate elasticity well below 1 will be hard won by sustained policy initiatives, especially while structural change continues to contribute to increased energy use.

The development strategy in the eleventh Five-Year Plan

While fully recognising the remarkable achievements of nearly three decades of rapid growth, it is acknowledged widely within China that some adjustment of the current development strategy is necessary. In his March 2006 *Report on the Work of the Government*, Premier Wen Jiabao said of the issues arising from the tenth Five-Year Plan period (2001–05):

> The main problems were an unbalanced economic structure, weak capacity
> for independent innovation, slow change in the pattern of economic growth,

excessive consumption of energy and resources, worsening environmental
pollution, serious unemployment, imbalance between investment and
consumption, widening gaps in development between urban and rural areas
and between regions, growing disparities between certain income groups,
and inadequate development of social programs. We need to work hard to
solve all these problems. (Wen 2006)

The eleventh Five-Year Plan outlined a vision of development that was
socially and environmentally sustainable and that contributed to maintaining
a harmonious society, and outlined programs to be implemented to address
these issues and to achieve such a form of development. As one observer
(Naughton 2006) wrote, the proposals in China's eleventh Five-Year Plan for
the period 2006–10 were striking:

There emerges from this Plan document a rich and comprehensive vision
of a sustainable development process in China, and a glimpse of the kind of
government role that would be required by this development process. The
vision is of a society that is more creative, more focused on human resource
development, and treads with a lighter and more environmentally benign
step.

It is one thing to outline a vision of a sustainable economy and a harmonious
society and quite another to define and implement a detailed set of programs
to give effect to this vision. This is especially so in such a diverse, vibrant and
internationally engaged society as contemporary China. The forces shaping
the current growth pattern—from the role of local governments and the limited
power of the central government, the strong influence of foreign companies and
investors and the level of the exchange rate to the popular desire for a strong
China and a better life—are complex and interrelated, and it will take a major
effort to realign them.

Achieving this vision will require sustained and effective delivery of new
policies in many different areas, especially given the momentum that has built
up behind the current development strategy since China's entry into the WTO.
For example, four important objectives of current government policy are:

- to make growth more sustainable and environmentally benign, to reduce
  the rate of energy and water use and reduce pollution

- to increase innovation within all sectors, including industry, and to shift
  the pattern of activity from low value-added output based on low labour
  costs towards higher value-added activities based on knowledge

- to change the structure of growth towards the service sector, and to accelerate the growth of particular service sectors that contribute directly to individual welfare
- to improve the position of people in the countryside, and to build structures to ensure that the benefits of growth also flow to people in rural areas and those regions that are lagging.

Achievement of each of these objectives would contribute, directly or indirectly, to reduced energy use and to environmental sustainability. This is particularly true of the intention to shift the pattern of economic activity from energy-intensive areas (such as specific forms of heavy industry) to industry and service sectors that are knowledge intensive and rely less on energy and other resource inputs; and to stimulate the adoption of advanced technologies, processes and practices that are energy efficient and more environmentally benign.

The attempt to change substantially the structure and technological base of a large, rapidly growing economy is without precedent and is likely to prove difficult, especially in a country still dealing with the transition from a planned to a market economy. Our question here is whether there are implementation paths within this general approach that could enable China to continue rapid growth while also stabilising emissions within 25–30 years.

A scoping model for China's energy use and carbon dioxide emissions

The analysis here is based on what we refer to as a scoping model: a simple modelling framework in which the key relationships and assumptions are spelled out so that the results are highly transparent. There is a wide range of global energy or integrated assessment models that now include China, as well as a number of models applied specifically to China. These more complex models have many theoretical advantages over such simple models, but require estimation of, or assumptions about, a large number of parameters, such as price and income elasticities of demand, elasticities of substitution, supply curves and the rate and structure of economic growth. The evidence available for many of these parameters is very limited in China, not only because of limited data but because of the sharp changes that have taken place in economic institutions and in the structure of the energy sector in the past decade. For example, for much of the period since the opening to the market in 1979, energy use was supply constrained; this means that there are difficulties in obtaining reasonable estimates of various demand elasticities.

The swings in the aggregate income elasticity of energy use discussed above also illustrate the level of uncertainty that prevails in this area, as well as the need to incorporate recent trends.

In such a situation, the use of a large, complex model might require heroic assumptions and disguise the key uncertainties, leading to either misleading or non-transparent results. Complex models also require more detailed data, which are often assembled only with long time lags. Sheehan (forthcoming) reviews the base cases in simulations with a wide range of integrated assessment models: those participating in the Energy Modelling Forum (EMF 21) modelling comparison project (Weyant et al. 2006) and the three models used in the Synthesis and Assessment Product 2.1a of the US Climate Change Science Program (CCSP) (Clarke et al. 2007). Sheehan found that none of the 10 models in the EMF 21 exercise that provided data on China and India or the three models used in the CCSP report provided a realistic assessment of recent and emerging trends in these two countries.

For the April 2008 edition of the *World Economic Outlook*, the International Monetary Fund (IMF) used the G-cubed model to analyse global emission trends and policy responses (IMF 2008), with surprising results. The baseline scenario shows global energy-related carbon dioxide emissions stagnating between 2002 and 2010 (a growth of only 0.5 per cent per annum), but then growing rapidly between 2010 and 2030 (3.1 per cent per annum). For China, the annual growth rates of emissions for these two periods are 1.8 per cent and 6 per cent, while for Non-Annex 1 countries as a whole the growth rates are 0.2 per cent and 5.1 per cent (IMF 2008:Table 4.4). In fact, we know that emissions in developing countries have surged since about 2001, especially in China, where the growth in energy use and energy-related carbon dioxide emissions for 2002–10 will be about 10 per cent per annum. The IMF figure for China in 2010 (3.8 giga-tonnes of carbon dioxide) is therefore less than half the likely outcome, given trends through to 2008. If global and Chinese emissions had stabilised since 2002, rather than surged, the international climate debate now would be much different. On the other hand, if the IMF's predicted baseline growth for 2010–30 is added to the real experience of this decade, the global problem will be dire indeed. These issues again illustrate the problems involved in using large models to analyse complex, rapidly changing situations about which there is limited knowledge.

A simple scoping model

Studies of the demand for energy frequently use a standard framework such as:

$$E_{it} = f(Y_{it}, P_{it}, Z_{it}),$$ (1)

in which E_{it} is the demand for energy in industry sector i in period t, Y_{it} is an income or output variable relevant to sector i, P_{it} is the relative price of energy in sector i in period t, and Z_{it} is a vector of other variables affecting energy demand in sector i, such as technological change and government policy initiatives related to energy conservation. Assuming that from 2006 onwards supply constraints on energy demand in China have been removed, so that real energy use can be treated as demand determined, we use this framework to construct the projection model for China, applying it to 11 sectors. In a log-linear specification, Equation 1 becomes

$$Ln\ (E_{it}) = \alpha_{it}\ ln\ Y_{it} + \beta_{it}\ ln\ P_{it} + \gamma_{it}\ ln\ Z_i$$ (2)

in which α_{it}, β_{it} and γ_{it} are the elasticities of energy use with respect to Y, P and Z, respectively. Partial differentiation of Equation 2 with respect to time and rearrangement gives

$$\delta E_{it} = \alpha_{it}\ \delta Y_{it} + \beta_{it}\ \delta P_{it} + \gamma_{it}\ \delta Z_{it}$$ (3)

in which in the projection model the change variables (δE and so on) represent rates of change with respect to time.

To enable examination of issues concerning the impact of the pattern of growth on energy use, we define two new elasticities relating the rate of growth of value added in sector i in period t to the growth rate of a higher-level variable, A_{it}; the ratio of the growth rate of value added in aggregate sector i in period t to growth in total GDP in that period; and I_{it}, the ratio of the growth rate of value added in industry sector i in period t to growth in total industry value added in that period (for the six sectors within industry). Therefore, A_{it} defines the pattern of growth across six aggregate sectors (in which industry is a single sector) for a given rate of growth of aggregate GDP, and I_{it} defines the pattern of growth across the six industry sectors for a given rate of growth of industry value added. That is

$$\delta Y_{it} = \delta Y_t\ A_{it}\ I_{it}$$ (4)

in which A_{it} takes a value of 1 if i is a disaggregated industry sector and I_{it} takes a value of 1 if i is an aggregate sector.

Substituting Equation 4 into Equation 3 and converting growth rates into levels, gives the following expression for total energy use in China in sector i in period t

$$E_{it} = E_{it-1} (1 + \alpha_{it} \, \delta Y_t \, A_{it} \, I_{it} + \beta_{it} \, \delta P_{it} + \gamma_{it} \, \delta Z_{it}) \tag{5}$$

Energy use involves different types of fuels (coal, oil, natural gas and various types of non-fossil and renewable fuels), and each of the fossil fuels has a different propensity to generate carbon dioxide emissions. The share of fuel type j in total energy use in sector i (s_{ij}) will vary over time, depending on availability, relative prices, investment patterns, policy initiatives and other factors. The energy use met by fuel j in year t in sector i can be denoted by $E_{ijt} = E_{it} \cdot s_{ijt}$, and total use of fuel j will be given by

$$E_{jt} = \sum s_{ijt} \, E_{it-1} (1 + \alpha_{it} \, \delta Y_t \, A_{it} \, I_{it} + \beta_{it} \, \delta P_{it} + \gamma_{it} \, \delta Z_{it}) \tag{6}$$

Finally, carbon dioxide emissions per unit of use of fuel j (m_{jt}) in China will also vary over time, depending, for example, on the quality of fuel used and the technological processes involved. Total carbon dioxide emissions from the use of fuel j in year t will then be given by

$$M_{jt} = m_{jt} \, E_{jt} = m_{jt} \, E_{jt} \tag{7}$$

with total emissions given by summing over fossil-fuel types. This simple model is used below to analyse and project China's future energy use and carbon dioxide emissions from energy use, given suitable projections or assumptions for the many parameters involved.

Data and the industrial structure of China's energy use

The data for the analysis are drawn from official Chinese statistics (from successive editions of the *China Statistical Yearbook*; NBS various years) and from the International Energy Agency (IEA) database and from the *World Energy Outlook 2007* (IEA 2007). Data for energy use and value added are taken from, or derived from, the *China Statistical Yearbook* (NBS various years), while data for fuel use by industry and emissions intensity are sourced from the IEA. Some issues arise in integrating the two data sources: for example, in terms of the different treatment of renewable energy sources. Further information on the issues involved in assembling and using these data can be found in Sheehan and Sun (2007).

As noted above, issues concerning the structure of energy demand are at the heart of the Chinese debate, and Table 18.1 provides some key statistics on China's energy use. Energy use is concentrated heavily in five industries: petroleum processing, chemicals, non-metallic minerals, ferrous metals and non-ferrous metals. These industries together have energy use per unit of value added more than four times the national average, and nearly six times that of all other industries. In 2006, these industries accounted for 44 per cent of national energy use but only 10 per cent of GDP. During 2001–06, real value added in this group grew by 15.2 per cent per annum, compared with the national rate of 10.2 per cent, and energy use grew by 15.7 per cent per annum, given an elasticity of 1.04. The five energy-intensive industries accounted for 54.8 per cent of the increase in total energy use, while providing only 13.6 per cent of the increase in China's GDP during this period.

Given the concentration of energy use in these industries, they are critical to the analysis, and reducing the rate of growth of these industries and their energy intensity is a key priority for government policy. On the other hand, the role of the services sector cannot be neglected. As Table 18.1 also shows, total energy use in services grew by 10.2 per cent during 2001–06—in line with the growth in value added—while the energy intensity of the transport sector was high and its energy use grew rapidly.

Policy intervention areas and simulation cases

The analysis undertaken below makes use of a base or unchanged policy case, and three alternative policy cases. In view of the adoption of the target of a 20 per cent reduction in aggregate energy intensity in the eleventh Five-Year Plan, the policy environment in China has been in flux in 2006 and 2007; new initiatives implemented in these years are included in the policy options cases. The policy stance in 2005 is interpreted as one of continued but modest efforts to contain energy use and carbon dioxide emissions from fuel use, through the introduction of market mechanisms, increased energy prices and programs to encourage energy conservation and the use of advanced technologies. These policies are incorporated in the base case.

In shaping these cases, six policy intervention areas are employed, and are included in the model

- the rate of growth in value added in industry relative to that of overall GDP
- the industry composition of the growth in industrial output

Table 18.1   Energy consumption and value added by industry in China, 2001–06

	Primary energy consumption		Real value added		Descriptive statistics	
	2006 (Mtce)*	Change 2001–06 (% p.a.)	2006 (100 billion 2000 yuan)	Change 2001–06 (% p.a.)	Elasticity of energy use, 2001–06	Energy use/value added (Mtce per 100 billion yuan)
Agriculture	8.4	5.6	19.0	4.4	1.27	0.44
Industry						
Mining	13.3	4.3	9.0	13.6	0.32	1.47
Petroleum processing	12.4	10.6	1.9	16.8	0.63	6.44
Chemicals	24.8	13.1	4.5	13.9	0.94	5.52
Non-metallic minerals	19.9	14.9	3.1	4.8	3.12	6.53
Ferrous metals	42.8	20.1	5.8	20.8	0.97	7.35
Non-ferrous metals	8.6	16.3	2.7	22.7	0.72	3.25
Other manufacturing	34.5	7.8	42.6	11.2	0.70	0.81
Electric power, gas and water	18.8	9.7	6.2	5.2	1.86	3.04
Total	175.1	12.3	75.8	11.7	1.04	2.31
Construction	3.7	10.7	10.0	11.0	0.97	0.37
Services						
Transportation	18.6	12.4	10.5	9.4	1.31	1.77
Wholesale, retail and hospitality	5.5	11.1	13.6	8.8	1.26	0.41
Other tertiary (including households)	34.9	9.0	45.5	10.9	0.83	0.77
Total	59.0	10.2	69.5	10.2	1.00	0.89
Total energy use	246.3	11.5	174.3	10.2	1.13	1.41
Memorandum item						
Five energy-intensive industries	108.5	15.7	17.9	15.2	1.04	6.05

* Mega-tonnes of coal equivalent

Note: The five energy-intensive industries included here are petroleum processing, chemicals, non-metallic minerals, ferrous metals and non-ferrous metals.

Sources: National Bureau of Statistics (NBS), 2007. *China Statistical Yearbook 2007*, China Statistics Press, Beijing; National Bureau of Statistics (NBS), various years. *China Statistical Yearbook*, China Statistics Press, Beijing.

- the rate of growth of energy prices relative to general output prices
- the reduction in energy use from programs to promote energy conservation and the adoption of new technologies
- the composition of energy supply by fuel
- the emissions intensity of the use of various fossil fuels.

We distinguish these policy intervention areas from the specific policy instruments (such as export tariffs, energy taxes, research and development subsidies or changes in relative prices) that might be used to achieve a given policy objective in a given area. Policy instruments are not addressed here, but the Chinese government is currently taking some action in each of these areas, some examples of which are provided below.

In taking action to achieve the eleventh Five-Year Plan's goal of reduced energy intensity, the government has relied mainly on 'command and control' measures, rather than price or tax measures, and has in part reverted to tried and true methods that were successful in very different circumstances before 1995. A number of measures have been taken to curtail the growth of energy-intensive industries, and the 'Top-1,000 Energy Consumers Enterprise' program has been established to reduce energy use in large firms. In terms of structural change, initiatives have centred on the imposition of export tariffs, on the removal of support at all levels of government for energy-intensive industries, and on closing down inefficient capacity and on expanding service industries. Since 2004, the government has progressively removed its export incentives for energy and resource-intensive exports, and replaced them with export tariffs. Export tariffs on 142 such products were increased from 1 June 2007, with some moving from a 5 per cent to a 10 per cent tariff and others moving from 10 per cent to 15 per cent (MOF 2007). The central government has also issued a serious warning to local governments that all policies that encourage energy-intensive developments must cease (NDRC 2007), and it is entering into agreements with provincial governments to phase out inefficient capacity in energy-intensive industries. For iron and steel, for example, the objective is to close 100 million tonnes of inefficient iron production capacity by 2010 and 50 million tonnes of steel capacity; in mid 2007, an agreement with 10 provinces to close 40 million tonnes of iron capacity and 42 million tonnes of steel capacity was announced (Central Government Portal 2007). Finally, the State Council has announced plans to accelerate the development of the service sector, and to increase its share of GDP by 3 percentage points, and its share of employment by 4 percentage points, by 2010.

Specification of the base and alternative policy cases

The three alternative policy cases, which are cumulative in the sense that additional policy objectives are added in the second and third cases, are as follows.

- **The new industrial structure case**, in which the relative role of industry in total output and the role of energy-intensive industries in industrial output are reduced rapidly.

- **The price and command measures case**, in which, in addition to the structural changes, there is a more rapid increase in relative energy prices and more aggressive action to promote energy conservation and the use of new technologies.

- **The additional measures case**, in which further measures are taken to reduce energy use in transport and other services and to reduce the level of emissions per unit of fossil-fuel use.

The full set of assumptions and parameter specifications is provided in Table 18.2 and is described briefly below.

GDP growth rate. A common rapid growth path for GDP is used in all of the cases. GDP growth is assumed to moderate progressively from the current high levels to 8 per cent per annum by 2010 and to 7 per cent per annum by 2020, maintaining that level until 2030.

Underlying elasticities of energy use with respect to value added. For the critical issue of elasticities of energy use with respect to value added by industry, we specify a pattern of underlying elasticities that remains unchanged during the projection period, even though the real elasticities of energy use change as a result of the policies that are in force—in all cases. Consistent with the argument above, for the eleven industries we use the average elasticity value for the industry for 2001–06 as the underlying rate, with an upper bound of 1.2. The upper bound is used to ensure that unusually high elasticity values in particular cases during 2001–06 do not distort the long-run picture. Even using this specification, the overall projected elasticities of energy use with respect to GDP are much less than 1 in the longer term, falling to below 0.8 by 2030 in the base case and to 0.5 in the fullest policy case.

Relative prices for energy. While information on overall energy prices in China is limited, there seems to have been a significant increase in weighted average energy prices relative to the general price level in the 1990s and again in recent years, as higher global prices have flowed through to some extent to Chinese producers and consumers. On the basis of limited historical data, we assume an increase in average relative energy prices of 6 per cent per annum for each of the three years from 2006 to 2008 in all cases, with a further 3 per cent per annum in all subsequent years in the base case. For the second and third policy cases, we assume that a much more aggressive pricing stance is adopted in the longer term, with the 6 per cent increase in relative energy prices being continued to 2030.

Price elasticity of energy demand. The question of the price elasticity of energy demand is an important and a vexed one. Estimating the price elasticity of demand is especially difficult when, as in China during much of the period from 1980 to 2005, real energy use was supply constrained and prices were partly responsive to the underlying supply–demand gap arising from supply constraints. In view of this fact, estimates of the price elasticity in China derived in the literature from demand equations estimated on historical data must be treated with caution for our purposes.

There is an extensive international literature on the estimation of energy price elasticities across countries. For example, Gately and Huntington (2002) found that the long-run price elasticity for the OECD region for 1971–97 was –0.24, and –0.08 for 14 developing countries (not including China) with above-average per capita income growth, with evidence in both cases of asymmetrical responses to rising and falling prices. Pesaran et al. (1998) found that the long-run elasticity was about –0.3 for 10 Asian countries (again, excluding China). For China, Fisher-Vanden et al. (2004) find an elasticity of –0.368 from their cross-sectional analysis of manufacturing firms in 1997–99; Shi and Polenske (2006) find a long-run price elasticity for the industrial sector of –0.78 for 1980–2002; Hang and Tu (2007) find an elasticity of –0.54 for 1985–95 and of –0.65 for 1995–2005; while Chen et al. (2007) use a range of elasticities from –0.1 to –0.5 in their MARKAL model analysis. In the light of the international literature and concerns about estimating demand elasticities for China in a supply-constrained market, we use an elasticity of –0.4 for the energy-intensive industries and –0.2 for other industries.

Sectoral growth elasticities. At the aggregate level, the key variable is the rate of growth of industry value added (excluding construction) relative to the overall GDP growth rate. In 2006, and during 2001–06, growth in industry value added was 16 per cent greater than in GDP (elasticity of 1.16). It is assumed, in the base case, that the elasticity of the growth of industry relative to GDP falls to 1.05 by 2015 and then declines gradually to 1 by 2030. Growth in agricultural value added is set at 40 per cent of GDP growth, and tertiary-sector growth is the residual. These assumptions imply a gradual retreat from industry-driven growth, with the growth in the tertiary sector exceeding that of industry after

Table 18.2 Model parameters and assumptions for alternative energy runs: specifications for the base case and policy options to 2030

Parameter/assumptions	Base case (2005 policies)	Industrial structure	Structure, price and command measures	Additional measures
Industry value added to GDP	Elasticity falls from 1.16 to 1.05 by 2015, and to 1 by 2030	Elasticity falls to 0.9 by 2015 and to 0.8 by 2030	As for industrial structure case	As for industrial structure case
Energy-intensive industries value added to total industry	Falling from 2001–06 average to 1.1 by 2015; then to 1 by 2030	Falling from 2001–06 average to 0.9 by 2015; then to 0.8 by 2030	As for industrial structure case	As for industrial structure case
Underlying income elasticity of energy use, by industry	Average elasticities for 2001–06 to 2015, then falling 10 per cent by 2030	As for base case	As for base case	As for base case
Average relative prices for energy	6 per cent per annum in 2006–08, then 3 per cent per annum increase to 2030	As for base case	Additional 3 per cent per annum increase over 2009–20	As for structure, price and command case
Price elasticity of energy use, by industry	Price elasticity of –0.4 in energy-intensive industries, otherwise –0.2	As for base case	As for base case	As for base case
Energy conservation and new technology effects	Reduction of growth in energy use by 0.5 percentage points for energy-intensive industries and by 0.25 points for others	As for base case	Reductions in growth in energy use twice that in base case	Additional reduction in transport and other services of 1 point phased in over 2011–14
Fuel use shares, by industry	Based on IEA (2007), adjusted for shift from oil to coal and more aggressive growth of non-fossil fuels	As for base case	As for base case	As for base case
Emission intensity of fuel use	Fixed at IEA (2007) values	As for base case	As for base case	Reduction of 15 per cent (coal), 10 per cent (oil) and 5 per cent (gas) during 2010–30

2015. The new industrial structure option involves a much more rapid move to a service-oriented economy, with the elasticity of industrial growth with respect to GDP being reduced rapidly to 0.9 by 2015 and to 0.8 by 2030.

The sectoral growth elasticities relate to relative rates of growth within the industrial sector (excluding construction). For 2001–06, real value added in the five energy-intensive industries taken as a whole grew by 15.2 per cent in comparison with overall industrial growth of 11.7 per cent—implying a growth elasticity of 1.30. Growth patterns were, however, variable across the industries. For individual energy-intensive industries, the elasticities of growth relative to all industry are set initially at the average for 2001–06, bounded at 1.5 where the historical figure exceeds that level. In the base case, these elasticities are assumed to decline gradually to 1.1 by 2015, and then to 1 by 2030. In the new industrial structure case, the pattern of change is more rapid, with the growth elasticities of these industries being reduced rapidly to 0.9 by 2015 and to 0.8 by 2030. The combination of these two sets of assumptions is that, in this case, the structure of the Chinese economy changes profoundly in the next decade or so; by 2020, the growth rate of the energy-intensive industries is only about 5 per cent in comparison with the assumed GDP growth rate of 7 per cent per annum at that time.

Technology and energy conservation policies. While there is clear evidence that, during the period of limited energy supplies and a command economy in the 1980s and 1990s, technology and energy conservation policies had a substantial impact on energy use, no quantitative measure of that impact is available. Some of the command and control measures that have recently been implemented in this area have been noted above. In the absence of more precise information, we assume that the more limited policies included in the base reduce energy use by 0.5 percentage points per annum in energy-intensive industries and by 0.25 points in other industries; and, in the price and command measures strategy, these effects are doubled in both industry types. Clearly, this is only a preliminary specification, and requires further work.

Fuel use by type and emissions intensities of fuel types. The values for China for the projection period of s_{ijt}, the industry-specific shares of various fuel types in total energy use, and of m_{jt}, the emissions intensity of different fuel types, are based on the values used in IEA (2007)—varying from those estimates only for fuel use by type, where later information and increased knowledge of the emerging energy-use path is available. Massive expansion of coal production and of coal-fired power stations has been under way in China in recent years, while there is clear evidence of fuel substitution away from oil and considerable attention being given to renewable energies and to nuclear energy. As the industry-specific fuel-use shares are held fixed but the pattern of value added shifts across different cases, the overall fuel-use shares also vary across the cases.

The IEA's projections of carbon dioxide emissions per unit of fuel type used are adopted in full, and held fixed for all scenarios, other than for the additional measures case. Here it is assumed that new technologies, such as advanced clean-coal generation methods and carbon geosequestration, are introduced gradually after 2010 and with increasing effect after 2020.

Reducing energy use and emissions with continued rapid development

The results of the model simulations with the four cases are summarised in Table 18.3 and Figure 18.2. In the base case, energy use is projected to increase by 6.5 per cent per annum during 2005–30 and emissions rise by 6.1 per cent to 6.8 giga-tonnes of carbon (GtC) by 2030. The main growth is in the period to 2015—with energy use growing by 7.7 per cent per annum and emissions by 7.6 per cent per annum during 2005–15—with annual emissions growth slowing to 5.1 per cent during 2015–30. Nevertheless, the scale of energy use and emissions implied in this projection is such as to have major implications for the environment within China, for the global climate and for world resource markets.

In the new industrial structure case, the growth of energy use and emissions slows, but only by about 0.5 per cent per annum. This is somewhat surprising in the light of the significant change in industrial structure that is assumed, but reflects the assumed continuing reductions in energy intensity in high energy-using industries (in the base case) and the importance of transport and the services sector (including households) in the longer term. The additional measures in the second alternative case—higher continuing increases in relative energy prices and more effective measures to increase efficiency and to support the use of new technologies—have more effect, reducing the rate of growth of emissions during 2015–30 to 3.3 per cent. Even these assumptions are not sufficient to approach stabilisation of emissions, because in this case the use of fossil fuels in the services sector is still growing strongly.

As a result, further measures to reduce energy use in transport and other services, together with gains from technologies that reduce the level of emissions per unit of fossil-fuel use, are necessary to make stabilisation of emissions achievable. In the additional measures case, emissions are held to 3.5 GtC in 2030, little more than half that in the base case, and grow by only 1.6

Table 18.3 Summary of unchanged policy and alternative policy projections

	Level (Mtoe* and GtC)				Annual rate of change (% p.a.)			
	1980	2005	2015	2030	1980–2005	2005–2030	2005–2015	2015–2030
Energy consumption[a]								
Base case	424	1,693	3,550	8,210	6.5	6.5	7.7	5.7
New industrial structure	424	1,693	3,431	7,522	6.1	6.1	7.3	5.4
Price and command measures	424	1,693	3,112	5,663	4.9	4.9	6.3	4.1
Additional measures	424	1,693	3,089	5,338	4.7	4.7	6.2	3.7
Carbon dioxide emissions								
Base case	0.38	1.56	3.23	6.84	6.1	6.1	7.6	5.1
New industrial structure	0.38	1.56	3.10	6.11	5.6	5.6	7.1	4.6
Price and command measures	0.38	1.56	2.80	4.54	4.4	4.4	6.0	3.3
Additional measures	0.38	1.56	2.73	3.47	3.2	3.2	5.7	1.6

* Mtoe = mega-tonnes oil equivalent
[a] excludes energy from traditional biomass and emissions from cement production
Sources: Authors' estimates and IEA database.

Figure 18.2 **Carbon dioxide emissions from fuel combustion in China, 2005–30: the base case and alternative policy cases** (GtC per annum)

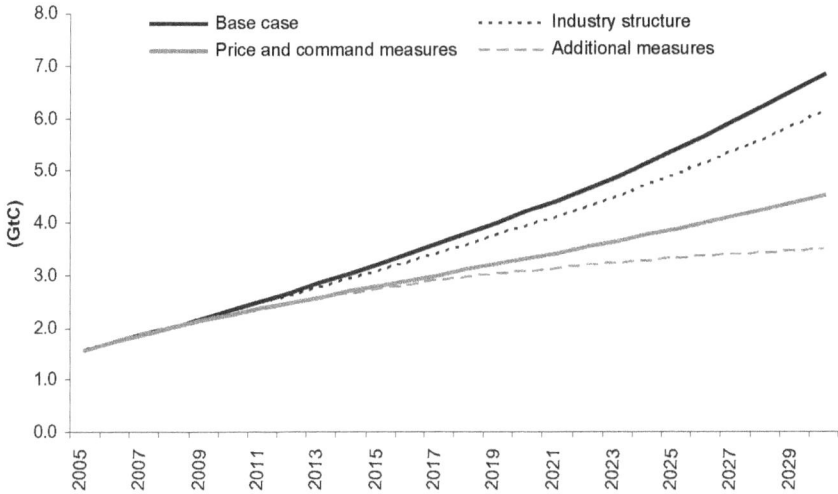

Source: Authors' estimates and IEA database.

per cent during 2015–30, with the annual growth rate approaching zero towards the end of the period. This is indeed a scenario in which emissions stabilise at less than 4 GtC not long after 2030.

Conclusion

There are many uncertainties about China's future energy use, and about the effectiveness of the policies currently being discussed and implemented to curtail it. What is not uncertain is that it is in China's interest to curtail the rapid growth in energy use, and hence curtail the human and environmental costs that it creates, and such an outcome is highly desirable in terms of global climate outcomes. Can this be achieved while continuing rapid economic development, or must China choose between rapid growth and acceptable environmental outcomes?

This scoping study offers only limited evidence about the answer to this question, but it does point to several tentative conclusions. First, there might be realistic options in which China does not have to choose; these forms of implementation of the alternative development path sketched in the eleventh Five-Year Plan certainly deserve further study. Second, such a task will be difficult, and will require extensive and sustained policy initiatives in each of the six policy areas discussed here. Finally, achieving such an outcome seems clearly to be in the interests of the Chinese people and the international community, so that strong measures by governments of industrialised countries to support this policy implementation process in China would be well justified.

Notes

1 For example, for eight countries in South East Asia (excluding China), the unweighted mean elasticity of energy use with respect to GDP for the period 1971–2002 was 1.12, while for another 61 developing countries for which data were available the unweighted mean elasticity during this period was 1.45 (IEA 2006b). These figures exclude traditional biomass energy, as do the figures used for China.

References

Andrews-Speed, P., 2004. *Energy Policy and Regulation in China*, Kluwer Law International, The Hague.

Central Government Portal, 2007. A total of 41.67 million tons of the steel production capacities with backward technologies in ten regions will be dumped, Beijing. Available from http://www.gov.cn/jrzg/2007-05/15/content_614511.htm

Chen, W., Wu, Z., He, J., Gao, P. and Xu, S., 2007. 'Carbon emission control strategies for China: a comparative study with partial and general equilibrium versions of the China MARKAL model', *Energy*, 32:59–72.

Clarke, L., Edmonds, J., Jacoby, H., Pitcher, H., Reilly, J. and Richels, R., 2007. *Scenarios of greenhouse gas emissions and atmospheric concentrations*, Sub-Report 2.1A of Synthesis and Assessment Product 2.1, US Climate Change Science Program and the Subcommittee on Global Change Research, Department of Energy, Office of Biological and Environmental Research, Washington, DC.

Fisher-Vanden, K., Jefferson, G.H., Hongmei, L. and Quan, T., 2004. 'What is driving China's decline in energy intensity?', *Resource and Energy Economics*, 26:77–97.

Garbaccio, R.F., Ho, M.S. and Jorgenson, D.W., 1999. 'Why has the energy-output ratio fallen in China?', *Energy Journal*, 20(3):63–91.

Gately, D. and Huntington, H.G., 2002. 'The asymmetric effects of changes in price and income on energy and oil demand', *Energy Journal*, 23(1):19–55.

Hang, L. and Tu, M. 2007. 'The impacts of energy prices on energy intensity: evidence from China', *Energy Policy*, 35:2978–88.

Huang, J., 1993. 'Industry energy use and structural change: a case study of the People's Republic of China', *Energy Economics*, 15(2):131–6.

International Energy Agency (IEA), 2006a. *World Energy Outlook 2006*, International Energy Agency, Paris.

——, 2006b. *World Energy Statistics and Balances*, International Energy Agency, Paris.

——, 2007. *World Energy Outlook 2007*, International Energy Agency, Paris.

International Monetary Fund (IMF), 2008. *World Economic Outlook*, International Monetary Fund, Washington, DC.

Lin, J., 2005. *Trends in energy efficiency investments in China and the US*, LBNL-57691, Lawrence Berkeley National Laboratory, University of California, Berkeley, California.

Lin, X. and Polenske, K.R., 1995. 'Input–output anatomy of China's energy use changes in the 1980s', *Economic Systems Research*, 7(1):67–84.

Ministry of Finance (MOF), 2007. China will adjust the tariff rate for some of its tradable goods, 21 May, News Office, Ministry of Finance, Beijing. Available from ://www.mof.gov.cn/news/20070521_1500_26483.htm (accessed 11 June 2008).

National Bureau of Statistics (NBS), 2007. *China Statistical Yearbook 2007*, China Statistics Press, Beijing.

——, various years. *China Statistical Yearbook*, China Statistics Press, Beijing.

National Development and Reform Commission (NDRC), 2007. *Urgent notice regarding the acceleration of industrial restructure and curbing the increasing expansion of high energy intensive industries*, National Development and Reform Commission Document No.933, Beijing. Available from http://www.ndrc.gov.cn/zcfb/zcfbtz/2007tongzhi/t20070525_137359.htm (accessed 11 June 2008).

Naughton, B., 2006. *The Chinese Economy: transitions and growth*, MIT Press, Boston, Mass.

Pesaran, M.H., Smith, R.P. and Akiyama, T., 1998. *Energy Demand in Asian Developing Economies*, Oxford University Press, New York.

Sheehan, P. and Sun, F., 2007. *Energy use and CO_2 emissions in China: interpreting changing trends and future directions*, CSES Climate Change Working Paper No.13, Centre for Strategic Economic Studies, Victoria University, Melbourne.

Sheehan, P. (forthcoming). 'The new global growth path: implications for climate change analysis and policy', *Climatic Change*.

Shi, X. and Polenske, K.R., 2006. *Energy prices and energy intensity in China: a structural decomposition analysis and econometrics study*, Working Paper No.0606, Center for Energy and Environmental Policy Research, Massachusetts Institute of Technology, Cambridge, Mass.

Sinton, J.E. and Levine, M.D., 1994. 'Changing energy intensity in Chinese industry: the relative importance of structural shift and intensity change', *Energy Policy*, 17:239–55.

Sinton, J.E., Levine, M.D. and Qingyi, W., 1998. 'Energy efficiency in China: accomplishments and challenges', *Energy Policy*, 26(11):813–29.

Wen, J., 2006. Report on the work of the government 2006, Delivered at the Fourth Session of the Tenth National People's Congress, 5 March 2006, Beijing, People's Republic of China. Available from http://english.gov.cn/official/2006-03/14/content_227248.htm (accessed 11 June 2008).

Weyant, J.P., de la Chesnaye, F.C. and Blanford, G., 2006. 'Overview of EMF–21: multi-gas mitigation and climate change', *Energy Journal. Special Issue Multi-Greenhouse Gas Mitigation*, 27:1–32.

World Bank, 2007. *2005 International Comparison Program: tables of final results*, World Bank, Washington, DC. Available from http://siteresources.worldbank.org/ICPINT/Resources/ICP_final-results.pdf (accessed 11 June 2008).

Acknowledgements

The authors gratefully acknowledge the exceptional research support provided by Alison Welsh, Stephen Parker and Margarita Kumnick, and funding from the Australian Research Council under a Linkage Grant and from the industry partners to that grant.

Index

www.ingramcontent.com/pod-product-compliance
Lightning Source LLC
Chambersburg PA
CBHW051441270326
41932CB00025B/3384